T5-CFQ-566

Walter D. and Deella Toms
Endowed Library Fund

Established
in loving memory
by

Dr. Esther C. Toms,
Class of 1943

H. D. Kumar · D.-P. Häder
Global Aquatic and Atmospheric Environment

Springer

Berlin
Heidelberg
New York
Barcelona
Hong Kong
London
Milan
Paris
Singapore
Tokyo

Har Darshan Kumar · Donat-P. Häder

Global Aquatic
and Atmospheric Environment

With 114 Figures

 Springer

LIBRARY
COLBY-SAWYER COLLEGE
NEW LONDON, NH 03257

QH
545
.W3
K86
1999
C.1

40821074

Professor emeritus HAR DARSHAN KUMAR
Banaras Hindu University
P.O. Box 5014
Varanasi – 221 005
India
hdkumar@banaras.ernet.in

Professor Dr. DONAT-P. HÄDER
Friedrich-Alexander-Universität
Institut für Botanik
und Pharmazeutische Biologie
Staudtstr. 5
D-91058 Erlangen
Germany
dphaeder@biologie.uni-erlangen.de

ISBN 3-540-65369-4 Springer-Verlag Berlin Heidelberg New York

Library of Congress Cataloging-in-Publication Data
Kumar, H. D., 1934- Global aquatic and atmospheric environment / Har Darshan Kumar, Donat-P.
Häder. p. cm. Includes bibliographical references and index.
  ISBN 3-540-65369-4 (hc.)
  1. Water – Pollution – Environmental aspects. 2. Air – Pollution – Environmental aspects. I.
Häder, Donat-Peter. II. Title. QH545.W3K86 1999   577.27–dc   21 99-19614   CIP

This work is subject to copyright. All rights are reserved, whether the whole or part of the material is concerned, specifically the rights of translation, reprinting, reuse of illustrations, recitation,
broadcasting, reproduction on microfilm or in any other way, and storage in data banks. Duplica-
tion of this publication or parts thereof is permitted only under the provisions of the German
Copyright Law of September 9, 1965, in its current version, and permission for use must always
be obtained from Springer-Verlag. Violations are liable for prosecution under the German Copyright Law.

© Springer-Verlag Berlin · Heidelberg 1999
Printed in Germany

The use of general descriptive names, registered names, trademarks, etc. in this publication does
not imply, even in the absence of a specific statement, that such names are exempt from the relevant protective laws and regulations and therefore free for general use.

Camera ready by the authors
Cover-Design by design & production GmbH, Heidelberg

SPIN 10693481   31/3137-5 4 3 2 1 0 – Printed on acid-free paper

Preface

During the last few decades anthropogenic activities in the industrially advanced countries have outcompeted nature in changing the global environment. This is best illustrated for example by the polluted lakes in Scandinavia and Canada, associated with acid deposition from fossil fuel combustion. One of the major challenges mankind is confronted with in the field of energy consumption is undoubtedly to ensure sustainability – a goal that requires improved management of natural resources and a substantial reduction of the noxious emissions which are dangerous to health and the environment. The threat of global climate change due to pollutant emissions causes serious concern to many nations, and reaching an international consensus is likely to take some time. Carbon dioxide emissions have slowed only marginally in industrialized countries during the last few years, but have increased significantly in most developing countries due to increases in energy demand and the increasing use of fossil fuels, which remain the most readily available energy sources today.

Unfortunately, far from learning lessons from the negative experiences of developed countries, many developing countries are taking the same path to development which has turned out to result in serious environmental consequences. In many developing countries, railway passengers already have to carry bottles of clean drinking water for their journey, and the day may not be far off when travellers may also have to carry breathable air! Looking across India, one is alarmed at the appalling state of most cities with respect to acute scarcity and bad quality of water which is often available for only an hour or two daily. Especially during the summer, the bad quality frequently results in epidemics of jaundice, hepatitis, cholera and other waterborne diseases which affect millions of people every year.

The pollution of poverty is perhaps the most threatening of all. We may not be able to save the purity of water, nor save the forests or the whales, unless we can save the people. This is a formidable task. Apart from controlling the rising population, the only other practical and feasible options to help improve the global environment is for people to make some sacrifice by curtailing their standard of living and by limiting their needs and requirements to levels that can be sustained – in keeping with the requirements of the future generations.

In this book we have aimed at drawing an overall picture of the general state of the world's aquatic ecosystems, rivers, lakes, wetlands, and oceans. The key environmental features of these water bodies are stated briefly, along with generalizations about their biotic and abiotic characteristics. Wherever possible, special focus is on how human activities have changed an aquatic ecosystem from its natural state. Remedial action to control or prevent further deterioration is suggested along with ways and means to restore the quality of some of the affected systems.

We have not gone into older, historical aspects of aquatic systems because many excellent books cover these in detail. Our focus has rather been on more recent data and on research carried out during the past two decades. Examples from developing countries are given wherever possible. The book is expected to be of interest to researchers, college and university students of botany, zoology, ecology, limnology, geography, environment, medicine and hydrology, and to all those who may be interested in the global environment.

Water, of course, is only part of the global environment. The other area of concern is the atmosphere which shows serious signs of global pollution. Two chapters are devoted to tropospheric and stratospheric ozone changes, and consequent UV-B radiation effects.

This book could not have been written without help and cooperation from a large number of authors, editors, societies, agencies, organizations and other professionals who have supplied reprints, reports, newsletters, monographs, journals, magazines and other scientific and technical literature. We have summarized the salient points of general interest from this literature and rewritten them suitably to increase comprehension and readability for the average reader. We sincerely thank all those who have helped with the genesis of this book. We also thank M. Barnett, F. Boggasch and M. Dautz for their help in converting the figures into an electronic form.

H.D. Kumar

D.-P. Häder

Contents

1 Aquatic Ecosystems

1.1
Introduction

Water is undoubtedly the lifeline of the environment. Some simple organisms can survive without air but none can survive without water. It constitutes over two thirds of the human body. A human being may be able to live without food for a month, but can live for no more than a few days without water. All living things need water to survive. Water possesses unique physical and chemical properties. It can be frozen, melted, evaporated, heated and combined. Normally, it is a liquid substance made of molecules (H_2O) containing one atom of oxygen and two of hydrogen. Pure water is colorless, tasteless and odorless. It turns into a solid form at 0 °C and to a vapor form at 100 °C. Its density is 1 g per cubic centimeter (1 g cm^{-3}). It is an extremely good solvent.

Global water tends to remain fairly constant in quantity and is continuously in circulation. Little has been added or lost over the millennia. The same water molecules have been transferred over and over again from the oceans into the atmosphere by evaporation, fallen upon the land as precipitation, then transferred back to the sea by rivers and groundwater. This endless circulation is known as the hydrologic cycle. At any given moment of time, about 5 l out of every 100 000 l are in motion. Regional short term fluctuations (days, months, to a few years) in the hydrologic cycle can result in droughts and floods as water and climate are intimately related.

Freshwater lakes, rivers, and underground aquifers hold only about 3.5% of the world's water. By comparison, saltwater oceans and seas contain about 95% of the world's water supply. Two thirds of the world's freshwater is found underground. Groundwater occurs in the tiny spaces between soil particles or in cracks in bedrock, very much like a sponge holds water. The underground areas of soil or rocks where large volumes of water are present are called aquifers and these are the sources of wells and springs. The top of the water in these aquifers forms what is called the "water table". Springs are created when groundwater naturally flows to the surface. groundwater becomes contaminated when man-made (anthropogenic) substances are

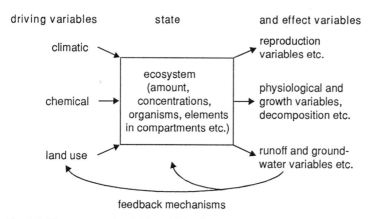

driving variables   state   and effect variables

climatic   reproduction
                                           variables etc.

                    ecosystem
                    (amount,   physiological and
chemical   concentrations,   growth variables,
                    organisms, elements   decomposition etc.
                    in compartments etc.)

land use   runoff and ground-
                                           water variables etc.

                  feedback mechanisms

Fig. 1.1. Three types of variables which affect any ecosystem

dissolved in those waters which recharge the groundwater zone. "Recharge" here refers to the replenishment of water in an aquifer. Much of the natural recharge of groundwater occurs in the spring or summer season as a result of the melting of snow or from streams in mountainous regions where the water table is often below the bottom of the stream bed. Recharge also occurs during local heavy rains. Quite often, groundwater discharges into a river or lake, maintaining its flow in dry seasons.

Some good examples of anthropogenic contaminants are petroleum products leaking from underground storage tanks, nitrates leaching from overfertilized agricultural areas, excessive applications of pesticides, leaching of fluids, and accidental spills of hazardous chemicals. Contamination can also result from overabundance of naturally occurring iron, manganese, or arsenic. Seawater can seep into groundwater in coastal areas, the seepage being referred to as saltwater intrusion.

Normally, groundwater is safer than surface water for drinking because of filtration and natural purification processes in the ground. However, these processes can be rendered ineffective due to sewage, fertilizers, toxicants and other substances seeping into the ground.

Water quality is defined in terms of physical, chemical and biological contents of water. The water quality of rivers and lakes changes with the seasons and geographic location, even when there is no significant pollution present. The quality can be affected by many diverse factors. A good quality drinking water should be free from pathogenic (disease causing) organisms, harmful chemicals and radioactive matter. It should be tasteless, aesthetically appealing, and free from objectionable color or smell.

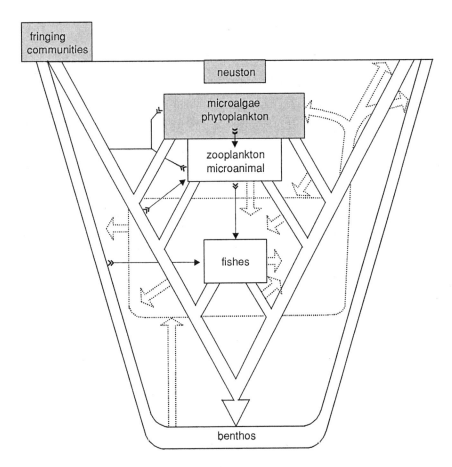

Fig. 1.2. Interrelationships among some semi-isolated subsystems or compartments into which aquatic ecosystems can be divided. Dotted open arrows nutrients, solid open arrows detritus, thin arrows food web relations, gray boxes photosynthesis

Bodies of water are inhabited by a group of interacting organisms collectively called the aquatic ecosystem. The individual organisms depend on one another and on their aqueous environment for nutrients (such as nitrogen and phosphorus) and shelter. Two familiar examples are lakes and rivers. Even a few drops of water can be considered to constitute an aquatic ecosystem since even this minute volume contains or is capable of supporting living organisms. In fact, aquatic biologists usually study drops of water brought from a lake or river to the laboratory in order to understand how these larger ecosystems work.

The complex nature of any ecosystem is illustrated in Fig. 1.1 indicating the driving, state and effect variables. The external driving variables usually act independently of the other variables. State variables are quantitative structural components of the ecosystem. Effect variables relate to those processes that are changed by the driving variables, usually causing long-term changes of the state variables and occasionally also of the driving variables by different feedback mechanisms.

Any aquatic ecosystem can be typically divided into three spatial compartments (Fig. 1.2) within which different processes are located: the pelagic community of the water mass, the benthic community in and on the bottom sediments or rocks, and in shallow regions the fringing communities dominated by emergent or submerged plants. The pelagic compartment may further be classified into the planktonic community suspended in the water, and the nektonic assemblage of larger, more mobile organisms which can swim through it. In some cases a neuston is formed by organisms which

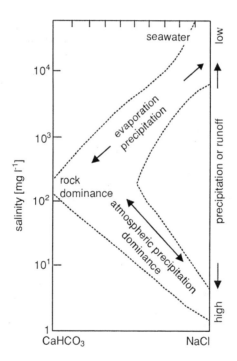

Fig. 1.3. Dependence of the salinity of surface waters on environmental factors and processes

Table 1.1. Ranges of phytoplankton biomass and productivity in relation to nutrient status and salinity in inland water bodies.. (After Likens 1975)

Characteristics	Oligotrophic	Mesotrophic	Eutrophic	Saline
Net primary productivity (g dry wt m^{-2} yr^{-1})	20–50	50–100	150–500	500–2500
Phytoplankton biomass (mg dry wt m^{-3})	20–200	200–600	600–10 000	1000–20 000
Total phosphorus (ppb)	< 1–5	5–10	10–30	30–100
Inorganic nitrogen (ppb)	< 1–200	200–400	300–650	400–5000
Total inorganic solutes (ppm)	2–20	10–200	100–500	1000–150 000

dwell in a layer right at the air/water interface. Aquatic habitats can be classified on the basis of the quantities of dissolved inorganic salts. The salinity of the water greatly affects the nature of the flora and fauna of any given habitat. More than twice as many classes of animals are represented in the sea as in freshwater even though the diversity in terms of individual species does not differ significantly in the two cases. The salinity of surface waters is controlled by several interacting processes shown in Fig. 1.3. Salinity is also an important variable in non-marine aquatic habitats as there it tends to correlate with levels of primary production (Table 1.1).

1.2
River Ecosystems

A river can be defined as flowing water in one or more channels having defined banks. Modern usage includes rivers that are intermittent or ephemeral in flow and channels that are practically bankless. The word "stream" emphasizes the fact of flow; as a noun it is synonymous with river and is often preferred in technical writing.

A river can have a wide range of physical characteristics from narrow, mountain torrents to vast expanses of silently flowing water. Rivers have played important roles in the development of civilization. When in flood some rivers can become highly hazardous for the inhabitants living along their banks. Rivers have always been transportation routes, both on the water and along their banks. Rivers are also suppliers of water and energy, and act as waste carriers for man and nature. Through their flood deposits they provide fertile level land. Supplying water can be regarded as the chief economic role of rivers. Water, as a renewable resource, is both the least expensive and the second most essential commodity (after air) for human use. Most rivers find their origin in rain that falls in the mountains or watersheds. Water flowing to lower levels makes its own path, forming gullies. Waters from these gullies run together, forming a channel. A small channel with flowing water is called a *rill*, a larger one a *brook* and still larger, a *creek*. Often, some small streams dry up in summer, then fill up and flow again during the rainy season; these are called intermittent streams (Baconguis 1995).

At the start, water flows in the channel only during or immediately after rain. Fast flowing water on slopes erodes the river bottom more than the sides, thereby lengthening and deepening the channel. The channel develops into a steep-sided V-shaped valley. This deepened channel can obtain water from underground. When water flows in the channel all year round, it becomes a perennial stream, and a new river is formed. After heavy rains, more water flows down the sides to form branching gullies and water from these gullies enters the river. Since they contribute water, the gullies become its tributaries. With continuing erosion, the river and its tributaries eventually become a branching system of water channels. Runoff water flows down to the tributaries of the small rivers. These drain into bigger rivers. This larger volume and faster flowing water further erodes the riverbank and the channel becomes wider and shallower, and its cross-section is U-shaped. The river valley profile becomes eroded into a gentle slope and the river matures.

Along its course, a river usually flows on almost level plains where it slows down. Any obstacle causes the river to change its course, go around the obstacle and form bends or curved channels, which is called "meandering". At a bend, it strikes and erodes the outer bank. Sediments eroded from the outer bank are carried downstream and deposited on the inner bank of the next bend. The bend deepens, loops, and becomes separated from the stream, forming an oxbow lake. Deep loops and oxbow lakes are generally formed on wide floodplains. The river at this stage is described as being old (Baconguis 1995).

Not all rivers flow on the Earth's surface. There are streams flowing beneath the ground. These are underground rivers. They are formed when the water seeps into the cracks and breaks in large limestone beds. CO_2 absorption by river water makes it acidic and so the limestone along the way dissolves. Over thousands of years, the cracks become cavities, then caves. Some of these caves look like long, vertical chimneys whereas others appear like horizontal chambers. These caves fill up with water and sometimes dry up. Water often flows from one cave to another. The flowing water laden with sediments erodes the caves, which further enlarge, eventually forming an underground river.

The elevated area where a river begins is its head. Its lowest point is its mouth which empties into a bay, lake or sea. The whole area drained by a river system consisting of the main river and the tributaries is the watershed, drainage basin or river basin (Baconguis 1995). The watershed acts like a catchment basin for water falling on land. A watershed should be thickly forested, otherwise water flowing down its slopes can take away much of the soil down the valley and little water will sink into the ground for storage. The ridge separating the land that drains into one river from the land that drains into another river is called a divide (Baconguis 1995).

1.2.1
Uses of River Water

1. Source of drinking water,
2. Source of water for irrigation, for different industries, for thermal control, and as a force to generate electricity,
3. As habitat for fish and other wildlife. Inland waters, rivers, natural and artificial lakes, swamps and marshes are good fishing grounds. Many of them are used for fish culture. Unfortunately, increasing pollution of rivers due to domestic, industrial, agricultural, and mining activities is destroying the natural spawning grounds of fish and other aquatic organisms. Wastewater, chemicals and toxic substances discharged into rivers kill fish and other organisms,
4. As waterways, several rivers are navigable, being used for transporting raw materials to factories along the river and finished products from factories to ports. Rivers are used for transporting agricultural and industrial products,
5. For recreation and sports, swimming, fishing or water skiing.

1.2.2
The River as a Biological Environment

Regardless of the size of a particle that is moved by a flood, it will be deposited when the flow slows down. A zone where the bottom has pebbles large enough for invertebrate animals to cling to is distinct from one where the bottom is sandy or muddy so that the larger invertebrates must burrow. If a river bottom is shallow, rooted plants will grow in it. Between the zone of erosion and the zone of deposition there is an intermediate zone where fine particles settle during low flow and are washed away by a flood, and the slope or river bottom grade is such that this intermediate zone is longer than either of the other two zones.

Water emerging from deep underground layers is usually rather cold with a small range of temperatures most of the time during the year, but may be warm if it drains from shallow soil. The smaller the volume of water, the warmer it will become when exposed to sunlight but it will also lose more heat at night. Larger water volumes have small daily temperature fluctuations.

A fast flowing stream is well oxygenated most of the time; a slow flowing river is well oxygenated during the day because of the activity of photosynthetic plants, but it becomes depleted of oxygen at night. Turbidity of the water increases gradually from the source to the mouth. Accumulated terrestrial organic matter washed into the river is the base of the food chain in running water. Therefore in hilly regions the quantum of material entering the stream that may become a source of food increases as area drained increases with distance from the source. The amount may decrease farther downstream where the carrying capacity of the river drops with decreased flow.

1.2.3
Stream Organisms

The slower the stream flow, the more diverse is its biotic community. Running water organisms have peculiar features that allow them to colonize fast flowing water. Stagnant waters fill and vanish but the watercourse must flow as long as there is precipitation to supply it. Running water therefore contains certain groups of organisms that have changed little over the years in which others have evolved and adapted. Most members of the order *Plecoptera* (stoneflies) and many of the order *Ephemeroptera* (mayflies) inhabit running water. Other primitive groups related to mayflies e.g., crustaceans and free-ranging caddis larvae of the insect family Rhyacophilidae also live in a similar body of water (Baconguis 1995).

Rapidly flowing water guarantees a good supply of oxygen and salts, at the same time removing the waste products of the inhabitants. It also carries food. The disadvantage of life in a moving medium is that any accidental displacement is always in one direction. Even the organisms that live in the substratum are occasionally subjected to this hazard when exceptional flow causes the substratum to shift (Baconguis 1995).

A variety of algae grow on stones and rocks. Algal communities show marked changes with season. Temperature, light, substratum, and dissolved substances in the water are factors that influence the composition of algal communities. Any stone or pebble that remains undisturbed and which is not scoured by sand and gravel becomes densely covered with algae. Macrophytes grow where substratum and water depth are suitable. This is partly due to the instability of the flowing water system. Once established, a tuft of vegetation affects the flow and facilitates the deposition of silt which in turn changes the substratum to the point it becomes favorable for other species to inhabit. Eventually the vegetation may present so much resistance to stream flow that the whole mass is washed away. On the other hand, copious plant growth can obstruct water flow to such an extent that it often becomes necessary to cut and remove it from the streams to prevent flooding.

1.2.4
Gravel-Bed River Mouths

A river's mouth is that region where energy from the river in the form of freshwater flow interacts with energy from the sea in the form of waves and tides. For most rivers this interaction is beneficial because it occurs over a fairly large area up the river. Thus, at the mouth the ocean dominates, but as we go upstream the river has an increasingly important effect until, at the limit of tidal influence, it is the river which dominates completely.

The river–sea transition is not always gradual. There are also some rivers where the interaction between river and ocean takes place in a narrow strip within a few hundred meters of the coast. For these rivers, the interaction is far from beneficial and in certain cases it is even catastrophic, at one extreme causing flooding and at the other extreme causing blockage of passage for those fish species which migrate between fresh and sea water. Such rivers are characterized by the following physical features:

1. Wide, steep and braided,
2. Existence of an elongated freshwater lagoon at the mouth,
3. A low, steep, gravel barrier beach (berm) separates the lagoon from the ocean,
4. A narrow, unstable outlet through the berm,
5. High rate of longshore transport (of sediment), and
6. The river experiences intermittent floods.

Our general understanding of the structure and function of large river ecosystems is based primarily on three riverine models:

1. The river continuum concept or RCC (Vannote et al. 1980),
2. The serial discontinuity concept (Ward and Stanford 1983), which integrates the effects of large dams and reservoirs on the RCC, and
3. The flood pulse concept in river–floodplain systems (Junk et al. 1989; Sedell et al. 1989).

These models stress the importance of nutrients derived from either headwater streams or seasonal floodplain pulses and virtually ignore the role of local instream primary production and riparian litter fall. They rely very heavily on data from either smaller streams, floodplain rivers (thereby excluding large rivers with constricted channels), or main channel habitats with their dominant collector feeding guild (thus de-emphasizing near-shore areas where species in many feeding guilds congregate; Thorp and Delong 1994).

To redress some of the weaknesses in the above models, Thorp and Delong (1994) have proposed a new model called the riverine productivity model (RPM) which brings out the importance to large river food webs of local autochthonous production and direct organic inputs from the riparian zone. The representation of ecosystem functions by the RPM differs most significantly from that of previous models for rivers characterized by constricted channels.

According to the RCC, streams should be viewed as longitudinally interlocked systems in which ecosystem dynamics of downstream stretches are strongly connected with processes occurring farther upstream as well as with local lithology and geomorphology. Biotic assemblages hypothetically follow an orderly longitudinal shift in response to upstream processes and changing stream morphology (Thorp and Delong 1994). With respect to large rivers, the RCC assumes:

1. The majority of fine particulate organic matter (FPOM) is derived from upstream processing and is the chief source of organic carbon for food webs,
2. The input of coarse particulate organic matter (CPOM) from adjacent riparian vegetation is insignificant, and
3. Instream primary production is greatly limited by depth and inorganic turbidity, so respiration typically exceeds production.

According to the RCC model, large rivers depend upon the inefficiencies ("leakage") of upstream processing of organic matter for their primary source of energy. Because these materials are processed and transported over a considerable distance, organic matter reaching large rivers would be rather low in nutrient content. The concept concludes that the community biomass of invertebrates in large rivers is dominated by collectors ("water column filter feeders" and "sediment burrowers") which rely on FPOM carried downstream from headwater streams through medium-sized rivers. This FPOM would be available to benthic heterotrophs only in deposition areas, and pelagic suspension feeders have relatively easy access to FPOM throughout the water column.

1.2.5
The Flood Pulse Concept

Allochthonous sources of organic matter other than those derived directly from upstream processing have been found to be important in certain large river ecosystems. In rivers having extensive floodplain interactions, productivity of riparian/floodplain vegetation and processing of organic matter within the floodplain can greatly modify the longitudinal patterns of ecosystem processes predicted by the RCC (Junk et al. 1989). The flood pulse concept proposed that long, predictable pulses in discharge are the principal force controlling biota in river floodplains. Most riverine animal biomass in some large river systems with floodplains is supposedly derived from production within floodplains and not from downstream transport of organic matter originating elsewhere in the basin. Organic matter from the floodplain should be somewhat more labile than FPOM transported from upstream (Thorp and Delong 1994).

Sedell et al. (1989) concluded that rivers with constricted channels tend to follow patterns predicted by the RCC whereas those with unconstrained channels (i.e. having extensive floodplains) should receive sufficient quantities of organic matter from lateral inputs to relegate downstream transport to a relatively minor role.

1.2.6
The Riverine Productivity Model (RPM)

The RPM postulates that a considerable part of the organic carbon assimilated by animals in certain types of large rivers comes from a combination of: (1) local autochthonous production (phytoplankton, benthic algae, aquatic vegetation), and (2) direct inputs from the riparian zone (e.g., abscised leaves, particulate organic carbon = POC, and dissolved organic carbon = DOC) during periods not limited to flood pulses. Autochthonous production and direct organic inputs from the riparian zone together constitute the main source of carbon driving the food webs of those large rivers which have a constricted channel and adequate firm substrate within the photic zone. In floodplain rivers, instream primary productivity (especially from phytoplankton) is a significant contributor to secondary productivity, though probably not the principal source of assimilated carbon (Thorp and Delong 1994).

According to the RCC, most macroinvertebrates in large rivers are collectors (filterers and gatherers) which exploit FPOM transported from upstream sources, while predators represent a small proportion of the community. The flood pulse concept only redefines the primary source of FPOM, but it agrees with RCC in predicting collectors to be the predominant functional feeding group.

1.3
Freshwater Resources and Water Quality

Aquatic ecosystems are usually divided into freshwater, estuarine, and marine systems, distinguished on the basis of their salt content. Aquatic ecosystems cover more than 70% of the Earth's surface and harbor a rich diversity of plant, microbial and animal species which interact among themselves and with their physicochemical environment.

Table 1.2. Aquatic productivity in different environments. (After Whittaker 1975)

Habitat	Net primary productivity (g dry wt. m^{-2} yr^{-1})	Secondary productivity (g dry wt. m^{-2} yr^{-1})	Plant/animal productivity ratio
Open ocean	2–400	125	1: 0.06
Upwelling areas	400–1000	500	1: 0.05
Continental shelf	200–600	360	1: 0.04
Algal beds/reefs	500–4000	2500	1: 0.02
Estuaries	200–3500	1500	1: 0.02
Lakes, rivers	100–1500	250	1: 0.02
Swamps, marshes	800–3500	2000	1: 0.008
Non-aqueous habitats	0–3500	760	1: 0.003–0.02

Of the water on the globe, 94% is salt water in the oceans and only 6% is represented by freshwater. Of the latter, 27% is bound in glaciers and 72% is underground. This leaves at any one time only about 1% of the freshwater in the atmosphere or in streams and lakes. This freshwater supply is continually replenished by precipitation as rain or snow. It has been estimated that the total annual runoff from the continents is about 41 000 km^3. Of these, 27 000 km^3 return to the sea as flood runoff, and another 5000 flows into the sea in uninhabited areas. Thus, only 9000 km^3 of water are readily available for human use world-wide. Since the world's population and usable water are unevenly distributed, the local availability of water varies widely. Much of the Middle East and North Africa, parts of Central America and the western United States are already short of water. By the turn of the century, many countries are likely to suffer acute scarcity of water due to increasing demand for water for agriculture, industry and domestic use.

The rising world population has imposed severe stress on global availability of freshwater resources. Figure 1.4 shows the global water supply and demands in km^3 per year. Freshwaters are usually richer in nutrients than seawater. Estuaries, lagoons, and coastal marine areas are also more productive than the open ocean (Table 1.2), at least partially due to nutrient inputs from rivers. Over 5 decades ago, Lindeman proposed his trophic dynamic hypothesis in an attempt to analyze the energy flow in lakes.

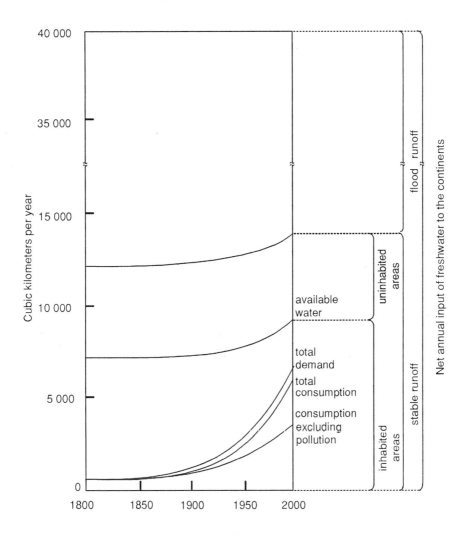

Fig. 1.4. Global water supply and demand in km^3 per year

In freshwaters, the production of biological matter usually terminates in the form of fish. Primary and secondary production processes are involved in fish production. He assessed the production of the three main trophic levels by measuring the sum of organic matter during the growing season.

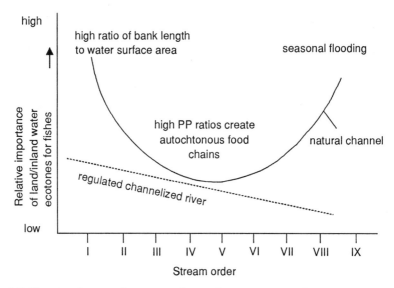

Fig. 1.5. Changes of ecotone importance along a river or stream continuum

However, his hypothesis is not universally applicable to all situations. For example, in Arctic lakes with a limited community the interrelationships between trophic levels are fairly difficult to understand; respiration introduces difficulties in precise estimation of primary production. In addition, animals of the secondary trophic level frequently have mixed food and change their feeding habits.

Rivers and streams constitute a continuum of flowing water. Figure 1.5 illustrates changes of ecotone importance along the river continuum. Here, the ecotone represents the boundary between water and land (i.e., the river bank).

Considering the enormous amounts of water in the hydrosphere, it is difficult to understand how it could become scarce. Nevertheless, the water available for human use, especially in the Earth's heavily populated regions, is very close to potential "needs" and the process of moderating those needs should begin now. The most comprehensive of the studies of world water balance appear to be those of L'vovich (Table 1.3). He estimates the total annual precipitation to be 525 100 km^3.

The demand for water varies markedly from one country to another and depends on population and on the prevailing level and pattern of socioeconomic development. Marked differences exist between developed and developing countries. The former use much more water than the latter, sometime manifold higher volumes. Freshwater ecosystems are characterized as having running water (lotic) or still water (lentic).

Table 1.3. Annual world water balance. (After L'vovich 1979)

Elements of water balance Peripheral land area (116.8 million km^2)	Volume (km^3)	Depth (mm)
Precipitation	106 000	910
Runoff	41 000	350
Evapotranspiration	65 000	560
Enclosed part of the land area (32.1 million km^2)		
Precipitation	7500	238
Evapotranspiration	7500	
Ocean (361.1 million km^2)		
Precipitation	411 600	1140
Inflow of river water	41 000	114
Evaporation	452 600	1254
The globe (510 million km^2)		
Precipitation	525 100	1030
Evapotranspiration	525 100	1030

Freshwater streams (including springs, rivulets, brooks and creeks) and many rivers are important parts of freshwater ecosystems; over their course they have a tendency to change from being narrow, shallow, and relatively rapid to increasingly broad, deep, and slow-flowing. Table 1.4 gives the catchment areas and other characteristics of some large rivers of the world, ranked according to their sediment loads.

1.3.1
Major Rivers of East Asia

Between the South China Sea and the bay of Bengal, six giant rivers dump large volumes of water and sediment into the sea. These are the Mekong, the Chao Phrao, the Salween, the Irrawady, and the Ganges-Brahmaputra (Table 1.4). All these have broad flood-plains and extensive deltas. The Irrawady and the Mekong originate in China, cut deep valleys and then flatten into plains and deltas during their flow to the sea. The floodplains of the Chao Phrao in Thailand are noted for their rice paddy cultures. Regular gentle floods from the river are conducive to rice farming. At the head of the bay of Bengal there are many mouths of the Ganges-Brahmaputra system.

Table 1.4. Catchment areas, water discharges and sediment loads of selected large rivers, ranked by sediment load. (UNESCO 1991)

River	Country	Catchment area (1000 km²)	Mean water discharge	Annual sediment load (1 000 000 tons)	Annual sediment load (tons km⁻²)
Yellow	China	752	1370	1640	2480
Ganges	India, Bangladesh	955	11 800	1450	1500
Amazon	Brazil	6100	17 200	850	139
Brahmaputra	India Bangladesh	666	12 200	730	1100
Yangtze	China	1807	29 200	480	280
Indus	Pakistan	969	5500	436	450
Irrawaddy	Myanmar	430	13 500	300	700
Mississippi	USA	3269	24 000	300	91
Kosi	India	62	-	172	2774
Mekong	VietNam	795	-	170	214
Red	VietNam	120	3900	130	1100
Parana	Argentina	2305	-	90	38
Yongding	China	51	-	81	1944
Congo	Zaire	4014	39 000	72	18
Danube	Romania	816	6200	65	80
Niger	Nigeria	1081	4900	21	19
Po	Italy	540	1550	15	280
Ob	USSR	2430	12 200	15	6
Rhine	Netherlands	160	2200	2.8	17

Over 40 000 m³ per second of muddy water pours from these shifting exits and carry what was once Himalayan ridge tops and canyon sides into the bay of Bengal. Year after year, the Ganges and the Brahmaputra bring down over 2×10^{15} g of soil into the sea. These large flows of the Ganges-Brahmaputra cause serious damage to property, livestock, and humans, especially in combination with storms and tidal surges into the bay. India has vast and varied aquatic resources. Besides the extensive coastline and an exclusive economic zone of over 2 million km², there is the large river system and the reservoirs, lakes and dams. These provide all types of geoclimatic zones as well as fresh and brackish water.

Table 1.5. Distribution of runoff between continents

Region	Volume of water (1000 km³)								
	1975			1974			1974		
	P[a]	E	R	P	E	R	P	E	R
Europe	6.6	3.8	2.8	8.3	5.3	3.0	7.2	4.1	3.1
Asia	30.7	18.5	12.2	32.2	118.1	14.1	32.7	19.5	13.2
Africa	20.7	17.3	3.4	22.3	17.7	4.6	20.8	16.6	4.2
Australia	7.1	4.7	2.4	7.1	4.6	2.5	6.4	4.4	2.0
N. America	15.6	9.7	5.9	18.3	10.1	8.2	13.9	7.9	6.0
S. America	28.0	16.9	11.1	28.4	16.2	12.2	29.4	19.0	10.4
Antarctica	2.4	0.4	2.0	2.3	0	2.3	-	-	-
Oceans	385	425	−40	458	505	−47	412	453	−41

[a]P = Precipitation, E = Evaporation, R = Runoff

Three fourths of Indian sea fish production comes from the west coast. The total Indian contribution is, however, only 3.5 million tons against a potential of 10 million tons, and the current world production of about 80 million tons. Edible crustaceans (such as prawn) also contribute significantly to the catch from marine and freshwater habitats. Some species of mollusks and clams also provide a cheap but nutritious food. The most important fishes currently exploited are sardines and mackerels.

The inland fishes of rivers and lakes are also a valuable resource of over 900 species belonging to over 300 genera. In rivers and lakes, fishes breed and there is a natural stock. In lakes and reservoirs fishes are stocked to forage for food and in smaller reservoirs they are fed and fertilized. The dominant species vary depending upon the geoclimate. The Indian reservoirs are, however, understocked, underexploited and badly managed. India has millions of hectares of derelict inland, saline and alkaline soils where marine and brackish water species of fish and prawns may be grown profitably.

The most meaningful measure of freshwater availability is average annual runoff from land areas since water use is a recurring activity dependent on continuing replenishment of supplies. Annual runoff from inhabited areas exclusive of the polar zones has been estimated at 39 000 km³ but potential use of runoff is substantially limited by distribution of supply, both geographically and temporally. The distribution of runoff between continents is highly variable (Table 1.5). Distribution of runoff is also nonuniform within many individual countries.

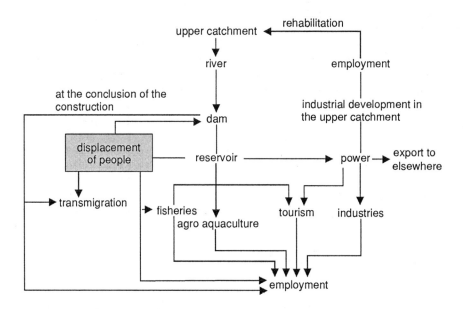

Fig. 1.6. The resettlement alternatives in a dam project. The power generated by the dam is partly utilized to improve the environmental quality of the area of the dam and the upper catchment

The establishment of a reservoir as a result of damming a river leads to the inundation of human settlements and extensive agricultural land, but on the other hand creates another resource in the form of a reservoir or a lake. That is why other strategies need to be developed to utilize new resources, i.e. lakes. Utilization and development of the lake can be used to provide new jobs for the people living in its surrounding. Opportunities include fishery, aquaculture and agri-aquaculture. If electricity is available and tourism encouraged, other sectors can also be developed without much problem. Figure 1.6 shows the resettlement options in a dam project.

1.4
World Water Use

Worldwide water use increased dramatically from about 1360 km³ in 1950 to 4130 in 1990 (Figs. 1.7, 1.8) and is expected to reach about 5190 by the year 2000.

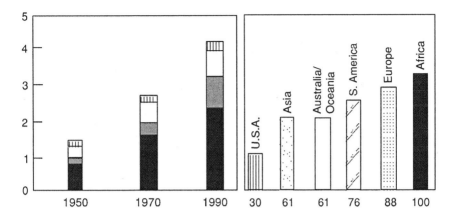

Fig. 1.7. Trends in water withdrawal (km³ per year)   **Fig. 1.8.** Increase in water withdrawal

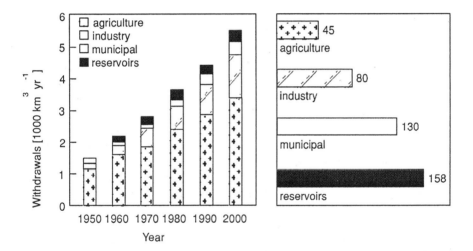

Fig. 1.9. Water by sector

Fig. 1.10 Increase of water use in different sectors (in percent compared to 1970–1990)

The uses to which water is put vary from country to country, but agriculture (69%) is the main drain on the water supply, with 23% for industry and 8% for domestic purposes (Figs. 1.9, 1.10). The percentage of river basins per continent is indicated in Fig. 1.11.

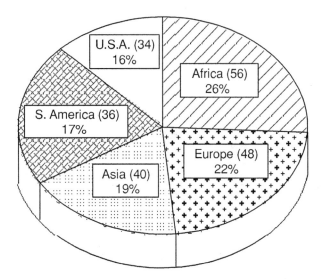

Fig. 1.11. International river basins (number and percentage of the total)

Apart from the acute scarcity of water, many countries have to worry about water quality, and concerns have grown since the 1960s. At first, attention focussed on surface water pollution from point sources, but more recently groundwater and sediment pollution and non-point sources have become equally serious problems.

Acidification of lakes by acidic deposition is common in some European countries, in North America, and now also in Asia. Wastes can also be carried to lakes and streams along indirect pathways – for example, when water leaches through contaminated soils and transports the contaminants to a lake or river. Dumps of toxic chemical waste on land become a serious source of groundwater and surface water pollution. In areas of intensive animal farming or where large amounts of nitrate fertilizers are used, nitrates in groundwater often reach concentrations that exceed permissible limits.

Many of the rivers which have been monitored are polluted, as they have a biological oxygen demand (BOD) of more than 6.5 mg l^{-1}. The average nitrate level in unpolluted rivers is 100 μg l^{-1}. Several European rivers have a mean value of 4500 μg l^{-1}. Some rivers outside Europe have a much lower mean value of 250 μg l^{-1}. The mean phosphate levels in several rivers monitored are 2.5 times the average for unpolluted rivers (the latter is about 10 μg l^{-1}). Organochlorine pesticides measured in some rivers from developing countries are higher than those recorded in European rivers.

1.4.1
Impacts of Mismanagement and Pollution

Water use has not been efficient in many countries. Overexploitation of groundwater has led to the depletion of resources in some areas and to increased encroachment of saline waters into aquifers along coastal zones in some countries. The rapid expansion of agriculture in desert areas may lead to overexploitation of groundwater for irrigation. Excessive irrigation has also caused waterlogging and salinization, thereby accelerating land degradation. The lack of maintenance of water delivery systems and overuse of water for domestic, commercial and industrial purposes, especially in the developing countries, has caused a host of socio-environmental and economic problems. Pools and ponds of water in rural areas and marginal settlements have become breeding grounds for various disease vectors.

The quality of freshwater depends not only on the quality of waste entering the water but also on the decontamination measures that have been put into effect. Even though organic waste is biodegradable, it nonetheless presents a serious problem in developing countries. Human excreta contains pathogenic microorganisms as waterborne agents of serious diseases. Industrial waste may include heavy metals and many other toxic and persistent chemicals not readily degraded under natural conditions or in conventional sewage-treatment plants. The high content of nutrients in rivers and lakes has created eutrophication. Eutrophication brings increasing difficulties and costs for water treatment works which have to produce safe, palatable drinking water. Acidification of freshwater lakes has adversely affected aquatic life. In most newly industrializing countries, both organic and industrial river pollution are on the increase, and decontamination efforts are often neglected. In these countries industrialization has had higher priority than reduction of pollution. As a consequence, in East Asia, degradation of water resources is now a grave environmental problem.

The traditional approach to water management involves building of dams and reservoirs both for flood control and for water storage. Thousands of dams and reservoirs have been built world-wide. Between 1950 and 1986 over 35 000 dams higher than 15 m were constructed, some being higher than 150 m. About half of these dams were constructed in China alone. Although these dams have provided several benefits, they have not been without environmental cost. The amount of stored water in man-made reservoirs in the world has been estimated at 3500 km^3, nearly equal to the total annual water withdrawal in the world. In addition to such water management schemes, various measures have been taken by several countries to improve the efficiency of water use. Despite these efforts, water use is still markedly inefficient, especially in developing countries, many of which provide water (especially for irrigation) either free or heavily subsidized. The past two decades have also witnessed increasing efforts to recycle water for use in industry and agriculture.

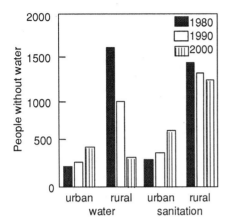

Fig. 1.12. Water supply and sanitation in developing countries

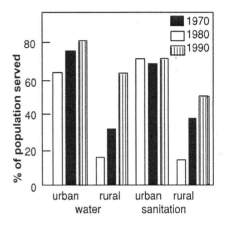

Fig. 1.13. Number of people without water supply and sanitation (in millions).

Table 1.6. Some odors in drinking water and reference chemicals for their identification

Odor	Reference compound
Almonds (sweet)	benzaldehyde
Bleach (sweet)	chloramine
Chlorinous	chlorine
Cucumber	nonenal
Earthy (beets)	geosmin
Grassy	3-hexen-1-ol
Solvent (sweet)	*m*-xylene
Musty (*peaty*)	2-methylisoborneol
Rubber hose, shoe polish	cumene
Green bell pepper	isobutyl-methoxypyrazine
Old grass, rotten nuts	hexanal
Moth balls, camphor	*p*-dichlorobenzene
Rancid nuts	heptanal
Decaying vegetation	grass
Septic	grass

Any discharge of effluent, waste, or pollutants in drinking water reservoirs can lead to the development of offensive smells. These odors resemble the odor of some chemicals (see Table 1.6).

In 1970, one third of the population in urban areas of the developing countries did not have access to safe clean water and 29% did not have access to sanitation services. At the same time, in those countries 86% of the population in rural areas did not have access to clean water and 89% did not have access to sanitation services (see Figs. 1.12, 1.13).

Aquifers near infiltration zones of polluted rivers or in the proximity of plumes of leached substances from landfills often change from aerobic to denitrifying, sulfate-reducing, and eventually methanogenic conditions. Some aromatic compounds (e.g. phenol and benzoate) are metabolized under both aerobic and anaerobic conditions, and it is believed that the presence of a functional group such as a hydroxy- or carboxy-substituent facilitates the anaerobic breakdown of the aromatic ring. Benzene, toluene, and xylene lack functional groups. The existence of anaerobic catabolic pathways for aromatic hydrocarbons has been a subject of debate and controversy. Some observations from polluted aquifers point to slow anaerobic degradation of aromatic hydrocarbons but conclusive evidence for a microbial metabolism of benzene, toluene, and xylene in the absence of molecular oxygen has emerged only recently (Zeyer et al. 1990). The mineralization of toluene in anaerobic aquatic sediments appears to be coupled to iron reduction. Toluene and m-xylene are rapidly mineralized under continuous flow conditions in the absence of molecular oxygen in a denitrifying aquifer column.

1.5
Nature of the Chemical Environment

About 15–20 elements are essential nutrients for organisms. In freshwater ecosystems, the major dissolved ions of calcium, magnesium, sodium, potassium, bicarbonate (or carbonate), sulfate and chloride are the chief factors affecting organisms; pH and other ions (e.g., silicate, ferrous, etc.) are also important. The types and numbers of organisms in a lake depend on several factors and not simply on the concentration of one ion. The presence of chelating substances influences ionic ratios. The algal growth rates and yields in a lake are controlled usually by those constituents which are present in low or limiting concentrations; these are commonly phosphate and nitrogen compounds, and rarely silicate and iron. As a result of many world-wide studies of the chemical composition of lakes, some broad generalizations about their important cations and anions have emerged (Table 1.7).

Table 1.7. Chemical constituents (rough averages) of freshwater. (Based on the data of several authors)

Major elements (mol%)[a]		Minor elements/ions, trace elements and organic compounds	Gases
Cations	Anions		
Ca^{2+} – 35	HCO_3^{2-} – 80	NH_4^+ NO_3^-, inorganic and organic phosphorus, SiO_2 $HSiO_3^-$ Fe, Co, Mo, Mn, B, Zn, vitamins, humic substances, metabolites	O_2, N_2,
Mg^{2+} – 33	SO_4^{2-} – 10		
Na^+ – 25	Cl^- – 10		
K^+ – 7			

[a]These figures include the data from Lake Tanganyika which is one of the largest freshwater lakes of the world, but is quite different from many smaller lakes. If the data of Lake Tanganyika are excluded, then the average concentration of Ca^{2+} and SO_4^{2-} ions will be much higher whereas those of Mg^{2+}, Na^+, K^+ will be much lower than the figures given here.

Though these values are highly variable for different lakes, a kind of "standard" or average picture is given in Table 1.7 as a rough approximation. Conductivity, total dissolved solids and calcium concentration are closely related to pH, but other chemical variables show complex relationships with other factors. On the global average, the bicarbonate-carbonate ion is the dominant anion, calcium and magnesium are roughly equal. In many African lakes, in contrast, sodium and magnesium (not calcium and bicarbonate) often dominate. In most European lakes, the ratio of (Ca^{2+} and Mg^{2+}) to (Na^+ and K^+) is greater than 1, but in most non–European lakes, this ratio is often less than 1.

Important interactions among calcium, carbonate, and phosphate occur in water bodies. In calcium-rich water, the calcium concentration is often regulated by the carbonate concentration. But supersaturation has also been observed in many lakes in which the product of $[Ca^{2+}]$ $[CO_3^{2-}]$ is greater than 10^{-8} (i.e., the solubility product of $CaCO_3$ at 20 °C). In some lakes, supersaturation increases with increasing pH. Furthermore, in eutrophic and calcium-rich waters the calcium concentration sometimes influences the phosphate concentration; calcium exerts a strong influence on phosphate concentration at high pH values in Lake George in Uganda.

Elements commonly enter natural waters by erosion. In oligotrophic lakes in the English Lake District the N:P ratio is considerably higher than 10, and phosphate becomes limiting. In certain cases, e.g. tropical lakes, in non-forested regions, the N:P ratio is less than 10 and then nitrogen rather than phosphorus becomes the limiting nutrient.

Dissolved organic material in marine and freshwater ecosystems constitutes one of the Earth's largest actively cycled reservoirs for organic matter. The bacterially mediated turnover of chemically identifiable low-molecular weight constituents of this pool has attracted detailed studies, but these compounds make up less than 20% of the total reservoir. In contrast, little is known about the fate of the larger, biologically more refractory molecules – including humic substances – which make up the bulk of dissolved organic matter. Humic substances are high molecular weight organic acids; when passed through hydrophobic resins they are retained. They are further categorized as humic acids or fulvic acids, based on their solubility at low pH. Humic substances make up the largest single class of dissolved organic matter (DOM), accounting for up to 60% of the DOM in most natural waters. Although abundant, they constitute that component of DOM from which all easily available energy has already been extracted, and so have been neglected by microbiologists. The accessibility of carbon in humic substances to bacteria may, however, be much higher than believed hitherto. Exposure to sunlight enhances the breakdown of humic carbon to lower molecular-mass compounds, some of which are assimilated rapidly by natural bacteria. There are recent indications that exposure to sunlight causes dissolved organic matter to release nitrogen-rich compounds that are biologically available, thus enhancing the bacterial degradation of humic substances (Bushaw et al. 1996). Bushaw et al. have shown that ammonium is among the nitrogenous compounds released and is produced most efficiently by ultraviolet wavelengths. Photochemical release of ammonium from dissolved organic matter has strong implications for nitrogen availability in many aquatic ecosystems, including nitrogen-limited high-latitude environments and coastal oceans, where inputs of terrestrial humic substances are high.

The general importance of photochemical nitrogen release from DOM (dissolved organic matter) is suggested by studies in which additions of riverine humic substances were found to increase nitrogen availability and stimulate rates of primary and secondary production. Worldwide, dissolved organic nitrogen (DON) accounts for almost 70% of the nitrogen that enters coastal oceans in rivers; future changes in global climate and precipitation patterns may increase the movement of riverine DON to the sea. Ultimately, the participation of DON in autotrophic and heterotrophic biomass production, N_2O production, denitrification, eutrophication and other fundamental biogeochemical processes requires conversion to biologically available forms. This up to now unrecognized mechanism for DON conversion to assimilable N by photochemistry changes our current understanding of the sources of biologically active nitrogen in aquatic ecosystems (Bushaw 1996). In lakes, characterized by slow sedimentation and rapid vertical mixing, phosphate accumulates in the water, whereas the reverse occurs where sedimentation is rapid but vertical mixing is slow. Rate of sedimentation depends on water current, calcium concentration, and mineralization rate in the epilimnion. Vertical mixing depends principally on local currents and eddy diffusion.

1.5.1
Input of Organic Matter from Terrestrial Ecosystems

Production of organic matter by terrestrial vegetation depends mainly on temperature and precipitation. Two regions can be distinguished where terrestrial production may be potentially limited: (1) by precipitation at latitudes less than 40° and (2) by temperature at higher latitudes. The highest annual production of about 3 kg dry matter per m^2 occurs in the wet tropics whereas the lowest (about 0.03 kg m^{-2}) occurs at cold and dry Arctic sites. How much of the terrestrial production reaches aquatic bodies? Small, flowing waters seem to receive most of their energy from terrestrial sources, but such contribution seems much smaller in larger water bodies. In large lakes, the energetic contribution of the organic matter from land is usually quite small or negligible.

1.5.2
Aquatic Humus

Aquatic humic substances are defined as colored polyelectrolytic acid particles isolated from water by sorption onto XAD resins at pH 2, or weak-base ion-exchange resins. Aquatic humic substances are formed by degradation of larger organic molecules or are polymerized through chemical and biological processes, from smaller precursor molecules, such as carbohydrates, proteins and simple compounds, to larger and more refractory condensates. Humic substances are thus polyectrolytes with carboxylic, hydroxyl, and phenolic functional groups. In natural waters they are the major class of organic compounds, comprising 50–75% of the dissolved organic carbon.

Acidification, liming and other anthropogenic effects alter the amounts and character of humic material and its diverse roles. Humic substances may buffer against acidification, but may also add acidity to surface waters. The humic material can affect biological processes through several different mechanisms, either directly through interfering with metabolic processes or indirectly by altering the bioavailability of nutrients or toxicants (Kullberg et al. 1993).

1.5.2.1
Effect on Biological Systems

The biotic effects of acidification are altered by the presence of dissolved humic compounds. The influence of humus includes reduced toxicity of dissolved metals, but both toxic and beneficial effects of the compounds occur.

In some respects, humus resembles anthropogenic surfactants. This suggests that a possible toxic mode of humus is interference with the structure of membranes, since this is an important toxic mode of synthetic surfactants. The function of gill membranes is critical for both invertebrates and fish in order for them to withstand low pH. As gills are the main site for ion uptake, their function is disturbed at low pH. At low concentrations, however, surfactants might stabilize membrane structure against acid and osmotic lysis. Under certain circumstances, humus could thus decrease pH stress by beneficial effects on gill membrane function (Kullberg et al. 1993).

Humus also alters the synthesis and activity of some enzymes. As the activity of ion-regulating enzymes in the gills depends on pH, humus may induce structural effects and thus change the pH dependence of these enzymes. In addition, the exchange of ions between organisms and water depends on the charge of the gill membrane surface.

1.6
Quantitative Pollutants

Nitrates, phosphates, radiation, heavy metals, petroleum etc. are serious pollutants when present in excessive quantities. Oil or petroleum pollution makes for headlines whenever there is a serious accidental spill from ships in the sea. Even without tanker collisions or accidents, normal shipping operations, coastal oil refining, submarines and other human activities often contribute to some oil pollution in the oceans.

Oil pollution destroys or injures marine life (fishes, oysters, shellfish, sea birds, etc.). Oil or its components can be toxic or carcinogenic to microbes, plants, animals and man. Oil pollution has caused declines in populations of diving birds (e.g., puffins, ducks) in the English Channel and elsewhere.

1.6.1
Pesticides

Effluent discharges from pesticide factories into rivers can cause mass mortality of birds and other animals (Koeman 1971). Discharge of pesticide wastewater into the Rhine estuary in Europe caused steep declines in the populations of *Sterna sandwichensis* in The Netherlands. Porpoises (*Phocaena phocaena*) virtually disappeared from Dutch coastal waters and Harbor seals (*Phoca vitulina*) decreased steeply. Most organisms in Dutch coastal waters were found to be contaminated with mercury, most of which originated in the Rhine. Bottom sediments of rivers contain significant levels of toxic compounds, heavy metals, biocides, etc. When these sediments (along with the absorbed biocides) reach lakes or the sea, the marine birds and fish become contaminated. Thus, cormorants fishing in some European lakes die because of heavy contamination with polychlorinated biphenyls. In Dutch lakes, eel (*Anguilla anguilla*) and perch (*Perca fluviatilis*) have been shown to contain much higher average amounts of mercury than the same species from other inland waters. After the closure of a pesticide factory the numbers of these species started increasing.

1.6.2
Mercury

Methylmercury is a highly dangerous pollutant that is concentrated in food chains in the same way as DDT. Mercury compounds (both organic and inorganic) can also be powerful pollutants in terrestrial ecosystems. Like DDT, mercury (mercuric chloride) causes eggshell thinning in birds. Tomatoes are particularly sensitive to mercury pollution, whereas beans and corn are comparatively less sensitive.

In Sweden alone, the total emission of mercury to air, water and soil was estimated at about 100 tons in 1967 (see Bjorklund et al. 1984), but it fell to around 30 tons annually by 1975. The mercury content of pike, *Esox lucius* is markedly elevated in lakes situated in forest areas in some parts of Sweden, with mean values ranging from 0.68–0.86 mg Hg per kg body weight. In these lakes, much of the mercury comes from airborne sources, e.g., emissions from a sulfide ore smelter in Ronnskar (Bjorklund et al. 1984). In acidic lakes, the Hg content in fish is higher than in neutral or alkaline lakes. This problem may be partly tackled by liming the lakes to increase their pH.

1.6.3
Thermal Pollution

Thermal pollution results from the discharge of heated water or effluents into water bodies. Thermal generation of electricity produces heated water as do some other industrial processes. When waste heat enters rivers, lakes and estuaries, serious problems can arise. The amount of dissolved oxygen falls in the aquatic body, thus affecting aquatic animals. Several changes can occur in the composition of plant and animal species in aquatic ecosystems, in relation to different temperatures.

1.6.4
Metals

The various types of metals that can occur in natural waters are shown in Table 1.8.

Table 1.8. Types of metal species in water. (After Wilson 1979; arranged in order of increasing size)

Metal species	Example	State
Free metal ion	$Cu (H_2O)_6^{2+}$	True solution
Simple radicals	VO_2^{2+}, VO_3^-	True solution
Inorganic complexes	$CdCl^+$, $Pb (CO_3)_2^{2-}$	True solution
Organic complexes,	$Cu\text{-}OOC\text{-}CH_3^+$,	True solution
Chelates and compounds	$Hg(CH_3)_2$	
Metals bound to organic matter of high mol. wt.	Polymers of Pb and fulvic acid	Transitional
Colloids	$Fe (OH)_n$	Transitional
Metals absorbed on colloids	Pb on clay	Transitional, particulate
Metals in living or dead organisms	$PbCO_3$, PbS	Transitional

1.6.4.1
Cycling of Iron and Manganese

Certain iron oxidizing bacteria e.g. *Thiobacillus* sp. can utilize ionic forms of iron as an energy source. Iron carbonates and sulfides act as substrates for *Leptothrix* at neutral pH and low oxygen levels. Some species of this genus can oxidize both iron and manganese salts. *Siderocarpa* and *Gallionella* are other examples of iron metabolizing bacteria, and *Pseudomonas manganoxidans* exemplifies a manganese oxidizing microbe.

The concentration of Mn appears to be an important factor controlling algal abundance in aquatic bodies: low concentrations favor cyanobacteria whereas concentrations exceeding about 40 µg l^{-1} are more favorable for diatoms. Some steps involved in the cycling of Fe in aquatic habitats are shown in Fig. 1.14.

1.6.4.2
Chromium Cycling in Natural Waters

Chromium is found in two thermodynamically stable oxidation states in natural water systems, depending on redox conditions, namely Cr(VI) and Cr(III).

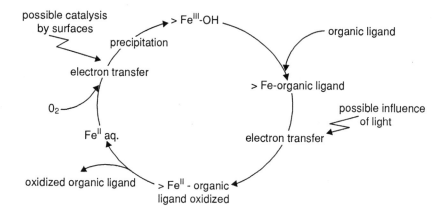

Fig. 1.14. The dissolution of Fe (III) (hydr.) oxides promoted by oxidizable organic ligands and the subsequent reoxidation of Fe (II) by O_2 plays an important role in soils, sediments and waters and causes relatively rapid cycling of electrons and of reactive elements at the oxic-anoxic boundary. This cycle can also be photochemically induced at oxic surfaces. The iron cycle mediates progressive oxidation of organic matter by oxygen. (After Hering and Stumm 1991)

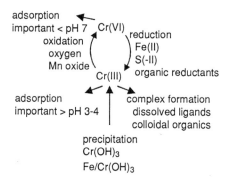

Fig. 1.15. The most important reactions of dissolved chromium in natural water systems

Figure 1.15 shows the important reaction pathways of these Cr species. The oxidized form, Cr(VI) exists as an anion and is relatively unreactive, being removed from the water column by adsorption on particulate material or by reduction to Cr(III). The adsorption reaction is important in acidic media. Chromate is reduced to Cr(III) in minutes to hours under sufficiently reducing conditions by Fe(II), S(-II) or organic reductants. The Cr(III) ion is readily hydrolyzed and binds strongly to particles and organic material. The oxidation of Cr(III) to Cr(IV) may occur by reaction with oxygen and Mn oxides, but the importance of oxygen as an oxidant is somewhat questionable and slow. Manganese oxides are faster oxidants.

1.7
Global Change and Freshwater Ecosystems

Expanding human populations and changes in global climate are expected to exacerbate already severe stress to freshwater resources in some regions (Ausubel 1991; NRC 1992; Walker 1991; Williams 1989). In others, increased availability of water will potentially mitigate stress (Ausubel 1991). The major consequences of global change for freshwater ecosystems can be viewed on temporal and spatial scales (Carpenter et al. 1992).

Over the past 25 000 years, freshwater ecosystems have undergone massive changes in spatial extent correlated with trends in regional climate, and future climate change may produce similar changes in the supply and distribution of freshwater. Greenhouse warming may increase both precipitation and evaporation in the future. Climate change may alter terrestrial vegetation, soils, and soil moisture, which will affect evapotranspiration (Overpeck et al. 1990).

Certain regions can become wetter as a result of global climate change. Increased stream flow, more frequent floods, and expansion of lakes and wetlands are also likely.

The belt of strong biogeochemical activity at the land-water interface may be predicted to expand and contract with fluctuations in the water supply, especially in streams, temporary ponds, and small lakes. Utility of terrestrial leaf litter to stream detritivores varies greatly among species and communities. Change in climate may alter the total biomass, productivity, and species composition of the riparian community and, in turn, the supply of organic matter and nutrients to freshwaters (Firth and Fisher 1991). Climate change may alter the composition of riparian vegetation.

Tolerance of temperature determines the distribution limits of freshwater fishes, which may be classified as cold water (e.g. Salmonidae), cool water (e.g. Percidae) or warm water (e.g. Centrarchidae and Cyprinidae) forms. Thermal limits will be altered by global warming. Distributions of aquatic species will change as some species invade higher-latitude habitats or disappear from the low latitude limits of their distribution (Shuter and Post 1990).

Warming should alter the stream habitat of cold water fishes such as trout and salmon. At low latitudes and altitudes immediate adverse effects may be expected on eggs and larvae. In contrast, at higher latitudes and altitudes, increased groundwater temperatures will increase the duration and extent of optimal temperatures; all life history stages will benefit accordingly (Meisner et al. 1988).

Warming of freshwater habitats at higher latitudes is more likely to open them to invasion. The constraint of size-dependent winter starvation would be abolished, allowing many species to expand their distributions (Carpenter et al. 1992).

Omnivorous warm water fishes having short life spans are expected to invade and flourish in habitats where impoundment, water quality degradation, and fishery exploitation have disturbed the environment. Fish species diversity, biological production rates, and fisheries yields are inversely related to latitude. The majority of freshwater fish species occupy low-to-mid-latitude environments. Most of the world's lakes are of glacial origin and occupy higher latitudes. With global warming, the extent of habitat available to fishes is likely to increase.

The most obvious and immediate effects of global climate change on stream ecosystems involve changes in hydrologic patterns (Carpenter et al. 1992). Complex relationships link climate to runoff. As streams dry, biotic interactions will undoubtedly intensify. Physiological or biotic stresses may differentially affect various species.

If climate becomes more variable, local populations will be more frequently exposed to harsh episodes that produce lethal conditions for short durations. Deletion of adult stocks by transient stress can extend throughout the trophic network (see Carpenter and Kitchell 1993) for long periods. Most vulnerable are the populations of cold water fishes in habitats near the low-latitude limits of their range. In rivers, analogous effects of variable temperatures and flows can affect the fishes and alter the network of interactions regulating both herbivores and the filamentous algae they consume (Power 1990; Power et al. 1985).

Global warming is likely to lead to a general increase in lake productivity at all trophic levels, and increased production will exacerbate hypolimnetic oxygen depletion in productive lakes (Carpenter et al. 1992). The global climate system redistributes energy from lower latitudes to higher latitudes. The energy surplus in the lower latitudes, whose boundaries correspond roughly to 35° N and S, results from the latitudinal gradient of the Earth's energy budget. The total energy distributed by ocean currents is greater than that distributed by the atmosphere.

The mean time during which a water molecule passes through a hydrological system is termed the mean hydrological residence time. The "memory" of a hydrological system increases with a longer residence time (Table 1.9). Global climate history can be understood by analyzing the stable oxygen isotope ratio in ice core samples taken from continental glaciers and ice sheets. The atmosphere has very short memory compared with the longer memory of the ocean. If we could block the energy supply to the atmosphere, its motion would cease within a month, but the ocean would continue its circulation much longer after such an energy cut-off.

Table 1.9. Global water inventory and the hydrological residence time

	Storage (km^3)	Mean residence time[a]
Ocean	135 000 000	3000 yr
Glacier	24 230 000	10 000 yr
Groundwater	10 100 000	1000 yr
Soil water	25 000	1 yr
Lake water	220 000	1 to 1000 yr
River water	1000	10 d
Atmospheric vapor	13 000	10 d

[a] yr = years, d = days

The total sum of heat contained in hydrological systems needs to be considered while assessing the future evolution of the global environment. A short memory means a small heat capacity. Thus the atmosphere contains insufficient heat to act as the source for future dynamic changes in the global climate. It can only respond to changes in solar irradiation or heat supply from the ocean. Future atmospheric behavior depends not only on increases in greenhouse gas concentrations and Earth orbital changes, but also depends greatly on changes in sea surface temperature, global ocean circulation, and the increased atmospheric turbidity (Kayane 1996).

The glacial and interglacial cycle is believed to have started around 800 000 years ago, when unknown feedback mechanisms began to operate. The amplitude of climate variation with time during the past three million years has increased, possibly as a result of increased regional differences in both continental-scale relief caused by crustal movement, and land surface wetness due to land use and cover change.

Vegetation has two roles in the hydrological cycle: the transportation of energy by evapotranspiration, and the storage of water in the soil-root zone. The former process includes a climate feedback mechanism by transporting energy and water to the atmosphere. The latter process affects the hydrological residence time in drainage basins.

Before the agricultural revolution in the present interglacial, the land was covered by "potential vegetation" which was determined only by physical factors and there was no anthropogenic intervention. The land surface then was wetter and the Earth transported more latent heat from the land surface to the atmosphere than it is doing now. Changes in land cover as a result of human activities are particularly large on the humid Indian subcontinent and on the north China plains.

One anticipated consequence of global warming is the change in the freshwater input into coastal oceans. Climate change influences extreme as well as mean climatic conditions, leading to anomalous conditions of wind direction and strength, freshwater

LIBRARY
COLBY-SAWYER COLLEGE
NEW LONDON, NH 03257

input and hence ocean dynamics. Coastal and oceanic current circulation may substantially change due to the abrupt variations in freshwater input from major river systems. The large scale circulation of the bay of Bengal is greatly influenced by the presence of large quantities of freshwater discharge from the Ganges-Brahmaputra-Meghana. The freshwater discharge from these rivers varies seasonally.

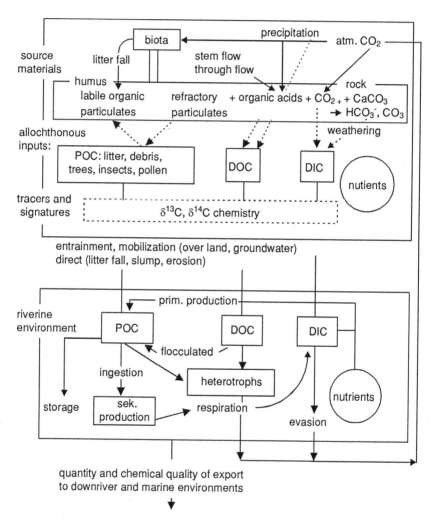

Fig. 1.16. Biogeochemistry of carbon in rivers

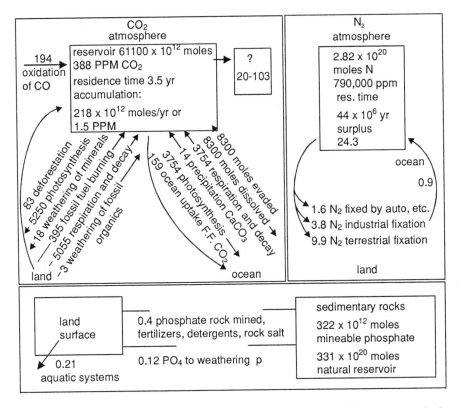

Fig. 1.17. Parts of global cycles of C, N and P as related to increased fluxes of organic C to sediments because of nutrient inputs to the surface environment. All fluxes are in units of 10^{12} moles yr^{-1}. (After Mackenzie 1975)

The greatest change in the surface circulation and salinity occurs during the period of maximum river drainage towards the end of the southwest monsoon. The distribution of low salinity waters at the head of the bay of Bengal varies with the direction of the currents. It may therefore be expected that any variation of the water transport by these rivers will affect the circulation pattern and the distribution of the physical oceanographic parameters in the bay of Bengal. These variations in the surface water may have important dynamic and climatic influences on the heat content of the oceanic surface layer. In this way, heat flow from the ocean to the atmosphere may also change some convective processes in the overlying atmosphere. Further, since the bay of Bengal is a breeding ground for cyclonic disturbances, these anticipatory changes may affect their frequency and strength (Dube et al. 1993).

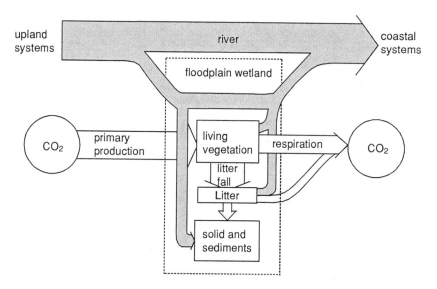

Fig. 1.18. Interactions between a floodplain wetland and its bordering river with regard to carbon flow

Figure 1.16 shows a general model of the biogeochemistry of carbon in rivers. Terrestrial source materials originate as shown, and with their characteristic tracers, enter a river via several pathways, representing land or floodplain inputs. Within the riverine environments, primary production sources also contribute carbon. The main carbon forms in the river are thus particulate organic C (POC: ranging in size from 0.45 µm to trees), dissolved organic C (DOC: < 0.45 µm), and dissolved inorganic C (DIC). The organic C in the river may be respired (in water, in the sediments), ingested, stored, exported, and flocculated. DIC may be lost through erosion. Figure 1.17 shows parts of global cycles of C, N and P as related to increased fluxes of organic C to sediments. Some interactions between a floodplain wetland and its bordering river with regard to organic carbon flow are illustrated in Fig. 1.18.

The west coast of India receives significant amounts of rainfall as a result of the southwest monsoon. Several processes seem to contribute to the variations in the magnitudes and the distribution of the rainfall:

1. The western ghats cause the orographic lifting of the air parcels (Sarkar et al. 1978),
2. Diurnal convection in which the temperature difference between land and sea, or heated mountain slopes produces mesoscale circulations (Prasad 1970),
3. Convective instability in which orografic lifting triggers deep convection,

4. Moisture flux from the Arabian sea,
5. Vertical wind shear can alter the location and the amount of rainfall,
6. Propagation of convective bands offshore (Benson and Rao 1987),
7. Local accelerations due to Somali jet, and
8. Presence of offshore vortices (Mukherjee et al. 1978).

Though there have been several quantitative studies, the effect of sea/land breeze circulations on the precipitation distribution in this region has not been investigated. These mesoscale circulations tend to influence regional climate and rainfall distribution in tropics (Reddy and Raman 1993).

1.8
Hydrological Cycle and Climate

The hydrological cycle involves the largest movement of any substance on Earth. The impact of human activities on climate cannot be assessed without including the role of water in all its phases. According to Chahine (1992), the uncertainties in assessing the effects of global-scale perturbations to the climate system are due primarily to an inadequate understanding of the cycling of water in the oceans, atmosphere and biosphere.

The exchanges of moisture and heat between the atmosphere and the Earth's surface strongly affect the thermodynamics of the climate system. In the forms of vapor, clouds, liquid, snow and ice, water plays opposing roles in heating and cooling the environment. Much surface cooling results from evaporation. Water vapor in the atmosphere acts as a powerful natural greenhouse gas and nearly doubles the effects of greenhouse warming caused by carbon dioxide, methane and other similar gases (Bolin et al. 1986). Clouds control climate by altering the Earth's radiation budget (Chahine 1992).

Figure 1.19 shows the main reservoirs and fluxes of water (see National Research Council 1986; Chahine 1992). The oceans are the dominant reservoir in the global water cycle, holding over 97% of the world's water. In contrast, the atmosphere holds only 0.001%, and the rest is locked up in ice caps, snow and underground storage. The hydrological cycle is indeed global, because continents and oceans exchange water. Over the oceans, evaporation exceeds precipitation and the difference contributes to precipitation over land. Over land, 35% of the rainfall comes from marine evaporation driven by winds, and 65% comes from evaporation from the land. As precipitation exceeds evaporation over land, the excess must return to the oceans as runoff (Chahine 1992).

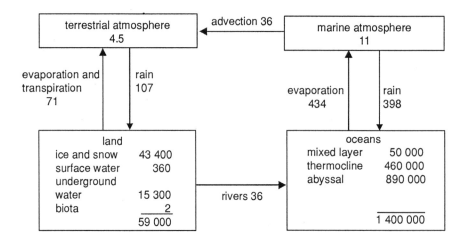

Fig. 1.19. Rough estimates of the global water cycle and its reservoirs. Boxes: reservoirs in 10^{15} kg, arrows: fluxes in 10^{15} kg yr^{-1}. (After National Research Council 1986)

The atmosphere recycles its entire water content 33 times per year (total yearly precipitation divided by atmospheric storage); this means that water vapor has a mean residence time in the atmosphere of about 10 days. In contrast, the mean residence time for the oceans as a whole is over 3000 years, but it is not the same at all ocean depths. In the ocean surface layers, it is only a few days or weeks, increasing to centuries and longer for the deep ocean levels. Over land, the mean residence time of water in vegetation and soil and in aquifers ranges from six years for the former to 10 000 years for the latter.

These two domains of residence time, days-to-weeks and decades-to-centuries, control the Earth's climate system in two distinct ways. The fast regime, consisting of the atmosphere, upper ocean layers and land surface, determines the amplitude and regional patterns of climate change. The slow regime, consisting of the bulk of the ocean, land, glaciers and ice caps, modulates the transient responses of the climate system and introduces considerable delay. The fast component of the hydrological cycle has a critical role in predicting climate change (Chahine 1992).

1.8.1
Atmospheric Circulation

Intensive heating at the equator expands the air which rises, becoming distributed towards the poles and is replaced by cooler air from adjacent areas at ground level.

Trade winds replacing the risen air move toward the equator. The risen air drops again at around 30°, spreading in both directions. During its heating, the tropical (equatorial) air expands and rises due to the lower pressure at higher altitudes. Expansion entails some loss of energy and hence the air cools at a rate called the adiabatic lapse rate. This rate varies around 6–8 °C per km. Cold air holds less moisture and the air which rises above the tropics, especially near mountains, loses water as precipitation. Mostly cold air lacking water is transported to latitudes 20–30°. This air heats up again during its descent, simultaneously absorbing (not releasing) water.

The above facts are also responsible for the observed scarcity of freshwater sites at latitudes between 20–30°. This is also the reason why artificial construction of reservoirs in these areas is ecologically unsuitable and in some cases even unsuccessful or unproductive.

The basic mineral composition of large rivers and lakes depends on the precipitation/evaporation ratio. With decreasing ratio, the content of total dissolved solids increases. Salinity and chloride of water bodies also depend on latitudinal factors and the distance of the water body from the sea. The concentration of suspended solids generally shows consistent drops with increasing annual precipitation.

1.9
Groundwater

About 95% of the Earth's useable freshwater is stored as groundwater. Groundwater is subsurface water which fills gaps in soils and permeable geological formations. The three primary groups of water-bearing formations or aquifers are illustrated in Fig. 1.20. Groundwater aquifers are periodically replenished by precipitation and by surface water percolating down through the soils. The degree of replenishment (recharge) depends on the climate, vegetation and geology of a given region. In humid areas with porous soils, over 25% of the annual rainfall may recharge the groundwater; aquifers in these areas often contain fossil groundwater which accumulated under entirely different climatic conditions.

Water stored in aquifers usually keeps flowing slowly downward by gravity, until it discharges into a spring, stream, lake, wetland or the ocean or is taken up by plants or extracted by wells. Flow rates are typically very slow, ranging from several meters to hundreds of meters per year.

Surface and groundwater systems are inextricably linked. In some cases, groundwater discharges into wetlands (Fig. 1.20), lakes and streams, maintaining water levels and sustaining aquatic ecosystems. In other cases, surface-water systems recharge the underlying aquifer.

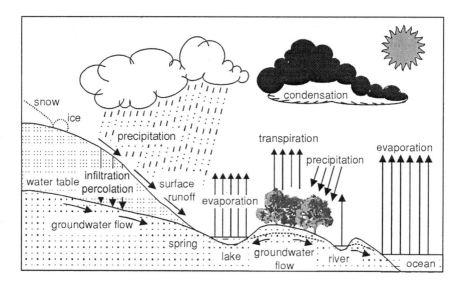

Fig. 1.20. Groundwater and hydrological cycle

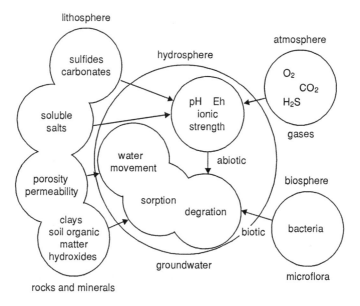

Fig. 1.21. Simplified sketch illustrating the interconnections in the geochemistry of groundwater. (After Hounslow 1985)

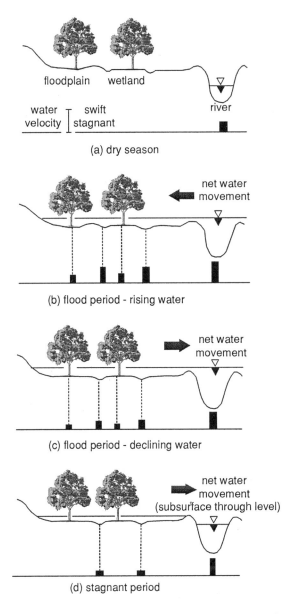

Fig. 1.22. Annual hydrological patterns in floodplain wetland ecosystems. Interactions between the river and floodplain wetland during various periods are shown

The direction of water flow in a given surface-water system often changes seasonally: during the wet season water flows from the surface to the subsurface while during dry periods the flow is reversed. Figure 1.21 indicates the interconnections between various facets of the geochemistry of groundwater and helps illustrate several problems that are still unresolved, at least in quantitative terms and Fig. 1.22 shows the annual hydrological patterns in floodplain wetland ecosystems.

The groundwater table follows the surface contours; it tends to be higher under hills and lower under valleys (Fig. 1.23). Groundwater problems can be divided into those caused by contamination and those caused by overexploitation (Fig. 1.24).

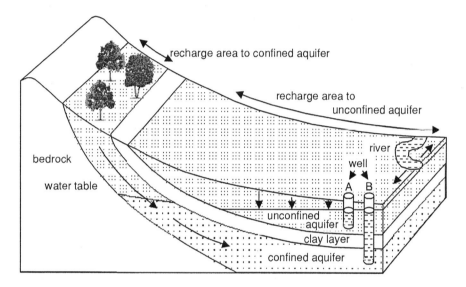

Fig. 1.23. Groundwater flow. The water table tends to be higher under hills and lower under valleys, roughly following contours of the land surface. Well *A* pumps water from an unconfined aquifer, whose recharge area occupies a large area around the well. Well *B* pumps water from a confined aquifer, whose recharge comes from a similar area located many kilometers away

Although groundwater is a renewable resource in most parts of the world, few aquifers can withstand enormous extraction rates indefinitely. Sustainable development requires that groundwater extraction from a given aquifer should not exceed its recharge rate (Fig. 1.25); else aquifers become depleted and the water table or water pressure begins to drop.

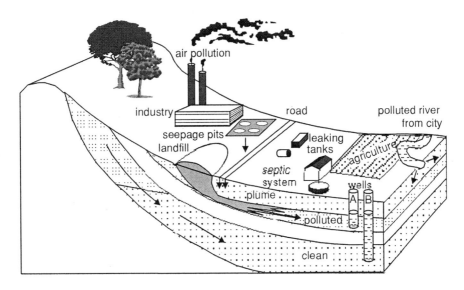

Fig. 1.24. Groundwater contamination. Various sources of contamination are shown. Well *A* is heavily polluted because its source is highly vulnerable to contamination. Water from Well *B* is much cleaner as it comes from a deeper, confined aquifer, whose recharge area is undeveloped

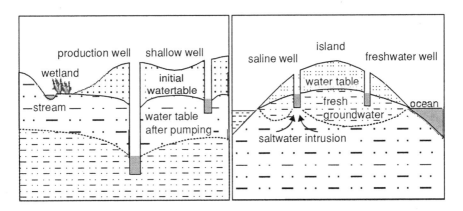

Fig. 1.25. Groundwater overexploitation occurs when the long-term average extraction rate exceeds the long-term average recharge rate

Fig. 1.26. Salt water intrusion into groundwater. In coastal aquifers, fresh groundwater is elegantly balanced on top of denser saline groundwater.

The following problems can then arise:

1. Shallow wells dry up,
2. Production wells have to be drilled to progressively greater depths,
3. Aquifers in coastal areas are contaminated by saltwater intrusion, or
4. Subsurface materials may gradually compact and cause land surface subsidence.

Saltwater contamination can persist for many years. Land subsidence is usually irreversible (see Vrba 1991). In extreme cases, the aquifer has to be abandoned as a source of water (Fig. 1.26). Polluting and over-exploiting groundwater can have the following serious consequences:

1. Contamination or loss of groundwater supplies can cause acute shortages, especially for islands, where desalination is often the only alternative source of freshwater. Where water supplies are inadequate for agricultural and industrial uses, the livelihood of entire sectors of the population can be at risk.
2. Contamination of drinking water supplies raises public health risk through exposure to pathogens, carcinogens and nitrates. Rural populations are particularly hard hit because of their greater dependence on groundwater (Vrba 1991). Where groundwater supplies fail altogether, people may be forced to drink untreated surface water, greatly increasing their exposure to waterborne diseases.
3. Aquatic systems can be devastated by groundwater problems. Nutrient-enriched groundwater discharging to lakes and reservoirs can induce algal blooms and cause eutrophication. Trace metals (see Fig. 1.28) and organic contaminants may enter the food chain, building up to toxic levels. Groundwater overexploitation can cause reduced base flows in rivers, declining water levels in lakes, loss of wetlands, and a reduction in soil moisture.
4. Where land subsidence is severe, buildings and infrastructure can be damaged, and low-lying coastal areas suffer from increased flooding.
5. The costs involved to extract and purify water to meet the most stringent water quality standards are high. Clean-up costs for a relatively minor gasoline spill into an aquifer can be extremely high. The depletion of a major aquifer can also lead to the permanent loss of agricultural and industrial productivity. The most effective and least expensive solution is to establish a program to protect groundwater.

Groundwater pollution is quite difficult to detect, and monitoring is costly, time consuming and a hit-or-miss affair. Contamination is often not detected until noxious substances appear in drinking water supplies, at which point the pollution has usually dispersed over a large area. The clean-up of subsurface pollution is notoriously time consuming and expensive, and can require advanced technological methods. Most groundwater contaminants are derived from agricultural, urban and industrial land uses (Fig. 1.24).

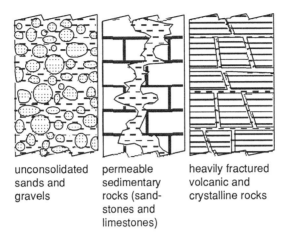

unconsolidated sands and gravels

permeable sedimentary rocks (sandstones and limestones)

heavily fractured volcanic and crystalline rocks

Fig. 1.27. Illustration of the concept of groundwater, and its three primary groups of water-bearing formations, called aquifers

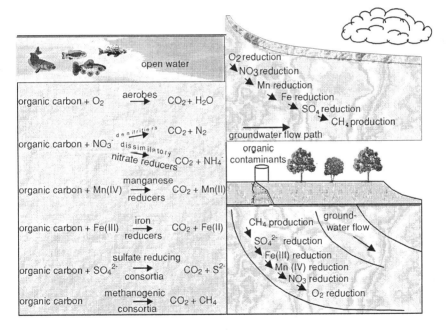

Fig. 1.28. Some interactions of trace metals and organic contaminants in aquatic sediments and groundwater. (After Lovley 1991)

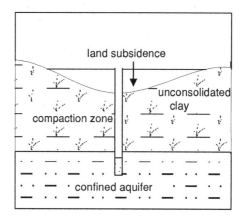

Fig. 1.29. Land subsidence occurs when groundwater is pumped from a confined sand and gravel aquifer overlain by highly compressible clay

In the past, most attention has focused on point sources: large pollution sources such as industrial spills, landfills, and subsurface injection of chemical and hazardous wastes. It is now known that sources of groundwater pollution are much more widespread and are related to diverse, typical activities at the surface. Groundwater pollution in most non-industrial areas can be attributed to such dispersed or non-point sources as fertilizers, pesticides, septic systems, street drainage, and air and surface water pollution (Fig. 1.28). The only effective method for the control of this type of pollution is by integrating land use and water management (see Jackson 1980; Vrba 1991). Figure 1.27 shows the three main groups of aquifers and Fig. 1.29 the causes of land subsidence when groundwater is removed from sand or gravel aquifers overlain by compressible clay.

The redox potential, oxidative (OXC), and reductive capacities (RDC) of groundwater are important parameters because the speciation, solubility and the sorption characteristics of substances are dependent on these quantities. The redox potential provides information on the driving force for redox reactions, while the OXC and RDC provide a measure of the redox capacity of the system, which is important both for measuring redox potentials and for determining the extent to which redox processes can take place in the system (Grenthe et al. 1992).

If we consider the redox intensity which separates oxic and anoxic environments (i.e. $PO_2 < 10^{-6}$ bar) as a reference point, O_2-MnO_2-$Fe(III)$-type minerals make up most of the oxidation capacity, whereas organic material, Fe(II) and S(-I, -II) components constitute most of the reduction capacity.

The OXC and RDC depend on the redox reactions that may take place in the system. The OXC/RDC of the bedrock is determined both by heterogeneous electron transfer processes involving pyrite, Fe(II) and Fe(III) in oxide and silicate minerals, and weathering which results in a release of Fe(II) and Mn(II) from bedrock minerals to the water (Grenthe et al. 1992).

The bedrock and the aqueous phases form two different compartments in the groundwater system, where the OXC/RDC in the latter tends to be more rapidly accessible for redox reactions than the OXC/RDC in the bedrock. The redox processes occurring at the interface between solid and solute depend both on sorption of the redox-active solutes and on the available wetted surface. The results of laboratory studies and field investigations of deep groundwater systems suggest that:

1. Stable and reproducible redox potential values can be obtained in anoxic deep groundwater systems. The measured potentials are consistent with redox reactions involving dissolved Fe(II) and hydrous Fe(III) oxide phases.
2. Redox potential (Eh) measurements are extremely sensitive towards dissolved oxygen and can be safely used as a sensitive indicator for the intrusion of oxygen.
3. There is often an extensive mixing of water of different origins, when using bore hole techniques. Hence, hydrochemical data can be used to obtain independent information on the hydrology of the system studied.
4. Reducing conditions, e.g., anoxic waters containing dissolved Fe(II) are usually encountered despite the fact that the water contains components of surface and drill water. This indicates a rapid reduction of intruding oxic waters on the surface of the Fe(II) minerals present in the bedrock system.
5. Intruding surface water at several sites has a net reducing capacity, due to the presence of dissolved organic matter from topsoil. The concentrations of both dissolved oxygen and organic matter decrease rapidly in the upper levels of the bedrock, and the main chemical effect of intruding surface water seems to be a redistribution of reduction capacity from the bedrock compartment to the more accessible water and fracture mineral compartments. This is of great importance for the rate of the redox reactions of dissolved species, e.g. oxygen or trace metals such as actinides (Grenthe et al. 1992).

The various types of ions and molecules commonly found in subsoil water are shown in Fig. 1.30. Sediments below streams play an important role in lotic ecosystems. Water flows not only across the surface of stream channels, but also through sediment interstices; consequently, surface and subsurface biogeochemical processes are linked.

Organisms have been found living in the saturated sediments below stream water and also in the saturated sediments at the side of streams. Many of these taxa are also found in the surface stream, but some are forms that have adapted to life in sediment interstices.

neutral molecules	cations	anions
O_2, N_2, CO_2	Na^+, K^+	Cl^-, F^-, (Br^-, I^-)
H_4SiO_4, H_3BO_3	Mg^{2+}, Ca^{2+}	HCO_3^-, NO_3^-, SO_4^{2-}
	Cu^{2+}, Mn^{2+}, Zn^{2+}	$H_2PO_4^-$, HPO_4^{2-}, $HMoO_4^-$
	Fe^{2+}, Fe^{3+}	

The transition metal ions are predominantly in complexed form

Fig. 1.30. Ions and molecules commonly found in groundwater

As most of these species are quite different from groundwater fauna, it appears that stream biota extend deep into alluvial sediments. Surface water often exchanges with subsurface water as it flows downstream, and these reciprocal interactions between surface and subsurface environments have been found to influence the chemistry and biology of the surface stream (Bakaloicz 1994).

The principal force in streams is flow. This flow transports organic matter, nutrients, oxygen and organisms. Water flows not only in the open stream channel but also through sediment interstices called the hyporheic zone (Fig. 1.31) and consequently the water flow links biological and chemical processes occurring on the surface of and within sediments. Flow occurs in three dimensions: longitudinally down the channel; vertically between the open channel and underlying sediments; and laterally to or from the riparian zone (Jones and Holmes 1996). Subsurface regions are categorized as either hyporheic or groundwater zones based on hydrologic exchange with surface waters. The division between the hyporheic zone and groundwater is the region where less than 10% of the subsurface water originated from the surface channel (Jones and Holmes 1996).

Being dark, the hyporheic zone prevents photoautotrophic production. Hence, biotic activity within sediments is supported by organic matter coming from either algal or macrophyte production on the surface of the sediment, or organic matter derived from terrestrial vegetation. Hyporheic dissolved O_2 levels result from an interplay between rates of O_2 consumption from detritus decomposition and supply from surface or groundwater zones. Water flows slowly through sediments, and this constrains the import of O_2. Anoxic regions often develop. Anoxia has two important effects on hyporheic processes: organic matter decomposition occurs via anaerobic pathways such as denitrification, sulfate reduction and methanogenesis, and the oxic-anoxic interface is the site of redox/precipitation reactions such as the oxidation of methane or ferrous iron. Anaerobiosis has important affects on nutrient cycling and may result in the production of methane and nitrous oxide (Jones and Holmes 1996).

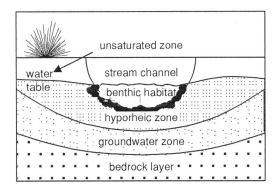

Fig. 1.31. Possible extent of the hyporheic zone in a gravel bed stream

The contribution of the hyporheic zone to stream ecosystem function depends on the rate of subsurface biogeochemical processes and the proportion of stream discharge flowing through hyporheic sediments. The interchange of water between the surface and hyporheic zone, and flow through sediments, is the product of hydraulic gradient and hydraulic conductivity (known as Darcy's Law). Hydraulic gradient is the difference in water pressure between two bodies of water, whereas hydraulic conductivity is a measure of resistance to flow imposed by a porous substrate such as stream sediments (Jones and Holmes 1996). The interchange of water between surface and subsurface environments in natural streams is complex owing to variations in hydraulic gradient and in conductivity; these variations are caused by heterogeneity of sediment porosity, and to the sorting of substrate particles. There are indications that groundwater can serve as a source of organic matter for hyporheic microbial activity (Fiebig and Lock 1991). But in streams low in organic matter, groundwater is not a major source of organic matter supporting hyporheic organisms.

1.9.1
Isotope Hydrology and Water Resources

Whereas global demand for freshwater doubles once every 21 years, pollution threatens this increasingly precious resource. Isotope hydrology techniques can help to identify new water sources and to manage water resources efficiently.

Isotope techniques can be applied in managing groundwater resources and in studying the origin and recharge of groundwater, groundwater dynamics, groundwater pollution, modeling approaches, and geothermal and palaeowater resources. They can also be used for examining surface water and sediments and unsaturated zones.

It is now possible to reasonably predict climate changes induced by human activities and study their impact on the global environment. One approach is the study of past climate changes through isotope investigations of climate archives. Isotope techniques are indispensable as proxy indicators of the climate and as a dating tool for past climatic events. Use of isotope techniques has generated new data on the concentration of noble gases dissolved in the groundwater of regional aquifer systems in North and South America, Europe and Africa. Isotope studies of lacustrine sediments from several lakes located in the Tibetan plateau have revealed large scale changes in the intensity of the Indian monsoon and associated rainfall during the Holocene period. This confirms the vulnerability of the climate in Southeast Asia to relatively small changes in the physical parameters of the atmosphere.

1.9.2
Biodiversity

Underground waters represent 97% of all global freshwaters. Groundwater is found in three main kinds of aquifer: karstic (limestone), fissured (granitic) and porous (alluvium; see Fig. 1.27). These habitats differ in their physical structure, especially in the size of the available space. This major characteristic determines the functioning of each system (such as the permeability of the transition zone with the adjacent surface system) and influences some biological characteristics of the groundwater fauna: the selection pressure differs in these different aquifers.

Groundwaters are used for drinking, agriculture and industry. Groundwaters have generally been regarded as having less habitat diversity than surface waters. A porous aquifer can be considered as a fairly homogeneous interstitial habitat; the seasonal cycles that so strongly influence the surface water environment have a much weaker effect on subterranean ecosystems (Marmonier et al. 1993). However, recent studies have revealed some local and regional heterogeneity of aquifers.

Groundwaters harbor all major invertebrate groups, the assemblages being composed of both narrowly specialized hypogean (subsurface) and generalist epigean (surface) fauna (Marmonier et al. 1993). For most of these invertebrate groups, abundance and biodiversity at local scales are low but they function actively as an "anticlogging" process. The groundwater animals are involved in bank filtration processes (filter effect) such as the transformation of organic matter. Because the groundwater food webs are short and the energy sources are weakly diversified, groundwater fauna and its biodiversity are highly sensitive to environmental changes and are perhaps a very efficient tool for understanding the functioning of the subterranean ecosystem (UNESCO 1986). In groundwater, animals are grouped as:

1. Stygoxens, typical epigean organisms that appear rarely, and at random, in groundwater (e.g. *Gammarus fossarum*),
2. Stygophiles, epigean organisms that occur in both surface water and groundwater, lacking adaptation to subterranean life, and
3. Stygobites, which are the true groundwater dwellers, are absent in surface waters, and are specialized to the subterranean environment (e.g. the amphipod *Salentinella delamarei*; Marmonier et al. 1993).

The occasional hyporheos have species which may live either in surface water or within the sediments, but which have an obligate epigean stage (e.g. most of the aquatic insects and surface benthos). Permanent hyporheos have species which live all their life either in surface water or within the sediments (without the obligate epigean stage; e.g. nematodes, tardigrades, copepods; Marmonier et al. 1993).

The stygobites are groundwater specialists with respect to morphology, physiology, and behavior. Ubiquitous stygobites are able to colonize all kinds of groundwater systems and phreatobites are restricted to porous (phreatic) aquifers (Marmonier et al. 1993).

The colonization of the groundwater environment by organisms has occurred from both marine and freshwater ancestors. Subterranean waters have been considered as a refuge for surface animals to avoid the environmental constraints of surface water environments (such as climatic variations).

Previously considered as "foreign elements" in subterranean waters, surface animals are now being viewed as an active component of the system. In the same way, groundwater biodiversity is now considered as a dynamic characteristic of the system. It can vary over time and not only during climatic crises (Marmonier et al. 1993).

One most striking feature of groundwater organisms is that they generally look alike. Despite the morphological convergence, groundwater organisms are not a homogeneous group. They can show a variety of different strategies to cope with oxygen deficiency, and a weak morphological diversity can mask a high diversity of potential physiological responses.

1.9.3
Pollution

Many people still depend on underground sources of drinking water. However, even groundwater is now threatened with pollution from refuse dumps, septic tanks, manure, transport accidents, and with diverse agricultural, chemical or biological pollutants.

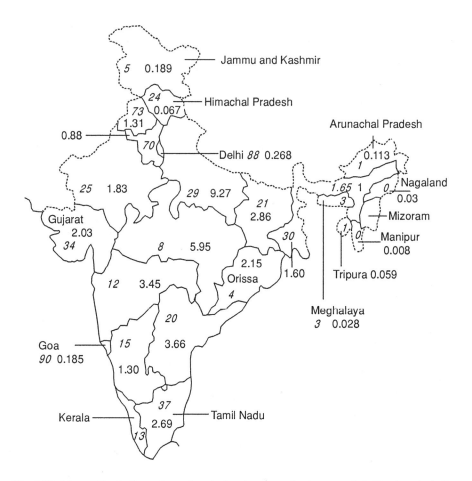

Fig. 1.32. Map of the Indian sub-continent, showing groundwater potential. Numbers in italics show percent usable water potential. Other figures show percent water actually being used in various states

Though the common belief is that most groundwater is free from microbes and safe for drinking, research has demonstrated the presence of viruses, enteroviruses, *Staphylococcus aureus, Klebsiella pneumoniae, Escherichia coli* and others in many well waters and undergroundwaters. Groundwater is also sometimes responsible for outbreaks of waterborne diseases. This happens not only in developing countries but also in advanced countries. In the USA many waterborne outbreaks of disease occurred in community and non-community water systems and in groundwater systems in the period 1970–1978. In these outbreaks, *Giardia lamblia* was the most commonly recorded pathogen. In India, several episodes of cholera, dracontiasis, giardiasis, hepatitis, shigellosis and jaundice have occurred from time to time and in most of these either groundwater or surface water have been the vehicles of disease transmission.

Groundwater pollution control requires accurate information on the distance between the aquifer and the ground surface, on the volume of the groundwater body and on the nature of the underlying stratum. A knowledge of the physico-chemical and biological properties of the pollutants is also necessary for controlling pollution. For groundwater control, observation pits dug at appropriate points are desirable and through these pits, samples of undergroundwater can be pumped from different depths. Fig. 1.32 shows the groundwater potential in India.

It is generally believed that background monitoring systems can be used to assess the impact of human activities on groundwater quality but this belief is a misconception since many years may elapse in some cases before a change in groundwater quality is observed (Fig. 1.33). Once degradation of the subsurface regime has occurred, it is very difficult or even impossible to remedy.

1.10
Population Increase and Water Management

The total world population is expected to rise from 4.5 billion in 1980 to about 6.5 billion in 2000, with the most rapid increases occurring in countries within the humid tropics and other warm humid regions; they will represent about a third of the total world population by the year 2000, and the proportion will continue to rise in the 21st century. All this will require proper, sustainable management of resources including water. Special attention is needed on the socio-economic and cultural factors involved in the control of water resources in rural areas of both flatlands and steep topography, and also their effects on erosion/water quality management.

Forests are a crucial factor in the preservation of water quality, and in climate control, but the growing population in the tropical areas is forcing continued destruction of these forests for agricultural expansion, the growing world timber trade and in domestic wood for fuel. Deforestation has already attained a level of 11 million ha per year. In Central America tropical forests have been reduced by 38% and in Africa by 23% during the last 3 decades. Besides population pressure, the major underlying causes of tropical deforestation are rural poverty, low agricultural productivity, inequalities in land tenure, ineffectiveness of forestry agencies and lack of integrated planning of forestry, agriculture and energy (UNESCO 1991).

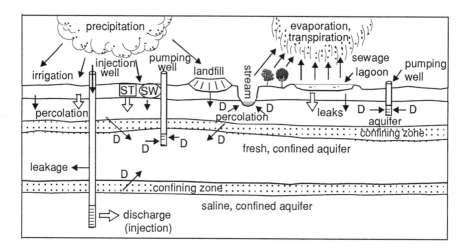

Fig. 1.33. Sketch to explain how waste disposal practices can contaminate the groundwater system. *ST* septic tank, *SW* sewer, *open arrows* intentional inputs, *dark arrows* unintentional input, *dark arrow with D* direction of groundwater movement

The impacts of large-scale deforestation for the timber trade and especially when that land is converted for the implementation of extensive cash crop estates, livestock farms or agriculture, need to be examined in terms of radiation and water balance changes and the role of atmospheric recycling of evaporated water vapor in assessing the effects of deforestation, and erosion/sedimentation processes and water quality changes resulting from forest burning and deforestation for conversion to agriculture.

Two centuries ago, only 1% of the world population lived in cities. By 1970, 37% of the entire world's population was urban (Table 1.10). By 2000, the urban population is expected to reach 51% of the total. In more developed countries, the urban population may grow from 717 million in 1970 to 1174 million by the year 2000, reflecting an increase of 457 million or 54%. In contrast, the corresponding urban population in the less developed areas is likely to increase from 635 million in 1970 to 2155 million in 2000. In 1970, about 37% of the global population lived in rural areas, most of them in the developing world. In 1980 only about 75% of urban communities in the developing world had access to a reasonable water supply and 50% to sanitary facilities. In rural communities only about 33% had access to good water and 13% to sanitary facilities, causing major health problems (Table 1.11). After ten years, the total number of people without these facilities is still staggering, and in spite of all the development and progress, most urban areas in the developing countries have no real treatment of their domestic and industrial wastes (UNESCO 1991).

Table 1.10. Urban and rural populations, 1970 to 2000. (UNESCO 1991, in millions)

Year	1970	1980	1990	2000
Urban Population				
World total	1352	1854	2517	3329
More developed regions	717	864	1021	1174
Less developed regions	635	990	1496	2155
Rural population				
World total	2284	2614	2939	3186
More developed regions	374	347	316	280
Less developed regions	1910	2267	2623	2906
Percentage of urban population				
World total	37.2	41.5	56.1	51.1
More developed regions	65.7	71.4	76.4	80.2
Less developed regions	25.0	30.4	36.3	42.6

Table 1.11. Health impacts of diseases related to water and sanitation. Figures for Africa, Asia and Latin America (1977–1978)

	Infection	Infections (thousands per year)	Deaths (thousands per year)
Waterborne	Amoebiasis	400 000	30
	Diarrheas	3–5 000 000	5–10 000
	Polio	80 000	10–20
	Typhoid	1000	25
	Ascariasis (roundworm)	800 000–1 000 000	20
Water-washed	Leprosy	1000	very low
	Trichuriasis (whipworm)	500 000	Low
Water-based	Schistosomiasis (bilharzia)	200 000	500–1000
	African trypanosomiasis (sleeping sickness)	1000	5
With water-related vectors	Malaria	800 000	1200
	Onchocerciasis (river blindness)	30 000	20–50
Fecal disposal	Hookworm	7–9 000 000	50–60

1. Waterborne diseases spread by drinking water or washing hands, food or utensils in contaminated water, which acts as a passive vehicle for the infecting agent.

2. Water-washed diseases spread by poor personal hygiene and insufficient water for washing. Lack of proper facilities for human waste disposal is another contributing factor.
3. Water-based diseases are transmitted by a vector which spends a part of its life cycle in water. Contact with water thus infected conveys the disease-causing parasite through the skin or mouth.
4. Diseases with water-related vectors are contracted through infection-carrying insects which breed in water and bite near it, especially when it is stagnant.
5. Fecal disposal related diseases are caused by organisms that breed in excreta when sanitation is defective.

Water will undoubtedly be one of the major issues confronting humanity at the turn of the century and beyond. We are already facing a crisis as regards the quantity (Tables 1.12, 1.13) and quality of water supply, though the social and political impact of that crisis is not yet being fully felt. The rising global population and the need for continued development lead to conflicting pressures on water resources which are the ultimate receptacle of pollution from various socio-economic activities associated with urbanization, agriculture, mining and clearing of native vegetation. Pollution originating from human waste, especially where adequate sanitation facilities are not available or are located too close to water supply sources in developing countries, affects both surface water and groundwater.

This makes water supply and health perhaps the most important issue for most of the global population. Paradoxically, the demands for "sustainable management" and increasing global population require more potable water from a declining available potable water base. Because of their limited size, small islands are particularly susceptible to these kinds of conflicting pressures on water resources. The following general principles for water management are noteworthy:

1. Freshwater is a finite and vulnerable resource, essential to sustain life, development and the environment,
2. Water development and management should be based on a participatory approach, involving users, planners and policy-makers at all levels,
3. Women play a central part in the provision, management and safeguarding of water, especially in developing countries, and
4. Water has an economic value in all its competing uses and should be recognized as an economic good.

Water-related challenges are facing several countries regarding availability, human health and safety, and environmental well-being. Since around 1975 in the USA, sewage treatment and drinking water supplies have improved, and grossly polluted systems such as the Cuyahoga River and Lake Erie are now much less polluted.

Table 1.12. Average annual availability of freshwater (surface and groundwater) in selected countries. (International Institute for Environment and Development and World Resources Institute)

	Total availability (km^3)	Per capita availability (x 10^3 km^3 per person)
Water-rich countries		
New Zealand	397	177.53
Canada	2901	111.74
Norway	405	97.40
Brazil	5190	36.69
Ecuador	314	31.64
Australia	343	21.30
Indonesia	2530	14.67
United States of America	2478	10.23
Water-poor countries		
Egypt	1.00	0.02
Saudi Arabia	2.20	0.18
Singapore	0.60	0.23
Kenya	14.80	0.66
Netherlands	10.00	0.68
South Africa	50.00	1.47
India	1850.00	2.35
China	2800.00	2.58

The chief focus has tended to be on point-source pollution; this has deflected emphasis away from other, no less harmful, forms of environmental degradation such as altered hydrological regimes, habitat destruction, invasion by exotic species, and spread sources of pollution (see Naiman et al. 1995 a,b; NRC 1992).

Some crucial issues meriting prompt consideration are how to improve protection and restoration of water resources and aquatic species and also how to integrate mankind's needs with protection and rehabilitation. Development of a workable plan for freshwaters depends on a proper understanding of freshwater ecosystems and on more effective and comprehensive laws and policies.

Table 1.13. Withdrawal of freshwater by some countries (percentage). (The World Resources Institute and Institute for Environment and Development)

	Domestic/ commercial	Industrial	Agricultural
Developed countries			
Austria	20	77	3
Belgium	11	88	2
Czechoslovakia	24	72	5
Finland	12	86	1
France	17	71	12
Netherlands	5	64	32
Poland	17	62	21
Switzerland	37	57	6
United Kingdom	21	79	1
Yugoslavia	17	75	8
Developing countries			
Algeria	23	5	72
China	6	7	87
Egypt	7	5	88
Ghana	44	3	54
India	4	3	93
Turkey	24	19	58
Uganda	43	0	57

Five high-priority areas for work on freshwater ecosystems have recently been identified on the basis of scientific significance, socio-political relevance, and the needs of decision makers (see Naiman et al. 1995b). These are summarized below:

1. **Ecological restoration and rehabilitation**. For effectively restoring and rehabilitating ecosystems degraded by human actions, it seems necessary to first find out how natural systems operate,
2. **Maintaining biodiversity**. Here the objective includes both individual species as well as the diversity of ecological processes and the integrity of ecological systems. Understanding relations between species and ecological processes as well as the consequences of exotic invasions is crucial,
3. **Modified hydrological flow patterns**. The hydrological regime in most bodies of freshwater has been modified by dams, diversions, and withdrawals. Hydrological changes have modified conditions for riparian and aquatic organisms: habitats for organisms adapted to natural discharge and water level patterns are reduced, rivers

become unable to act as migratory and material transport corridors, and riparian zones fail to serve as filters between upland and aquatic systems (see Glieck 1993),

4. **Ecosystem goods and services**. Modifications have severely changed the resources provided by freshwater ecosystems: water quantity and quality, biological productivity and other living functions, and aesthetics and recreation (see Glieck 1993; Likens 1992), and

5. **Predictive management**. Three types of uncertainty perpetuate resource management failures. Data are needed on disturbance regimes and their physical and biological legacies for being able to predict the consequences of cumulative and synergistic impacts (Naiman et al. 1995).

1.11
Water and the Environment

Judicious water administration is a critical element of sustainable development in many countries. Indeed, water is an essential component in many productive activities, of which one of the most important is the production of food by irrigation. This activity globally accounts for two thirds of the water resources used by humanity. Supply of drinking water and sanitation in urban centers is crucial for preserving human health in developed countries and for improving it in developing countries. Several diseases in developing countries come from lack of clean drinking water and hygiene.

The misuse of water resources is responsible for many important environmental problems. In many industrialized countries both surface water and groundwater are seriously contaminated as a result of a range of human activities, sometimes in isolation, sometimes over a large area or a long time. One example of the latter is modern agriculture, whether it uses irrigation or not, as a result of the intensive use made of mineral fertilizers and pesticides. In some cases the transfer of water from one basin to another has given rise to considerable ecological or social problems.

Some of the above problems have created the impression that water shortage will be one of humanity's big problems in the coming decades. Sometimes this feeling is due to genuinely manipulative publicity campaigns to justify hydraulic megaprojects. But except for a few specific cases, no serious problems of water shortage are to be found almost anywhere. On the other hand, cases of bad water management are very widespread.

Over the last few years, a standard figure of 1000 m^3 per person per day has frequently been given as an indication of the amount of water any country needs to reach normal development, but indiscriminate use of this figure makes little sense: Israel has a supply of less than 500 m^3 per person per day. In Europe, average water use as a proportion of gross water resources is in the order of 16%. Israel uses 110% of its resources each year; this is explained by the fact that part of the water used for urban needs is recycled for agricultural use.

In some dry countries, for example, in the Middle East or in the Persian Gulf, water use has been or could be the cause of conflicts. However, water problems in these countries can be suitably and fairly resolved were it not for the existence of political problems.

Water shortage is rarely a serious problem, but the contamination of surface and groundwater tends to be a problem which rarely receives adequate treatment. Successful water management should be based on three basic principles: solidarity, subsidiarity and participation. The specific way in which these principles are applied will vary from one country to another, but the effectiveness of water management will depend to a large degree on the hydrological education of the general public.

Rain is the universal source of freshwater river systems which contain the excess rain water. On the one hand, rainwater penetrates permeable soils, saturates them and accumulates to form groundwater reservoirs, or aquifers, which can come to the surface in the form of springs. On the other hand, water is absorbed by vegetation, which uses it for pumping minerals and then evaporates it by transpiration. Some rainwater is lost because it evaporates immediately on falling on impermeable surfaces like the asphalt of roads and cities. Running water courses finally flow over saturated soils, shaping the complex systems of the watersheds or river basins.

All water supplies are of variable volume. Both the discharge of rivers and the level of lakes and aquifers depend on rainfall. As all these resources are components of a larger river basin, a reasonable policy would be to manage water resources according to the characteristics of each basin. This requires a proper understanding of the system so as to match use and consumption to the existing supply. Conserving river systems as much as possible in their natural state is the best guarantee for the preservation of the landscape and of a constant supply. The following points should be borne in mind:

1. Hydraulic works and infrastructures, if not planned with great care, can cause alteration in the ecological systems of rivers and wetlands.
2. The canalization of river beds leads to a loss of flood surfaces and therefore a drop in the level of aquifers.
3. Damming modifies sediment loads to differing degrees and causes the gradual loss of river deltas which are extraordinarily rich ecosystems.
4. Excessive freshwater extraction from aquifers near the coast can cause salinity problems.
5. The intensive use of water does not give it sufficient time to purify itself by natural processes and necessitates the construction of industrial purifying plants.

Groundwater reservoirs are more like lakes, so that pollution leads to the build-up of a debt which continues to be paid in years to come. Water consumption has increased globally in recent years as a result of population growth and increase in living standards. The introduction of new farming methods, the spread of irrigation and the excessive use of fertilizers and pesticides leads to high consumption. More than two thirds of the world water consumption is used in irrigation. Agricultural pollution also endangers both surface water and aquifers, which receive water full of chemical products. Many cases of eutrophication, the enrichment of water by nutrients that accelerate the growth of algae, emanate from the runoff of fertilizers. The practice of intensive stock-raising on farms with large numbers of animals also exacerbates the problems of overconsumption and pollution. Cleaning the stockyards requires large amounts of water which is then released into the environment with high concentrations of nitrogen.

Industries have taken little care of water consumption and dumping, and in many countries the need for proper attention comes as something new. One rational approach would be to make industry take its water at a point down-river from where it returns it or, better still, generalize the use of closed circuit systems based on the constant recycling and reusing of the same water.

As regards human consumption, the general attitude to cleanliness is based on diluting the waste products. Those countries whose annual rainfall is below the world average (the average being just below 1000 mm or $1000 \, \mathrm{l \, m^{-2}}$ per year) are in a critical position. While it is possible for them to survive with a supplementary energy expenditure either for purification or desalination, survival will probably always call for water-saving measures.

Purification techniques should be based especially on the natural processes that include biological activity. Use of physico-chemical methods entails side-effects such as an excess of mud or sediments. The strategy to follow is to optimize operations in our use of water according to the discharge and to the distribution of contamination. A system in the form of a conduit or channel, such as a river, can respond relatively quickly. On the other hand, lakes and dams can only do so up to a point, because they show more inertia and irreversibility and take much longer to clean.

1.12
Freshwater Augmentation Technologies

Insufficient protection of the quality and the supply of freshwater limits sustainable development. Many health hazards in developing countries are related to poor water quality and limited water availability. Many of these countries use more than their annual freshwater renewal rate by satisfying demands from non-renewable sources. These water shortages are likely to worsen from causes such as continued rural-urban migration, compounding population growth, pollution of surface water sources, and rising living standards warranting increased demand. Governments respond to extra demand by increasing water supplies for urban dwellers, but this practice is becoming untenable as sources of good quality water become more and more distant, making them more expensive to explore and develop. Overpumping of groundwater cannot make up the difference either, as overpumping leads to salt water intrusion and a lowering of water tables. Planners should therefore consider making better use of both conventional and non-conventional technologies for augmenting and maximizing the use of freshwater resources.

Perhaps the most important development issue we will face in the coming years will be that relating to freshwater, and it is high time that concerted efforts are made now to identify and better understand suitable technologies for augmenting freshwater resources.

Some recent surveys have made an inventory of local conventional and non-conventional technologies for maximizing and augmenting freshwater resources, and wherever possible, these are being applied in the field. These surveys have revealed that:

1. Both conventional and non-conventional technologies exist to address freshwater problems, and
2. Several of the non-conventional or alternative technologies have applications not only in a specific region, but also in other regions and could be shared through international and regional technical cooperation projects. One basic method to increase water supply is through rainwater catchment.

One of the strongest challenges in ecology is predicting responses of ecosystems to perturbation. Wide-ranging predators or large-scale geochemical processes just cannot be included in small experiments. Microbial metabolism or plankton populations quickly become unrealistically unnatural when isolated in containers. Some of the resulting differences can be overcome by direct experimental manipulations of entire ecosystems. These experiments involve large areas, such as catchments or the natural ranges of mobile predators, for quite long periods of time.

Ecosystem experiments have already been conducted is several countries on lake eutrophication, lake biomanipulation, ecosystem acidification, and effects of forest management on catchment hydrology and biogeochemistry.

Catchments constitute basic units of the landscape. As they have distinct boundaries, fluxes of water and chemicals can be measured into and out of the ecosystems. Gases, particles, water, and dissolved chemicals are some of the atmospheric inputs. When water passes through a terrestrial ecosystem, its chemical composition changes because of biological and geochemical processes. Runoff from catchments integrates the net effect of terrestrial processes. Changes in the runoff result from changes in atmospheric inputs or in the terrestrial ecosystem (Carpenter et al. 1995).

The paired catchment concept is central to whole-catchment research. Typically, runoff is compared from two ecosystems. After a pretreatment study, one of the two catchments is manipulated whereas the other serves as untreated control (Carpenter et al. 1995). Using paired catchments (or paired forest stands), the effects of acid deposition on surface water have been studied. In Norway, exclusion of acid rain from a catchment by means of a roof showed that the effects of acid inputs were reversible. A parallel experiment with increased inputs of acidic materials to a pristine catchment demonstrated that surface waters could be acidified after only a few years of acid deposition. Effects of elevated CO_2 concentration and temperature on catchments are also being studied now.

In freshwater ecosystems, the responses of both communities and biogeochemical processes to a wide variety of environmental stresses can be studied. Lakes have been manipulated to study the effects of nutrient enrichment, acidification, alkalization, methane addition, reservoir formation and linkages between acidification, climatic warming, and exposure to ultraviolet radiation (see Carpenter et al. 1995). Data from whole-lake experiments have refuted the view that eutrophication could be prevented by low carbon concentrations, as carbon can invade lakes from the atmosphere at rates high enough to support algal blooms. This gas flux is prevented in bottle-scale experiments and has proved crucial in ecosystem experiments. Whole-lake experiments showed that N also could be drawn from the atmosphere, after establishment of nitrogen-fixing cyanobacterial blooms. These experiments have clearly demonstrated that eutrophication can be controlled by managing P alone (see Likens 1992).

Whole-lake acidifications have revealed that impacts are transmitted through food webs to affect populations of fishes and aquatic birds. Ecosystem manipulation can change foraging and predator-avoidance behaviors of fishes within days, causing certain indirect effects that cascade through food webs to alter productivity and nutrient cycling (Carpenter et al. 1995).

An open ocean ecosystem experiment has been conducted to answer a puzzling question: why do vast regions of the oceans have high concentrations of N and P and yet support low phytoplankton biomass? One hypothesis is that phytoplankton of these regions is limited by iron. Bottle experiments are not suitable for testing this hypothesis because corrections for zooplankton grazing cannot be made properly and also there is a high probability of contaminating the bottles with traces of iron. With a view to overcoming these limitations, the first open ocean fertilization experiment, called IRONEX, has been carried out in the equatorial Pacific (see Carpenter et al. 1995). Dissolved iron was added to an unenclosed 64 km^2 patch of water so as to increase the ambient concentration 100-fold. It was found that photosynthetic efficiency increased within the first 24 h, and by the third day, chlorophyll concentration as well as primary productivity had increased three times. Increased photosynthesis caused a small though significant decrease in CO_2 in the surface waters but was not sufficient enough to produce detectable declines in nutrients. This short-term experiment did not indicate whether the modest productivity increase and negligible impact on N and P levels resulted from the disappearance of available Fe, increased zooplankton grazing, or the short duration of the experiment. In the next experiment, a patch is intended to be repeatedly fertilized with Fe and sampled for a longer time period. Also, zooplankton abundance and grazing will be monitored, thus determining whether herbivores mitigate the response of phytoplankton to Fe (see Carpenter et al. 1995).

Some have suggested that fertilization of the entire Southern ocean with Fe could sequester substantial quantities of atmospheric CO_2, delaying predicted greenhouse warming of the Earth to a modest degree. Debates about the feasibility, wisdom, and ethics of such climate engineering are greatly affected by our limited knowledge of the response of oceans to nutrient enrichment. Experiments like IRONEX can contribute towards this effort.

An important issue in aquatic ecology research is that of scale. Smaller scales allow greater replicability and recent trends have been towards smaller and more homogeneous study plots. Conversely, studies of large-scale processes have to contend with the twin problems of difficult replication in space and of having characteristic time scales of change that are too long.

The ability to scale-up local estimates of biogeochemical fluxes to the regional scale is essential in the context of climate change. The need to make observations and formulate hypotheses on a range of spatial and temporal scales and to be able to see how events at one scale affect those at another, has been felt in recent years (see Giller et al. 1994).

There is a strong relationship between pattern and scale. Sea birds and acoustically detected fish can be recorded simultaneously along a transect and the data can be presented graphically at several different scales of aggregation from 0.25 to 3.0 km.

There are many examples of the way in which physical parameters impose patterns on organismal distribution and on ecosystem processes. In rivers and streams, the gradient has a broad overall effect on community organization. Extreme current speeds, e.g., flash floods, create disturbances which vary in intensity between highland streams and lowland meanders and between rivers according to the landscape. The disturbances create physical patchiness; the adaptation of the organisms to patchy conditions leads to very subtle and complex situations on the small scale.

In the open water and benthic boundaries of lakes and the ocean, physical heterogeneity can be seen on many scales. Vertically, the downwelling of light contrasts with the upwelling of nutrients, producing regularities in vertical profiles of phytoplankton and zooplankton. Wind stress at the surface and fluctuations in surface temperature create vertical turbulence. Horizontally, the range of scales is much greater, starting with the major ocean-basin gyres, which produce turbulent eddies on scales of hundreds of kilometers. Their energy is transferred to smaller eddies on an ever-decreasing scale until the energy of centimeter scale turbulence is spent in the viscosity of the water. Similar processes operate in lakes, albeit at a more limited scale. These physical aspects impose patterns on the distribution of organisms when the characteristic time and space scales of the organisms correspond.

Processes on large spatial and temporal scales have some hierarchical control of those on smaller scales. At the biological level, several cycles of fluctuation in phytoplankton production can be integrated within the lifetime of a member of the meso- or macrozooplankton, and several cycles of zooplankton production can be integrated into the lifetime production of a fish. Moreover, the larger organisms usually travel longer distances and are capable of integrating the production of several geographical areas within their range of migration. Similarly, on the physical level, an interannual climatic fluctuation such as the El Niño Southern Oscillation (ENSO) which operates on the ocean-basin scale, alters seasonal and annual temperature fluctuations over a large part of the Pacific Ocean.

1.13
Water and Development

It is not easy to show how much water input can influence the development of society. Social changes are not easily characterized in quantitative terms alone. Thus, one would wish to be able to measure the outcome of a water project in monetary units but the qualitative effects must not be overlooked.

Many attempts have been made to state the influence of water in qualitative terms. A beginning is usually made describing the initial fundamental stage where a man will need one or 2 l per day for his survival, depending on the climatic conditions. Thereafter the following three stages follow, all being affected by water use:

1. A supply-oriented society will prevail as long as the water supply is abundant. Small-scale activities will occur in the drainage basins. Water is looked upon as free.
2. Water demand will increase in the next stage. Watercourses will find multiple uses and will be regulated with reservoirs. Transfers from one river basin to another may occur. Artificial infiltration of surface water may even be undertaken in order to increase groundwater storage (Lindh 1985).
3. The water-courses are exploited in such a way that there is a rapid increase in the marginal costs in the development of the water resources and in water supply. The society becomes water demanding in nature and cost becomes an important consid-

eration. The concept of basin-wide management of the river occurs in its most developed form.

The last stage has already been reached in many parts of the world. At this stage, basically, nothing more remains than to direct one's interest toward developing unconventional means of producing water and energy resources.

Water may be used for different purposes. The first thing to note is that more than 80% of the world's available water resources is used to irrigate food crops, and that occurs too inefficiently. Large quantities are lost by evaporation, leakage and excessive irrigation.

One aspect where the positive influence of clean water is a factor in societal development is in the struggle against water-related diseases now being carried on intensively in the developing countries.

Mankind has never been too far from a good source of water, except for those rare tribes which could manage to live in some truly forbidding deserts. Some of the earliest concentrations of mankind were located in the delta of the Nile river and in ancient Mesopotamia where the Tigris and Euphrates rivers meander their way toward the Persian Gulf. Other major gatherings of humans took place along the Indus river and the Huang Ho in the Far East.

This assembling in the river valleys and the subsequent development of irrigation methods by the riverine dwellers was not, of course, the only factor in the centuries-long transition from the nomadic mode to fixed settlements. The use of fire and the ability to work metals and to devise tools that extended human muscle power were also very important (Lindh 1985).

The terms "hydraulic" and "fluvial" have been used for these early cultures which came into being along the Nile, the Indus, and the Huang Ho. In fact, the last location may have been one of the first river-based cultures, for it is known that river irrigation was practiced in China earlier than 5000 BC.

The poorest countries are mostly situated between the two Tropics (Capricorn and Cancer). They also have the lowest level of water management and service coverage (Table 1.14), and their populations are the most susceptible to water-related diseases. But there are several tropical countries, all rather small, with good service coverage. These are Brunei, Costa Rica, Hong Kong, Singapore, Trinidad and Tobago (Table 1.15).

Almost everywhere rivers and lakes are being contaminated by chemicals (from industries, municipal sewage discharge and agriculture). Surface water is becoming unfit for human consumption unless it is treated. Groundwater is often under-utilized in the humid tropics because rainwater and surface water are normally abundant, but its use is growing and its protection will become increasingly important in the future.

Table 1.14. Urban population served with water (percent) (Prost 1989)

Countries	Served	Not served
High income	100	0
Upper middle income	90	10
Lower middle income	77	23
China and India	69	31
Low income	65	35
World total	86	14

Table 1.15. Access to water supply services (Prost 1989). The table only includes countries and territories with a population of over 0.1 million. South Africa, Namibia, Comoro, Equatorial Guinea, Cambodia, Macao and Taiwan are also excluded as no data are available (total population excluded is about 55 million)

Per capita income categories	Total population (millions)			Population served (millions)			Percentage of total population		
	Urban	Rural	Total	Urban	Rural	Total	Urban	Rural	Total
High income (over $ 5000) 37 countries	807	283	1090	806	280	1086	100	99	100
Upper middle income ($ 1800 to 5000) 24 countries	347	196	543	312	125	437	90	64	80
Lower middle income ($ 500 to 1800) 46 countries	282	444	725	218	200	418	77	45	58
China and India ($ 300)	427	1408	1835	295	858	1153	69	61	63
Low income (less than $ 500) 42 countries	144	518	662	93	168	261	65	32	39
Total 151 countries	2007	2849	4855	1724	1631	3355	86	57	69

Table 1.16. Status of sewage treatment facilities and their satisfactory operation in developing countries (WHO 1991)

Developing countries in	Percentage of cities with sewage treatment facilities	Percentage of facilities operating satisfactorily
Sub-Saharan Africa	2	30
Asia Pacific	5	50
Arab, moslem countries	10	40
Latin America	25	45

In the tropics, there is intense solar energy which exerts a more seasonally uniform impact on surface waters. Not only is the quantity of available water per capita now shrinking, but the quality of freshwater is also constantly deteriorating. In tropical areas, the health hazards caused by polluted water supplies are more numerous and more serious than those in the temperate and developed areas. The WHO and UNEP have estimated that, in 1990, around 90% of sewage collected in industrialized countries was adequately treated as opposed to only 2% in developing countries. Only a few large urban centers are being partly served by wastewater treatment schemes in Africa and Asia. However, as is apparent from Table 1.16, having treatment facilities and operating them satisfactorily are two different things.

The lack of adequate sewage treatment in tropical countries leads the people to use the same water for the irrigation of vegetable gardens and for drinking, cooking and washing. This constitutes an ideal breeding ground for diarrheal diseases (more particularly, cholera) and malaria.

Water influences and is influenced by socio-economic development in several ways. It makes essential contributions to all human activities but can also create obstacles to development. Water project construction can facilitate development by enhancing these positive contributions and mitigating negative effects; but project construction can also produce adverse effects.

The term "socio-economic development" refers to advancement of a given society to a higher level of welfare or well-being. While the "economic" aspect of the term refers to goods and services related to material welfare, the "socio" part includes the full range of socio-cultural aspects of welfare.

The objectives of development include, at a minimum, a lessening of problems associated with inadequate nutrition and unhealthy living conditions. Development objectives also include such other components as the widespread diffusion of certain goods and services above essential needs and the fulfillment of nonmaterial needs to allow individuals the opportunity for full lives and fulfillment of human potentials.

Socio-economic development implies advancement of society on a broad front and includes both material and nonmaterial elements, and its scope is broader than that of economic growth. Economic growth is seen as the main means for achieving broad development objectives. Indeed, social progress is usually regarded as an inevitable outcome of an expanding economy, and economic growth itself sometimes is viewed as the chief social objective.

It is now being realized that economic growth alone may not reduce poverty among low income groups. An approach to development that concentrates directly on the needs of the poor has arisen. This strategy aims to meeting the minimum needs of all members of a population in the areas of nutrition, water supply, sanitation, health, housing, and education. Unlike the traditional approach, the "basic needs" approach emphasizes direct satisfaction of these needs rather than stimulation of economic growth as an intermediate means to satisfy such needs (Cox 1989).

Natural resource inputs to productive activities are essential to economic growth and development. These inputs include agricultural land, industrial raw material, power and water supplies, and a suitable climate and terrain for the activities involved. The existence of abundant high-value resources can be the main factor in the development of a particular nation or region as exemplified by the case of the petroleum-producing countries. However, in spite of the potential role of natural resources in growth and development, the actual contribution of natural resources depends, inter alia, upon the availability of capital and human resource capabilities for resource utilization (Cox 1989).

Water is a unique resource and some water is required for essentially all development activities. But many applications of water are not based solely on absolute needs. Above some minimum requirement for domestic needs of workers and certain other essential uses, the volume of water used in many productive activities is very often a discretionary matter. Modification in processes can significantly reduce the amount of water traditionally used although the feasibility of such changes may be limited by associated costs. The role of water in socio-economic development cannot be isolated and defined independently of other development factors. Its specific roles in development will vary among individual situations e.g. initial resource endowments and other natural conditions of the country involved. The existing level of development will also influence the role of water. Expansion of agricultural and industrial water supply tends to receive greater emphasis in a developing nation while preservation of natural water environments receives greater attention in a nation having already achieved a relatively high level of material welfare or affluence. Limitations of resources dictate a relatively low level of water supply service in many developing countries. A basic application of water to productive activities is agricultural water use for irrigation. One major use of water bodies is aquaculture. Figure 1.34 compares aquaculture production in developed and developing countries.

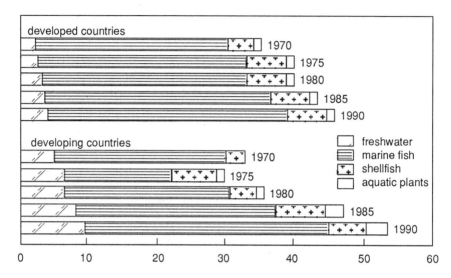

Fig. 1.34. Comparison of aquaculture production in developed and developing countries (in million tonnes, values are approximate)

Use of water to improve agriculture is important because of the overarching role of the agriculture in the economies of developing nations. A large portion of the population of these countries is directly employed in agriculture, and agriculture has a variety of linkages with the rest of the economies of these nations. Growth in agriculture usually parallels industrialization and general economic growth.

The role of water in a wide range of industrial and commercial activities is also important for the process of development. Industrialization is a principal means of increasing the output of goods and services. The magnitude of industrial withdrawals in the developed countries points to the need for large supplies of water as a prerequisite for industrialization in the developing countries.

Electricity generation is an important industrial application of water. Electricity is an input into most productive activities as well as a direct component of human welfare and use in everyday life. Inadequate supply limits industrialization and is an obstacle to expansion in agricultural output. The role of water in power production is most direct in hydroelectric power generation, but water also plays a significant role in steam-electric generation, primarily as a cooling medium. Water is withdrawn in many developed countries for cooling purposes and the quantities needed for cooling can be dramatically reduced by recycling, a change involving substitution of other inputs, primarily capital investment, for the large quantities of water traditionally used.

Water has contributed significantly to socio-economic development through its role in transportation. Marine navigation constitutes the main form of transporting raw materials and products in international commerce. Inland navigation plays an important, though selective, role in international and national transportation systems (Cox 1989).

Another traditional role of water has been in waste disposal services. There are many water-using activities based on a flow-through approach in which water is discharged after a single passage through an industrial or other process; this has generated large quantities of wastewater from growth in population and economic activity (Cox 1989). Use of natural bodies of water for discharge of wastewater has been in vogue since times immemorial.

Water, as it naturally moves through the phases of the hydrologic cycle, has the potential to constrain socio-economic development in several ways. Naturally occurring periods of excess cause flooding and the disruptions associated with such events. Another negative aspect is the role in disease transmission.

Although water in some minimum amount is a necessary condition for growth, its availability, even in abundant quantities, is not sufficient to guarantee growth. At least for industrial development, other considerations such as labor and markets can be the more important constraints. The fact that availability of abundant water supplies cannot ensure growth is illustrated by the existence of many water-rich areas without significant economic activity (Cox 1989). One very significant, and even catastrophic, socio-cultural disruption is the displacement of human populations by the construction of large water development projects such as dams and reservoirs. This can also inundate or force relocation of culturally significant sites and objects.

Some environmental disruptions from water projects include adverse effects on natural ecosystems and aesthetic values. An environmental impact often merges into a socio-cultural or economic impact. Some environmental effects of water resources development can become substantial and assume many forms. Water projects alter the physical, chemical, and biological characteristics of water and, in some cases, totally transform an existing ecosystem into a different form (Cox 1989).

Alterations in discharge and/or water levels are one important class of commonly occurring environmental changes. Changes in lake levels produce important ecological consequences since shore areas are important in the life cycle of many aquatic organisms. Water quality alteration is a second class of environmental impact associated with many water uses and water development activities. A most obvious source of quality alteration resulting from water use is the discharge of municipal and industrial wastewater and the addition of toxic substances to natural waters. Major ecosystem transformation occurs with the draining and filling of wetlands; this can replace an aquatic environment with a terrestrial one. Natural patterns of surface water and groundwater movement become modified when land is covered by impervious surfaces such as buildings and pavements.

Chronology of management of water resources with respect to human health (Condensed from UNESCO 1992)

1875	World population about 1.75 billion.
1876	Koch: birth of bacteriology.
1877	Sir Patrick Manson: microfilariae of *Wuchereria bancrofti* (filariasis) are taken up by mosquitoes with a blood meal and develop larval stages in the mosquitoes' tissues. He wrongly concluded that the disease was transmitted by drinking water in which mosquitoes had drowned.
1880	Alphonse Laveran: the malaria parasite. Evans: discovery of the first pathogenic trypanosome.
1881	Pasteur's first vaccine against anthrax.
1886	Camilio Golgi observes the stages of the malaria parasite in human blood and that paroxysms coincide with the multiplication of blood parasites.
1894	Alexandre Yersin and Shibasaburo: the plague bacillus.
1998	Paul-Louis Simond: fleas as possible vectors of bubonic plague.
1898	Sir Ronald Ross works on malaria transmission by mosquitoes. Also filariasis transmitted through mosquito bites (*Culex* mosquitoes). Grassi: human malaria shows same stages in *Anopheles* mosquitoes as bird malaria does in *Culex*.
1900	Sir William Leishman: the leishmania parasite.
1902	Forde: trypanosomes in the blood of sleeping sickness patients.
1906	Leiper draws attention to the guinea worm infection as a major tropical disease "but singularly neglected" (although first described under the name of dracunculosis by Linnaeus in his Systema Naturae, in 1758).
1907	P.M. Ashburn and F. Craig: Dengue virus in Philippines.
1914	Opening of the Panama Canal: first large-scale environmental engineering related to water and health, due to Gorgas. Its construction had been interrupted in the late 1880s because of the death of 52 000 workers, out of a total workforce of 85 000, due to yellow fever.
1925	Blacklock, Sierra Leone: simulids as vectors of onchocerciasis.
1928	Sir Alexander Fleming: penicillin
1940	Muller: DDT
1948	First World Health Assembly: creation of the World Health Organization (WHO).
1951	World Population: 2.5 billion
1955	Start of global DDT campaign against malaria under WHO auspices: eradication in many countries of the temperate zone and high levels of control in countries of tropical Asia.
1969	Man lands on the moon.
1974	Start of the WHO Onchocerciasis Control Programme.
1975	Introduction of *Bacillus thuringiensis* to control mosquito and blackfly larvae.
1978	WHO: global eradication of smallpox achieved.

1981   The U.N. Water Supply and Sanitation Decade launched.
1987   World Population: 5 billion

1.14
Health and the Water Cycle

Since 1945, there has been a 50% decline in infant and child mortality as a result of the improvement of public health services. In 1990, the proportion of people 15 years of age and under formed 32.4% of the population worldwide. The current dramatic population growth in developing countries puts pressure on natural resources. It is rapidly putting pressure, too, on institutions that have not been designed to cope with this change and of the resulting age structure of the population. But much remains to be achieved in tropical areas where most diseases associated with water are endemic. Even when they do not lead to early death, they inhibit social and economic development in view of their debilitating and disabling effect on the population. In some cases, disease becomes a way of life for people who may host several different infectious agents and, yet, must work or their family will starve. Much illness can be prevented by ensuring access to safe drinking water and adequate sanitation. In most tropical regions, domestic sewage remains a major problem, especially in urban areas. In the poor outskirts of cities, water is often without any form of pretreatment, and shows traces of pollution. Water also often becomes polluted during transportation.

The modern approach to water resource management is based on the system theory according to which any system that performs work dissipates free energy and, unless it replenishes its energy stores, will tend to run down. The living systems in nature are no longer considered as separate entities but as interactive components. The relationship between humans and nature must involve proper care for the living systems, or biotopes, that constitute nature. Biotopes may be animals, plants, forests, rivers or lakes and their conservation in good condition is vital to the future of life on this planet. The modern trend is a shifting away from the century-old exclusive concern for the quality of water to the notion of the importance to health of water quantity, whatever its quality. It has become apparent in recent years that stringent quality standards can be counterproductive since they may reduce the quantities of water that, otherwise, would be available. Severe quality standards could also delay the supply and increase the cost of water development and distribution (UNESCO 1992).

Table 1.17. Global estimates of vectoral diseases associated with water. (WHO 1987)

Diseases	Vectors	Population affected
African trypanosomiasis	tsetse flies	50 million people at risk 25 thousand infected
Dengue	*Aedes* mosquitoes	Over 1 million people infected per year
Dracunculiasis	*Cyclops* (crustacean)	Over 1 million per year in Africa alone
Filariasis	Mosquitoes (several species)	905 million people at risk
Malaria	*Anopheles* mosquitos	2210 million people at risk
Onchocerciasis	*Simulium.* black flies	86 million people at risk about 18 million infected per year
Schistosomiasis	Freshwater snails	500 million people at risk about 200 million infected
Yellow fever	*Aedes* mosquitoes	Epidemics in tropical zones of Africa and America

The concept of water-washed diseases refers to those infections that decrease as a result of increasing the volume of available water. Diminishing health risks thus involve a focus on increasing the quantity of water. Lack of water is associated with a greatly increased risk of trachoma, a blinding condition (UNESCO 1992). Water quantity indeed also influences diarrhea and other conditions that were previously considered to be exclusively dependent on the water quality. Table 1.17. shows global estimates of vectoral diseases associated with water, and Table 1.18 summarizes the cases of cholera incidence in selected countries.

1.15
Toxicity Testing in the Aquatic Environment

Thousands of different chemicals are being used daily, for purposes ranging from household cleaning to large-scale industrial processes. At some stage many of these chemicals find their way back into the environment in one form or another. Those not released into the atmosphere may be discharged into streams or rivers or into the sea, perhaps with some kind of "treatment" (e.g. in oxidation ponds). Some gradually leach through the soil into water courses and are deposited in marine or freshwater sediments.

Table 1.18. Cumulative figures of cholera cases and deaths for 1991 as reported by selected countries to WHO (i imported)

	Cases	Deaths
Africa		
Angola	8412	247
Benin	7474	259
Chad	13409	1313
Ghana	13095	409
Malawi	8088	245
Mozambique	6624	273
Nigeria	56352	7289
Rwanda	466	28
Togo	2396	81
Uganda	279	28
Zambia	11789	996
The Americas		
Bolivia	186	12
Brazil	1299	20
Canada	2 i	0
Chile	41	2
Colombia	11979	207
Ecuador	46284	697
Guatemala	3530	47
Honduras	11	0
Mexico	2690	34
Nicaragua	1	0
Peru	321034	2896
USA	25 i	0
South east Asia		
Bhutan	422	19
India	4262	79
Indonesia	6202	55
Nepal	31120	875
Sri Lanka	68	2
Western Pacific		
Cambodia	770	97
Japan	93	0 (66 i)
Malaysia	201	2
Republic of Korea	112	4
Singapore	34	0 (4 i)

The effect of these chemicals on marine and fresh-water ecosystems depends on their concentration, their toxicity to organisms at this concentration, and how long the toxicity persists. High doses of toxic chemicals result in fish kills and extensive destruction of aquatic biota. The challenge is to assess the effects and fate of low levels of contaminants which are often associated with organic and nutrient enrichment.

1.15.1
Assessing Toxicity

It is now possible to identify and measure chemical contaminants even in extremely low concentrations, but the only way to measure toxicity is to test the effects of a substance on living organisms. Two approaches are available to assess the potential toxicity of a water or sediment sample. A detailed chemical analysis of the sample can be made, followed by reference to published water quality criteria which give "safe" concentrations for the contaminants identified. Or, toxicity may be assessed directly using a biological toxicity test.

A toxicity test is a simple laboratory bioassay procedure. For effluents, a series of dilutions of the solution to be tested are made, the control usually being the water, which is also used to dilute the test sample. The aim is to simulate the chemical interactions which may occur when an effluent is discharged. Test organisms are placed in each dilution and the control and are incubated for a set time, under controlled temperature and lighting conditions. At the end of the exposure period the toxic effect is measured.

The test end point varies. It can be mortality, reduced fertility, lowered fecundity or reduced growth. By examining responses in the dilution series, one can quantify the toxicity and determine the lowest concentration which causes a measurable toxic effect. It is also possible to calculate the EC_{50} – the effective concentration resulting in 50% of the organisms showing a response of the end point being considered. This term includes LC_{50} which is the lethal concentration resulting in 50% of the organisms being killed after the chosen time.

For sediments, dilutions can also be made by adding clean sediment to the test sample. Commonly, however, toxicity is determined by comparing the survival or responses of test organisms placed in the test sediment with those in a reference (or control) sediment.

Most tests rely on **acute** responses (short term relative to the life span of the organism), usually 48 or 96 h. A **chronic** (long term relative to the life span of the organism) test generally measures as its end point reproductive success or failure or development and/or growth of embryonic or larval forms and usually lasts for up to 28 days.

A comprehensive assessment of potential contaminant effects involves chemical, toxicity and biological monitoring, the so-called "triad" approach. The relative merits of toxicity testing and environmental bioassessment for detecting contaminant effects are described below. The relative importance of each of the triad components differs among situations, and emphasis may be dictated by the potential environmental risk involved (Hickey and Roper 1994).

1.16
Biological Toxicity Testing Versus Environmental Monitoring of Impacts

Problems arise in attributing cause-effect relationships in complex receiving water environments with many potentially deleterious inputs. Environmental monitoring helps establish that significant impacts are occurring, but may not be able to identify the major contributor(s) or show whether there are other important sources of contamination as well. Toxicity tests on effluent samples can help to quantify the potential impacts of contaminants on marine organisms. The biological toxicity testing approach has the following advantages and disadvantages compared with environmental impact monitoring (Hickey and Roper 1994).

1.16.1
Toxicity Test Species

Several physiological responses of test organisms can be used to monitor toxicity in water samples including growth, photosynthesis, pigmentation, motility etc. Recently, an early warning system, called ECOTOX (Fig. 1.35) has been developed using different movement parameters of the motile unicellular flagellate *Euglena gracilis* as end points (Häder et al. 1997). The orienting in the gravity field, the motility and the velocity of the cells are monitored by a real time image analysis system. ECOTOX can be employed to detect quality changes in effluents and aquatic ecosystems and to indicate whether aquatic life may be endangered. Effluent testing often involves three test species each representing a different phylogenetic level. No single species is expected to be sensitive to all chemical contaminants. Therefore the range of test species provides a greater level of ecosystem protection. The standard test species are *Daphnia* and *Selenastrum (freshwater)* and *Chaetocorophium* (marine). Table 1.19 compares the advantages and disadvantages of chemical measurements and bioassays. Organic and inorganic contaminants in water have various origins (Table 1.20).

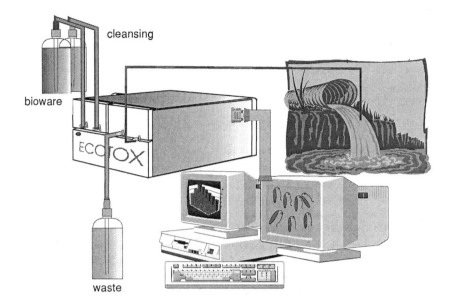

Fig. 1.35. ECOTOX. Fully automatic biomonitoring instrument to assay toxicity in water. First the control sample containing the flagellate *Euglena gracilis* is pumped into the chamber and mixed with water. Cell tracks are followed and movement parameters (percent of motile cells, average velocity and orientation in the gravity vector of the earth are calculated). Subsequently, a new sample of cells is pumped into the cuvette and mixed with the test water sample. The analysis of movement parameters is repeated. If one of the parameters deviates from the control values by more than a predefined threshold an alarm is elicited (Häder et al. 1997)

Advantages of biological toxicity testing:

1. It deals with a more concentrated contaminant so it is more likely to detect effects,
2. It establishes extent or degree of toxic risk,
3. A rigorous testing protocol can form the basis of routine compliance monitoring,

Disadvantages of biological toxicity testing:

1. Lack of measurable toxicity does not necessarily mean no environmental impact because laboratory tests use a few species only,
2. Laboratory tests may not integrate other stress factors operating on local communities, and
3. Laboratory tests cannot easily be used to predict community level ecological impacts

Table 1.19. Comparison of chemical measurements versus bioassays for assessing toxicity

Chemical measurements *Advantages*	Toxicity bioassays *Advantages*
Treatment systems can be more easily designed to meet chemical requirements as the procedures are well established.	The aggregate toxicity of all constituents in a complex effluent or contaminated sediment is measured.
The fate of a pollutant can be predicted through modeling.	The bioavailability of the toxic constituents and interactions of constituents are assessed.
Chemical analyses are usually less expensive than toxicity testing.	It is easily understood by the public, and provides tangible evidence of presence or absence of environmental impact.
Disadvantages	*Disadvantages*
In complex samples, not all potential toxicants may be identified.	Properties of specific chemicals (such as potential for bioaccumulation) are not assessed.
It is not always clear which compounds are causing toxicity.	Wastewater engineers cannot identify specific toxic components and have only limited ability to design or manipulate treatment systems.
Measurement of individual toxicants in complex samples is expensive.	If chemical/physical conditions are present that act on toxicants in such a way as to "release" toxicity downstream (or away from the discharge point), such toxicity may not be measured.
The bioavailability of the toxicants at the discharge site is not assessed, and the interactions between toxicants (e.g., additivity, antagonism) cannot be accounted for.	

Table 1.20. Principal sources and types of contaminants

Source	Contaminants
Agriculture	Ammonia, hydrogen sulfide, cadmium, pesticides
Mining	Heavy metals, suspensoids
Forestry	Resin acids, chlorinated organics, doxins, copper, chromium, arsenic, PCPs, chlordane
Geothermal	Mercury, arsenic, boron
Storm waters	Heavy metals, PAH, suspensoids
Harbor dredging	Heavy metals, organics
Municipal wastes	Ammonia, hydrogen sulfide, pesticides etc.

1.17
Diseases Associated with Water and Poverty

The population living in the developing countries (excluding China) is estimated at about 2485 million. In the mid 1980s, only about 1425 million of these had access to a reasonably safe water supply. The situation is worse in the least developed countries (Table 1.21).

Table 1.21. Coverage in water supply (percentage) and sanitation (percentage). (World Health Organization 1991)

Region	Water supply		Sanitation	
	Urban	Rural	Urban	Rural
Africa	78	30	54	21
Central and South America	90	62	79	37
Western Pacific including China	91	82	99	97
South east Asia	79	72	50	13
Eastern Mediterranean	98	49	62	20

Many people still live without water piped directly into the house. This has direct and important implications on health since it restricts the use of water for washing clothes and utensils, for cooking and for personal hygiene. People instead have to carry water for long distances and often queue at the tap. In addition, public water supplies are often contaminated by sewage in poorly maintained water distribution systems. Moreover, sewers are not always equipped for hygienic removal of fecal matter; they often run directly into the rivers, lakes and coastal waters from which many poor people draw their water.

Diarrheal diseases which often exacerbate or cause malnutrition, are highly prevalent and responsible for mortality in children under two years of age. The scarcity of clean water, in addition to poor housing conditions that attract a high density of insects and rodents, poor personal hygiene and general poor hygienic conditions are typical of the urban slums.

1.18
Restoration of Water Quality in Some Rivers

In some European countries such as the Netherlands, the quality of surface waters has improved during the last three decades. Less pollution is brought downstream by the Rhine and the Maas. Levels of mercury, cadmium, copper, lead and zinc have decreased by between 50 and 80%. Quantities of certain organic micropollutants, such as hexachlorobenzene and polyaromatic hydrocarbons have been more than halved. Discharges of oxygen-demanding substances have been reduced to such an extent that there are no longer serious problems with oxygen levels in Dutch waters.

The Rhine was described in the 1970s as the largest open sewer in Europe; can we now justifiably brand it as clean? Unfortunately not. If we look at levels of pesticides, nitrogen and phosphate, we get a completely different picture. The Rhine Action Program (RAP), launched by the five countries through which the Rhine flows (Switzerland, Germany, France, Luxembourg and the Netherlands), aims to reduce concentrations of these substances. The specific aim of the program is to reduce emissions of a large number of substances by between 50 and 70% and to restore the Rhine to its natural state so that salmon and other fish can return to its waters.

1.18.1
Diffuse sources

Each country is responsible for the section of the river that runs through its territory. Timetables have been drawn up for reductions by industry, domestic households and from diffuse sources. Though some progress has occurred, the targets for mercury, copper, lead and 1,1,1-trichloroethane have not been reached. It has also proved difficult to regulate the diffuse sources. The targets for chromium and cadmium have, however, been achieved.

But even the water that meets the standard requirements is not 100% reliable. Proving the presence of pesticides in water is very similar to the dope tests carried out in various sports. Many substances slip through the net simply because there are no analytical methods capable of identifying them. For example, a method of identifying bentazone, a pesticide which dissolves easily and is therefore very difficult to trace, was only developed in 1988. Once the drinking water companies in the Netherlands put the method into practice, they found that bentazone was present in practically all surface and drinking water in the country. In Amsterdam, people were receiving more than ten times the permitted safe dose. The Rhine and Maas water supply companies worked together to track down the source of the bentazone by taking samples of river water every twenty kilometers. The culprit was discovered to be the BASF factory in Ludwigshafen. The company has since stopped the discharges.

Agriculture in many developed countries uses large quantities of pesticides. From a half to three quarters of 50 pesticides investigated have been found to be present in groundwater in some European countries. If pesticide use continues at 1987–1989 levels, 35 of the 150 most-used crop protection agents will exceed the norm for clean groundwater in the Netherlands.

And this is only part of the pesticide problem. Many crop protection agents and other organic micropollutants do not dissolve and, like many heavy metals, do not end up in the groundwater but adhere to silt particles suspended in the water. Many river and lake beds are contaminated with polyaromatic hydrocarbons and polychlorobenzene. Approximately half of the sediments that have been tested so far are so polluted that they cannot be dredged for use on land. It is estimated that 130 million m^3 of polluted sediment will be created in the next 20 years, of which only 2 million m^3 will be able to be treated and reused in the near future. The technology is simply not available, with the consequence that the sediment is left where it is.

In recent years, companies processing drinking water from the rivers have become little more than chemical plants. Although those which draw their water from aquifers are better off, groundwater pollution is increasing, particularly nitrate pollution resulting from the excessive use of fertilizers.

The excess of phosphorus is even greater. Of the phosphorus spread on the land in the form of fertilizers, 60% remains in the soil. Half of all agricultural land with sandy soil is already saturated. This means that the land can no longer absorb the phosphorus that is still being spread on it, causing it to leach through to the groundwater. Most drinking water is drawn from the deeper layers which are less polluted. Water can take up to 10 000 years to reach these lower levels. Fertilizers have only been used intensively for the past twenty years and have therefore not yet reached the deeper groundwater. But it takes far less time – approximately 25 years – to reach the upper layers. Thus, the first problems have been felt only since the 1980s – and they will get worse because there is more pollution on its way.

Restoring the water table, cleaning up our surface waters and lake and river beds, restricting the use of fertilizers and wasting less water are all obviously connected and form part of the water cycle. Measures aimed at individual stages in this cycle are doomed to failure. The water, lake and river beds, banks and shores, and the flora and fauna that live in these surroundings all form part of a cohesive ecosystem, and it is this whole that needs to be restored.

1.19
Rural Water Treatment for Developing Countries

In developing countries, many people are forced to use surface water, water drawn from polluted rivers, irrigation canals, ponds, and lakes. This surface water, however, is a carrier of many infectious and tropical diseases and therefore should be treated before consumption. The main target of any water treatment is the removal or inactivation of pathogens, such as harmful bacteria, viruses, protozoal cysts, and worm eggs. Disinfection – usually by application of chlorine and slow sand filters – is the most widely used treatment process for bacteriological water quality improvement.

Slow sand filtration (SSF) applied as surface water treatment is particularly effective in improving the bacteriological water quality. SSF offers the great advantage of being safe and stable, simple and reliable. SSF is essentially a self-reliant technology largely reproducible with local resources. Operation is easily handled by unskilled personnel and does not depend on the external inputs. However, efficient application of the treatment process requires water of low turbidity, hence pretreatment of the surface water is usually necessary. Chemical flocculation combined with sedimentation is often inapplicable, because rural water supplies in developing countries generally face serious operational problems with chemical water treatment, e.g. since chemical flocculation is fairly sensitive to water quality changes. Prefiltration is a simple and efficient alternative treatment process. Disinfection efficiency and SSF performance are strongly influenced by the turbidity of the water to be treated. Turbidity mainly arises from the solids and colloids present in the water. A large number of microorganisms remain attached to the surface of these solids. Adequate water disinfection is only possible with water of low turbidity or water virtually free of solid and organic matter. SSF also requires a relatively clear water. Pretreatment of surface water for the reduction of turbidity or solid matter concentration is therefore required for both chlorination and SSF. Sedimentation, possibly preceded by chemical flocculation, will remove the settleable matter and part of the suspended solids. Similarly, a safe and reliable chlorination of the water remains a target difficult to attain under rural conditions in developing countries (Wegelin et al. 1991).

Filtration is a more effective process for solid removal due to the large filter surface area available. Sound water treatment concepts start with the separation of coarse matter, usually easy to achieve, and end with the removal or inactivation of small solids and microorganisms generally more difficult to separate. Figure 1.36 illustrates the relationship between particle size and solids classification. It also shows the main applications for various treatment processes.

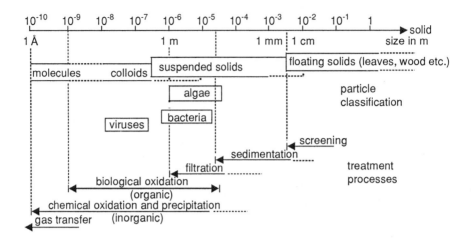

Fig. 1.36. Particle size and treatment processes. (After Wegelin et al. 1991)

The bulk of solids is separated by the coarse filter medium, while the subsequent medium and fine filter media have more of a polishing function. Roughing filters reduce turbidity to a level that allows a sound and efficient SSF operation. SSF as the final treatment stage substantially reduces the microbial populations in the water (Wegelin et al. 1991).

1.19.1
Roughing Filters

Roughing filters combined with slow sand filters present a reliable and sustainable treatment process, particularly appropriate for developing countries (Wegelin et al. 1991). Studies on the design and performance of prefilters to cope with highly turbid surface water have been made in different institutes.

A roughing filter is composed of layers of filter material that successively decrease in size ranging from 25 mm to 4 mm in size. Gravel is usually used as filter material. The design and mode of application of roughing filters varies considerably. The different filter types can be classified according to their location within the water supply scheme, and with respect to the flow direction. There is therefore a distinction between intake and dynamic filters, which form part of the water intake structure, and the actual roughing filters, which are further subdivided into down, up, and horizontal flow filters. Figure 1.37 shows one type of roughing filter (Wegelin et al. 1991).

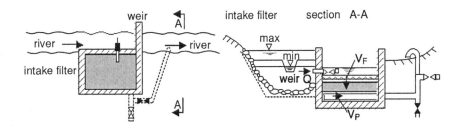

Fig. 1.37. Layout of intake filters. (After Wegelin et al. 1991)

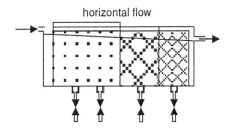

Fig. 1.38. A horizontal-flow roughing filter. (After Wegelin et al. 1991)

Intake and dynamic filters are installed next to small rivers. Part of river water, impounded by a small weir, is diverted either into a filter box. The river water is filtered through different gravel layers, the top layer being of coarsest size. These two filters therefore act mainly as surface filters. Intake and dynamic filters are not only designed for continuous solids separation, but also to protect the treatment plants from high solid concentration. In contrast to intake and dynamic filters, roughing filters possess a large silt storage capacity. In vertical-flow roughing filters, the available height of the filter medium is limited to about 1.0–1.5 m. The filter length of horizontal-flow roughing filters varies in general between 5 and 9 m (Fig. 1.38). Roughing filters are operated at filtration rates ranging from 0.3 to 1.5 m h^{-1}. During the last few years, horizontal-flow roughing filtration (HRF) has received greater attention than any other prefiltration technique. A treatment scheme, illustrated in Fig. 1.39, makes utmost use of local resources. 20 l of gravel and 10 l of sand used as filter material are able to transform contaminated turbid surface water into 30 l of quality drinking water per day (Wegelin et al. 1991).

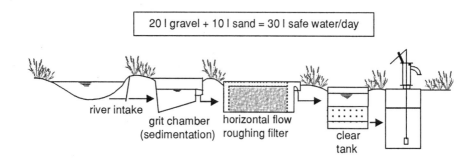

Fig. 1.39. A self-reliant water treatment plant. (After Wegelin et al. 1991)

References

Ausubel JHA (1991) Second look at the impacts of climate change. Am Sci 79: 210–221

Baconguis SR (1995) River ecosystem. Canopy International (Laguna, Philippines) 21:1–9

Bakaloicz M (1994) Water geochemistry: water quality and dynamics. In: Gilbert J, Danielopol DL, Stanford JA (eds) Groundwater ecology. Academic Press, London, pp 97–127

Benson CL Jr, Rao GV (1987) Convective bands as structural components of an Arabian sea convective cloud cluster. Mon Wea Rev 115: 3013–3023

Bjorklund I, Borg H, Johansson K (1984) Mercury in Swedish lakes: its regional distribution and causes. Ambio 13:118–121

Bolin B, Doos BR, Jager J, Warrick RA (1986) The greenhouse effect, climate change, and ecosystems. Wiley, Chichester

Bushaw KL, Zepp RG, Tarr MA, Schulz-Jander D, Bourbonniere RA, Hodson RE, Miller WL, Bronk DA, Moran MA (1996) Photochemical release of biologically available nitrogen from aquatic dissolved organic matter. Nature 381:404–407

Carpenter SR, Fisher SG, Grimm NB, Kitchell JF (1992) Global change and freshwater ecosystems. Ann Rev Ecol Syst 23:119–139

Carpenter SR, Chisholm SW, Krebs CJ, Schindler DW, Wright RF (1995) Ecosystem experiments. Science 269:324–327

Carpenter SR, Kitchell JF (1993) The trophic cascade in lake ecosystems. Cambridge University Press, London

Chahine MT (1992) The hydrological cycle and its influence on climate. Nature 359:373–80

Cox WE (1989) Water and development. Managing the relationship. UNESCO, Paris

Dube SK, Rao AD, Sinha PC, Jain I (Jan. 25–30, 1993) Implications of climatic variations in the freshwater outflow on the wind induced circulation of the bay of Bengal. Abst Internat Conf Sustainable Development Strategies and Global/Regional/Local Impacts on Atmospheric Composition and Climate, Ind Inst Technol, New Delhi, pp 180–181

Fiebig DM, Lock MA (1991) Immobilization of dissolved organic matter from groundwater discharge through the stream bed. Freshwater Biol 26:45–55

Firth P, Fisher SG (1991) Climate change and freshwater ecosystems. Springer-Verlag, New York

Giller PS, Hildrew AG, Raffaelli DG (1994) Aquatic ecology: scale, pattern and process. Blackwell, Oxford

Glieck PH (1993) Water in crisis: a guide to the world's freshwater resources. Oxford University Press, New York

Grenthe I, Stumm W, Laakusharju M, Nilsson A-C, Wikberg P (1992) Redox potentials and redox reactions in deep groundwater systems. Chemical Geology 98:131–150

Häder D-P, Lebert M, Tahedl H, Richter P (1997) The Erlanger flagellate test (EFT): photosynthetic flagellates in biological dosimeters. J Photochem Photobiol B: Biol 40:23–28

Hering JG, Stumm W (1991) Oxidative and reductive dissolution of minerals. In: Hochella MF Jr, White AF (eds) Reviews in Microbiology, Vol. 23. Mineral water interface geochemistry, pp 427–465

Hickey C, Roper D (1994) Testing for toxicity in our aquatic environment. Water and Atmosphere (NIWA, New Zealand) 2 (4):11–14

Hounslow AW (1985) Strategy for subsurface characterization research. In: Ward C, Giger W, McCarty PL (eds) Groundwater quality. Wiley, New York, pp 356–369

Jackson RE (1980) Aquifer contamination and protection studies and reports in hydrology series, No. 30. UNESCO, Paris

Jones JB, Holmes RM (1996) Surface-subsurface interactions in stream ecosystems. TREE 11:239–242

Junk WJ, Bayley PB, Sparks RE (1989) The flood pulse concept in river-floodplain systems. In: Dodge DP (ed) Proceedings of the International Large River Symposium. Canad Sp Publ Fish Aquat Sci 106:110–127

Kayane I (1996) Interactions between the biospheric aspects of the hydrological cycle and land use/cover change. IGPB NEWSLETTER 25:8

Koeman JH (1971) Het voorkomen ende toxicologische beteknis van enkele chloorkool water stoffen an de Nederlandse kust in de period van 1965 tot 1970 (In Dutch, with English summary). Thesis, Utrecht Univ, p 136

Kullberg A, Bishop KH, Hargeby A, Jansson M, Peterson RC Jr (1993) The ecological significance of dissolved organic carbon in acidified waters. Ambio 22:331–337

Likens GE (1975) Primary production of inland aquatic systems. In: Lieth H, Whittaker RH (eds) Primary productivity of the biosphere. Springer Verlag, Berlin pp 185–202

Likens GE (1992) The ecosystem approach: its use and abuse. Ecology Institute, Oldendorf, Germany

Lindh G Water and the city. UNESCO, Paris (1985)

Lovley DR (1991) Dissimilatory Fe(III), Mn(IV) reduction. Microbiol Rev 55:259–287

Mackenzie FT (1975) Man's impact on global element cycles. Amer Chem Soc Abst Papers of 169th Annual Meeting, Port City Press, Baltimore

Marmonier P, Vervier P, Gilbert J, Dole-Olivier MJ (1993) Biodiversity in groundwaters. TREE 8:392–395

Meisner JD, Rosenfeld JS, Regier HA (1988) The role of groundwater in the impact of climate warming on stream salmonines. Fisheries 13:2–8

Mukherjee AK, Rao MK, Shah KC (1978) Vortices embedded in the trough of low pressure off Maharashtra Goa coasts during the month of July. Indian J Meteor Hydrol and Geoph. 29:61–65

Naiman RJ, Magnuson JJ, McKnight DM, Stanford JA (1995a) The freshwater imperative: a research agenda. Island Press, Washington DC

Naiman RJ, Magnuson JJ, McKnight DM, Stanford JA, Karr JR (1995b) Freshwater ecosystems and their management: a national initiative. Science 270:584–585

National Research Council (1986) Global change in the geosphere. Biosphere 91. National Academy Press, Washington DC

NRC National Research Council (1992) Restoration of aquatic systems: science, technology, and public policy. Natl Acad Press, Washington, DC

Overpeck JT, Rind D, Goldberg R (1990) Climate-induced changes in forest disturbance and vegetation. Nature 343:51–53

Power ME (1990) Effect of fish in river food webs. Science 250:811–814

Power ME, Matthews WJ, Stewart AJ (1985) Grazing minnows, piscivorous bass and stream algae: dynamics of a strong interaction. Ecology 66 1448–56

Prasad BC (1970) Diurnal convection of rainfall in India. Indian J Meteor Hydrol and Geoph 21:443–450

Prost A (1989) The management of water resources, development and human health. UNESCO International Colloquium on the Development of Hydrologic and Water Management Strategies in the Humid Tropics. UNESCO, Paris

Reddy NC, Raman S (25–30 Jan 1993) The effect of mesoscale circulations on precipitation distribution over the west coast of India. Abst Internat Conf Sustainable Development Strategies and Global/Regional/Local Impacts of Atmospheric Composition and Climate. Ind Inst Technol, New Delhi, pp 182–185

Sarkar RP, Ray KC, De US (1978) Dynamics of orographic rainfall. Indian J Meteor Hydrol and Geophy 29:335–348

Sedell JR, Richey JE, Swanson FJ (1989) The river continuum concept: a basis for the expected ecosystem behavior of very large rivers? In: Dodge DP (ed) Proceedings of the International Large River Symposium. Canad Sp Publ Fish Aquat Sci 106:49–55

Shuter BJ, Post JR (1990) Climate, population variability, and the zoogeography of temperate fishes. Trans Am Fish Soc 119:314–336

Thorp JH, Delong MD (1994) The riverine productivity model: an heuristic view of carbon sources and organic processing in large river ecosystems. Oikos 70:306–308

UNESCO (1986) Pollution et protection des aquiferes. UNESCO, Paris

UNESCO (1991) Water related issues and problems of the humid tropics and other warm humid regions. Internat Hydrological Programme. UNESCO, Paris

UNESCO (1992) Water and Health. IHP Humid Tropics Programme Series No 3. UNESCO, Paris

Vannote RL, Minshall G, Cummins KW, Sedell JR, Cushing CE (1980) The river continuum concept. Can J Fish Aquat Sci 37:130–137

Vrba J (1991) Integrated land-use planning and groundwater protection in rural areas, Technical Documents in Hydrology Series. UNESCO, Paris

Walker BH (1991) Ecological consequences of atmospheric and climate change. Climate Change 18:301–16

Wegelin M, Schertenleib R, Boller M (1991) The decade of roughing filters-development of a rural water treatment process for developing countries. J Water SRT Aqua 40:304–316

Whittaker RH (1975) Communities and ecosystems. 2 ed Macmillan, New York

WHO (1987) Technology for water supply and sanitation in developing countries. Technical Report Series No 742. WHO, Geneva

WHO (1991) Surface water drainage for low-income communities. World Health Organization, Geneva

Wilson AL (1979) Trace metals in waters. Philos Trans R Soc Lond 228B:25–39

Williams P (1989) Adapting water resources management to global climate change. *Climate Change* 15:83–93

Zeyer J, Eicher P, Dolfing J, Schwarzenbach RP (1990) Anaerobic degradation of aromatic hydrocarbons. In: Kamely D (ed) Biotechnology and Biodegradation. Portfolio Publ Co, The Woodlands, pp 33–40

2 Lakes and Wetlands

2.1
Introduction

Lakes and reservoirs, as integral components of our planet's life-support systems, are essential for the maintenance of human life. They are resources and assets of enormous economic, cultural, aesthetic and recreational importance. They are subject to many demands, to significant loss of biodiversity and to an increasing degree of pollution and degradation. These processes are accelerating because of rising human pressure. It is being recognized that the main issues for the management and conservation of lakes include those of regional and global concern and have great current and future significance.

Since there exists a strong interdependence between human populations and aquatic ecosystems, and since lake deterioration and loss of biodiversity are intimately linked to population pressure and human activity, there is need to support efforts (1) to limit human population pressures, to develop ecological sustainability, (2) to maintain the biodiversity of our planet and (3) to bring about such changes to individual life-styles as can minimize the ecological impact of human activities on lakes.

The developments in water quality management, environmental protection and nature conservation are characterized by interacting physical, social and institutional responses. Resilience towards change and time lags in responses occur both in the ecosystem and in society. Both delays contribute to frequently observed inadequate management. Prevention or curtailment of environmental deterioration requires awareness of the lengthy and hidden degradation in the ecosystem and a proactive decision-making structure in society.

The functions of lakes and reservoirs are well defined and diverse. Their benefits can often be expressed in numbers and measures and their control can be done by engineers and scientists. The benefits can be expressed in volumes of drinking water produced, megawatts, crop yields, fish catch, etc. The benefits can also be expressed in monetary units as there is a market for these goods and a price associated with their benefits (Lijklema 1995).

Water supply in many parts of the world is rapidly becoming of major concern for dependent human populations. Large-scale water resource projects have become increasingly significant. These large projects should be subjected to environmental impact assessments and careful monitoring of ongoing activities and extensive post-evaluation.

It is well known that impacts on lakes are the consequence chiefly of activities in catchments. Therefore integrated catchment management needs to be supported and encouraged. This involves detailed planning of all components accompanied by environmental impact assessment for those projects which significantly modify the landscape, land-use or drainage characteristics. Catchment management must also include a greater sensitivity to environmental habitat change, lead to lifestyle adaptations, protect the natural cleansing properties of wetlands and adopt a holistic approach to catchment management.

Ecosystems and social systems share several common features: complexity, non-linear behavior and some inertia and resilience towards change. Ecosystem behavior is essentially deterministic but the environmental forcing functions are highly stochastic. The reactions of society are only mildly rational – both over-reaction and apathy occur. The behavior of ecosystems tends to be adaptive and evolutionary and strategically oriented toward suicide and self-destruction. Yet, when there is a price to pay, society will react. Very often this reaction comes too late, after much damage has been done. Irreversible changes are exemplified by the extinction of several terrestrial and aquatic species.

Understanding of ecosystem behavior and the risks of diverse human activities in modern society must be well analyzed so as to enhance the capabilities for prevention-oriented behavior. This means the continuation of such ecosystem studies as ecotoxicology, effects of changing land-use patterns, propagation of risks in water systems, development of restoration techniques, and biomanipulation. A more systematic and critical exploration of risk assessment techniques is desirable.

There are more than a million lakes and reservoirs in the world. Lakes are not only a necessary water resource for humans but also have an important role in the natural environment. The International Lake Environment Committee (ILEC) has recently surveyed the world's lakes and published the findings in five volumes titled *Survey of the State of World Lakes*, covering 217 large and small lakes and reservoirs from 73 countries (see Seko 1994). The following aspects concerning the lakes and reservoirs have been investigated: location, general description of the water body, physical dimensions, physiografic features, water quality, biological features, socio-economic conditions, water utilization, deterioration of the environment and hazards, wastewater treatments, improvement works, development plans, legislative and institutional measures for upgrading and sources of information. Table 2.1 shows the morphometric and hydrological differences between lakes and oceans. It was concluded that lakes of the world face six major problems, outlined below (see Fig. 2.1):

1. **Eutrophication**. Eutrophication deprives lakeside residents of good water quality in many densely populated areas. The percentages of lakes and reservoirs with eutrophication problems are, by region, as follows: Asia and the Pacific 54%, Europe 53%, Africa 28%, North America 48%, and South America 41%.
2. **Water level fluctuations**. Withdrawing too much water from a lake or its tributary sources results in a decrease in lake area and volume, temporary eutrophication symptoms and concentration of minerals in the water.
3. **Siltation**. This occurs mainly because traditional forms of sustainable land use are rapidly disappearing with increases in population and the invasion of cash-based economies, even in remote rural areas. Siltation decreases flood control capacity and increases flood damage.
4. **Toxic chemicals**. The problem of toxic chemical contamination of lake water, sediments and organisms chiefly plagues industrialized countries, but similar contamination has now spread also to many parts of the developing world.
5. **Acidification**. Acidification and the resulting death of fish and other animals has so far been confined to the lakes of northwestern Europe, the northeastern United States, and some parts of Canada. These symptoms have not yet appeared in many other parts of the world which receive rainwater of similar acidity.
6. **Disruption of lake ecosystem**. Lake ecosystems, and the fauna and flora they contain, normally evolve slowly over time. The sudden introduction of new species, however, can quickly upset a stable ecosystem. Equally disruptive effects can occur when lake conditions change as a result of excess nitrogen or phosphorus in agricultural runoff, causing a sudden explosion in the population of one or more species that were previously in balance with other members of the ecosystem.

The above problems are mainly the direct result of careless human activities. It takes very long time for the environment to recover when lakes and reservoirs are damaged (Seko 1994).

The structure of the water body in any lake is determined primarily by the weather. From late winter to spring, it loses the heat remaining in it from the previous year to its environment. In heavy storms the water can undergo mixing down to the lake bottom. The dissolved oxygen concentration rises up to the surface saturation value and volatile substances escape. With the increase in incident solar radiation in late spring, the upper water layer warms up, thereby decreasing in density and forming a cover over the colder, deeper water, preventing further downward mixing (Ambuhl and Buhrer 1993). The lake has now changed from winter "turnover" to summer "stagnation". Winds, waves and convection currents agitate the water.

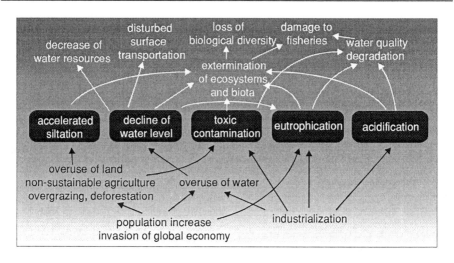

Fig. 2.1. Six major environmental problems in world lakes/reservoirs and the ways they are related to one another. (After Seko 1994)

The pelagic or free-water zone (usually at the upper layer, the epilimnion) is populated with plankton. Algae and animals which form this community are constructed simply, but are morphologically, physiologically and ethologically highly adapted to their environment and to their continually floating or swimming mode of life. The plankton form a typical (classical) ecological community of simple structure. Two decades ago, many European lakes were overloaded with phosphate. Today the phosphorus content is decreasing, and has attained minimum values in some lakes, whereas in others, it is still falling. In a eutrophic (nutrient-rich) lake, essentially the same processes occur as in an oligotrophic (nutrient-poor) lake, but with different weightings and different distributions of nutrients and processes within the water body. Putrefaction processes are mostly absent in oligotrophic lakes but are conspicuous in eutrophic ones. Table 2.2 compares the conditions for life in pelagic waters and on land.

Figure 2.2 shows the seasonal variation of phytoplankton development and its dependence on environmental factors and Fig. 2.3 shows the ideal and perturbed seasonal successions of phytoplankton in temperate water. Winter diatoms (W) are replaced by r flagellates as stratification occurs in spring. Large K species succeed the r types in summer. W species may recur in autumn. The upper limit of biomass (the envelope) is set by the availability of light and the stocks of N and P. In the perturbed condition W, r and K can occur at various times as the stratification comes and goes.

Table 2.1. Morphometric and hydrological differences between oceans and lakes (typical values or order of magnitude). (After Imboden 1990)

	Oceans (total)	Lakes Large	Lakes Small
Volume (km^3)	1.4×10^9	2×10^2	10^{-2}
Surface (km^2)	3.6×10^8	10^3	1
Mean depth (m)	4×10^3	2×10^2	10
Surface area per boundary length (km)	10^6	10	10^{-1}
Residence time (yr)			
Surface water	10^2	1 to 10	10^{-1} to 1
Total water body	4×10^4	10 to 10^2	10^{-1} to 10

Table 2.2. Conditions for life in pelagic waters and on land. (After Imboden 1990)

Factor	Open water	Land
Control over position	*Limited.* Control by buoyancy or self-propulsion	*Normal.* By roots, attachment etc.
Gravity	*Small* or absent (buoyancy)	*Large.* Supporting structures needed (trees)
Light	Low transparency of water	High transparency of air
Vertical attenuation of light	in the order of 0.1 m^{-1}	Negligible
Temporal variation of light for individual	Large and unpredictable (depending on position)	Slow and regular
Temperature variation	Moderate	Often large or even extreme
Water	Abundant	Limits the spreading of life
Nutrients	Usually growth limiting, but easily distributed within system	Local inhomogeneities co-exist (fertile/barren land)
Spreading of species	Easy by water currents	Usually difficult: easy by "tricks" only (seeds transported by animals, through air, etc.)
Ecological niches	Few	Many
Number of plant species in total ecosystem	~ 40 000	~ 400 000
Average standing crop of living biomass	2 g C m^{-2}	6000 g C m^{-2}
Total primary production (approximately)	50×10^{15} g C yr^{-1}	40×10^{15} g C yr^{-1}
Productivity per biomass	75 yr^{-1}	0.04 yr^{-1}

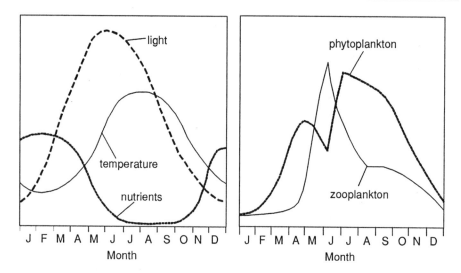

Fig. 2.2. Seasonal variations of factors important for plankton (top) and a schematic diagram of plankton development with time (bottom). (After Ambuhl and Buhrer 1993)

The upper few meters of lake water inhabited by plankton form a bioreactor. Phytoplankton photosynthesis occurs in this euphotic zone, and the infrared component of light is responsible for building up the thermal stratification. New algal production, either living or in the form of detritus, sinks to the lake bottom; a part enters the food chain and is sedimented in the form of fecal pellets. Physico-chemical phenomena which express themselves here include the adsorption and desorption of dissolved substances onto and from particles on their way to the sediment at various temperature-dependent exchange rates. The production of new biomass in oligotrophic lakes is so low that the subsequent aerobic degradation of this new biomass does not lead to a lack of oxygen (Ambuhl and Buhrer 1993). In eutrophic lakes there are more algae, and hence more zooplankton and more fish than in oligotrophic lakes. Less tolerant species are replaced by more tolerant species.

In deep water, oxygen is depleted; aerobic degradation is replaced by anaerobic decay, possibly accompanied by sulfate reduction and the production of H_2S. With the onset of autumnal cooling, the water body gradually turns over. Circulation brings oxygen into the depths, and also brings nutrients up to the surface along with the upwelling deep water. These nutrients form the "starting capital" for the new spring production (Figs. 2.3, 2.4). This goes on for some time, whereafter production is stimulated by some external nutrient supply or by nutrient recycling. Algal growth is available for exploitation by zooplankton which graze on algae.

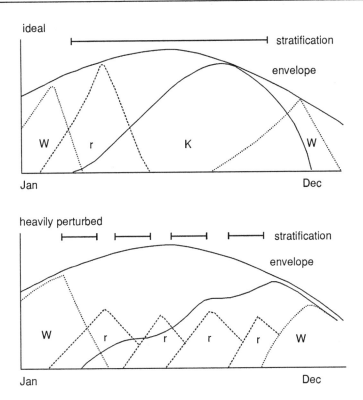

Fig. 2.3. General patterns of seasonal successions of phytoplankton in temperate water. For explanation, see text

Since zooplankton development requires higher temperatures, however, their development lags behind that of the phytoplankton. When water temperatures are high, the zooplankton can undergo explosive development. Later in the season, the algal biomass recuperates rapidly, but its species composition changes: now those algae which the zooplanktom was not able to graze on earlier because of their size or unwieldy shape, can develop. Despite the presence of a higher (but less edible) algal biomass, the zooplankton starve and just survive. In most lakes, phosphate is a key factor governing biological production. The general relationship between total phosphorus and depth of lakes is shown in Fig. 2.4. Phosphate is often regarded as an unpleasant or even damaging substance. However, out of all the nutrients, phosphate is the limiting factor for algal growth. It did not gain attention until it began to enter lakes in large amounts unintentionally in wastewater and consequently ceased to be limiting.

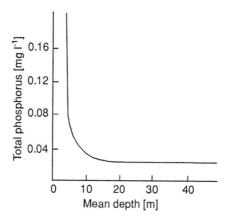

Fig. 2.4. Relation between total phosphorus in lakes of different depths. (After Sakamoto 1966)

Because the production of biomass requires relatively little phosphorus (about 1/100 of the total biomass weight), it takes very little phosphorus to remove the limitation and restart growth. The only way of regaining control is by cutting back the external phosphate input and interrupting the internal lake cycles.

Dead or living organic material sinks down into the lake depths, undergoes further degradation and transformation, and finally becomes sediment. At this point, the nutrients are partly stored in the permanent sediment and partly go back into aqueous solution and are returned to the nutrient-hungry epilimnion by the mixing processes.

2.2
Trophic State

The evaluation of the trophic state of a lake is based on many types of data such as the lake basin morphometry, physicochemical parameters, biological parameters, and the rates of lake metabolism including nutrient inputs. Table 2.3 shows the OECD boundary values for an open trophic classification system. Issues relating to predator and resource controls of trophic-level biomass and concerning predator dependence versus resources dependence of trophic-level biomass are currently being debated (Hunter and Matson 1992). If trophic-level biomass is predator dependent, increasing resources should increase predator biomass, but not the prey biomass (Fretwell 1977). In contrast richer resources would be expected to produce proportional increases in both predator and prey biomass (Arditi and Ginzberg 1989; Slobodkin 1992).

Table 2.3. OECD boundary values (annual mean values) for an open trophic classification system. (After OECD 1982)[a]

Parameter		Oligotrophic	Mesotrophic	Eutrophic	Hypertrophic
Total phosphorus	x	8.0	26.7	84.8	
(μg l^{-1})	x ± 1 SD	4.9 – 13.3	14.5 – 49	48–189	
	x ± 2 SD	2.9 – 22.1	7.9–90.8	16.8–424	
	range	3.0–17.7	10.9–95.6	16.2–386	750–1200
	n	21	19 (21)	71 (72)	2
Total nitrogen	x	661	753	1875	
(μg l^{-1})	x ± 1 SD	371–1180	485–1170	861–4081	
	x ± 2 SD	208–2103	313–1816	395–8913	
	range	307–1630	361–1387	393–6100	
	n	11	8	37 (38)	
Chlorophyll *a*	x	1.7	4.7	14.3	
(μg l^{-1})	x ± 1 SD	0.8–3.4	3.0–7.4	6.7–31	
	x ± 2 SD	0.4–7.1	1.9–11.6	3.1–66	
	range	0.3–4.5	3.0–11	2.7–78	100–150
	n	22	16 (17)	70 (72)	2
Chlorophyll *a*	x	4.2	16.1	42.6	
Peak value	x ± 1 SD	2.6–7.6	8.9–29	16.9–107	
(μg l^{-1})	x ± 2 SD	1.5–13	4.9–52.5	6.7–270	
	range	1.3–10.6	4.9–49.5	9.5–275	
	n	16	12	46	
Secchi depth	x	9.9	4.2	2.45	
(m)	x ± 1 SD	5.9–16.5	2.4–7.4	1.5–4.0	
	x ± 2 SD	3.6–27.5	1.4–13	0.9–6.7	
	range	5.4–28.3	1.5–8.1	0.8–7.0	0.4–0.5
	n	13	20	70 (72)	

[a]The geometric means (after being transformed to base 10 logarithms) were calculated after removing values which were greater than two times the standard deviation obtained (where applicable) in the first calculation.
x = geometric mean, SD = standard deviation, () = the value in brackets refers to the number of variables (n) used in the first calculation.

Although many hypothetical models have been proposed to explain how predation and resources may shape the structure of communities, the models have received little field verification. Food chain models need to be verified to determine when and how much of the observed variation in trophic-level biomass is produced by predation vs. resources. According to some of these models, plant biomass should increase in response to increasing resource availability in odd-link ecosystems, but not in even-link ecosys-

tems. For lakes, it is difficult to find odd-linked and even-linked lakes because om-
nivory obscures the delineation of organisms into well-defined trophic levels (Mazumder 1994). Hansson (1992) attempted to test the predictions of the food chain models
with data from small lakes containing odd- and even-link trophic levels. An alternative
approach is to describe functionally dominant trophic interactions rather than to desig-
nate the lake ecosystems by the number of trophic levels present (Mazumder 1994). In
many North temperate lakes, *Daphnia* is a keystone filter-feeding species and is an efficient consumer of small, edible algae. It is in turn a prey of choice for planktivorous
fish (Hansson 1992; Sarnelle 1992). The presence of the large *Daphnia* (e.g., *D. pulex,
D. magna, D. longispina*) in any lake indicates that these primary consumers are released from predators, because these *Daphnia* species are rarely encountered when the
biomass of predators (planktivorous fish) exceeds 20 kg ha^{-1} (McQueen and Post
1988). Conversely, smaller species of *Daphnia* (e.g., *D. cuculatta, D. ambigua, D.
cristata*) often coexist with a high biomass of planktivorous fish (Gulati 1990). Commonly, zooplankton communities include small cladocerans (*Bosmina* 0.2–0.4 mm,
Ceriodaphnia 0.3–0.5 mm), rotifers (0.06–0.2 mm), cyclopoid copepods (0.4–
1.0 mm), and nauplii (0.1–0.3 mm) in high planktivore systems (Brooks and Dodson
1965; Mazumder et al. 1992); these smaller species are less efficient consumers of
algae. Although the absence or presence of large *Daphnia* and a biomass of
planktivorous fish <20 kg ha^{-1} are not good indicators of the number of trophic levels
present or the length of the food chain, they do point to the dominant trophic link in a
lake (Mazumder 1994). Based on the results from several temperate lakes covering a
wide gradient of potential productivity, Mazumder (1994) has shown that algal
biomass response to phosphorus is stronger in those lakes where nutrient availability is
the primary determinant of algal biomass because here the grazer control is less
important. This observation is consistent with the predictions of the food-chain models.
However, the view that herbivore pressure may be most severe in relatively
unproductive environments (McQueen et al. 1986) is not supported by his data.
According to Mazumder (1994), algal biomass appears to be both predator- and ratio-
dependent because grazers can often modify algal response to nutrients, and algae
show a positive response to nutrients even in the presence of large grazers.

According to the cascading trophic interaction hypothesis of Carpenter et al. (1985)
and the biomanipulation hypothesis of Shapiro and Wright (1984), an increase in the
biomass of top predators (piscivores) in a lake leads to decreased planktivore biomass.
An alternative hypothesis is the so-called bottom-up: top-down model of McQueen
et al. (1986). It combines the predicted influences of both predators (top-down) and resource availability (bottom-up) and postulates that top-down forces should be strong at
the top of the food web and weaken towards the bottom. In eutrophic lakes particularly, the impacts of changes in piscivore biomass tend to weaken as they cascade
down through the food web so that there is little, if any, effect on chlorophyll biomass.

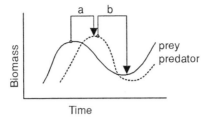

Fig. 2.5. Population dynamics of a hypothetical predator-prey system. **a** delayed bottom-up impact; **b** delayed top-down impact. (After Matveev 1995)

Trophic cascade theory also postulates that fluctuations in biomass of the higher trophic levels do not cascade instantaneously through lake food webs; rather there are lags in ecosystem response (Carpenter et al. 1985); the lags would be expected to be longer in organisms of higher trophic levels in view of the fact that they have longer generation times (Fig. 2.5; Matveev 1995).

Matveev (1995) analyzed a previously described system consisting of three trophic levels (phytoplankton, zooplankton and an invertebrate predator in a subtropical, northern Argentinian eutrophic lake) to test the generality of the above two hypotheses of food web theory, i.e. (1) top-down or predator-on-prey impact weakens the food web and (2) time lags in the trophic level interactions decrease in duration from top to bottom of food webs. In the lake studied, a spring pulse of phytoplankton had a delayed impact on the biomass of upper trophic levels – herbivorous zooplankton and the predatory water mite, *Piona* sp. Later in the season, the top-down control by *Piona* cascading down the food web showed that the top-down effect did not weaken downwards; rather, the impact of zooplankton biomass on phytoplankton biomass was stronger than that of the invertebrate predator on zooplankton, despite the fact that the predator caused a decline in zooplankton biomass. Contrary to the theory, the time lags did not differ significantly between the impacts of trophic levels of a given route direction. However, they did differ between bottom-up versus top-down routes in accordance with the prediction of the theory. Thus, it appears that some predictions of the food web theory may be restricted to fish-impacted systems, for which it was originally developed. An alternative explanation may be that the strength of top-down impact is associated with the width of consumers' feeding niches; the impact is stronger when niches are wider (Matveev 1995). Thus the findings of Matveev do not support the idea of McQueen et al. (1986), that top-down forces should weaken the food web. In the Argentinian eutrophic Laguna lake, the effect of zooplankton biomass on phytoplankton biomass was stronger than the effect of invertebrate predators' biomass on zooplankton biomass. Similarly, bottom-up impacts became stronger up the food web, contrary to the dictates of the bottom-up:top-down view.

According to Matveev, the observed failure of the bottom-up:top-down model to predict the events in Laguna lake may be found in the comparison of the feeding niches of zooplankton on invertebrate predators. The impact of zooplankton on phytoplankton was stronger than that of *Piona* on zooplankton because *Daphnia*, which dominated in zooplankton, consumed a wider diversity of food items than the water mite, and had a wider feeding niche than that of *Piona*. Hence, *Daphnia* grazing affected all components of the lower trophic level, while *Piona* predation affected only a small part, because *Piona* virtually did not consume copepods. Thus, the predators' top-down control was less efficient than grazers' control.

2.3
Interactions Between Physical and Chemical Processes

Few other ecosystems are better suited for the study of the basic principles governing the interactions between physical, chemical and biological processes than a lake. Lake ecosystems have distinct boundaries, are relatively homogeneous and are subject to external influences to only a limited degree.

Let us examine the aquatic ecosystem by considering a mass of water contained in a depression in the Earth devoid of life. It receives some inflows, which renew the water forming the water body, and also add kinetic energy into the lake, which creates mixing. Wind is another important and obvious mixing agent. Together with the heat exchange between the lake and its environment, it is the wind which mainly dominates the physical events occurring in the lake. A lake is an energy reservoir. Over short periods of time it takes up thermal and mechanical energy and later releases it to its environment. In the long term, however, the energy input and output tend to balance out. Mechanical energy shows itself as the kinetic energy of currents or as potential energy stored in the so-called density stratification, by which lighter water overlies heavier, thereby greatly limiting the vertical exchange of water (Imboden et al. 1993).

Seasonal fluctuations in the thermal energy flux result in corresponding variations in the density stratification, and consequently in the ratio of the intensity of vertical mixing to that of horizontal mixing. In the horizontal direction, mixing times are much shorter. For complete horizontal mixing, a medium-sized lake requires a period of several weeks; for complete vertical mixing, however, a period of several years may be needed. The density of water depends on its temperature and on its chemical composition, and the chemical substances dissolved in the water influence the physical processes occurring in the lake.

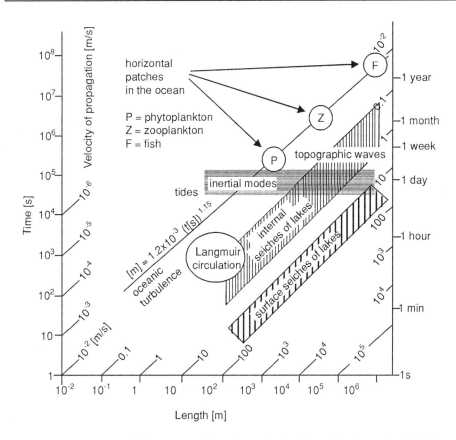

Fig. 2.6. Typical length and time scales of mixing modes in lakes. The diagonals with slopes equal to one represent lines of constant velocity of propagation (e.g., surface seiches moving with velocities between about 10 to 100 m s^{-1}). Typical differences among scales of biological patchiness may be interpreted as the direct consequence of the influence of the turbulence scales on species of different growth rates. (After Imboden 1990)

Mixing in lakes is influenced not only by inflows and climate but also by geochemical factors. This is well known in the oceans, with their much higher salt concentrations: the so-called thermohaline circulation, driven by the differing salt concentrations in the various oceans, is ultimately responsible for the global ocean circulation and for the renewal of the oceanic deep water, a process which occurs on a time scale of centuries (see Imboden 1990).

Along with chemical and biological processes, mixing processes in lakes (Fig. 2.6) are responsible for the spatial and temporal distributions of various substances dissolved in the lake water. The extent of heterogeneity or homogeneity of distribution of such substances is determined by the relative rates of mixing and reaction by chemical or biological processes. Substances which react only slowly are usually homogeneously distributed whereas those which react rapidly are more heterogeneously distributed in a lake. Because lakes are mostly horizontally stratified (Fig. 2.6), concentration variations of dissolved substances with depth are generally much more pronounced than variations along the horizontal direction. This is particularly true of those substances involved in the redox reactions associated with the photosynthesis/respiration cycle. Chemical and biological events in lakes are thus characterized by redox conditions which exhibit a typical temporal and vertical spatial structure. The limnetic or pelagic zone of a temperate lake is divided into three regions:

1. The trophogenic zone in the well-illuminated warm upper layer rich in photosynthetic algae. It is defined largely by the epilimnion, but is not identical to it. A high absorption of light occurs by the phytoplankton, and the photosynthetically available radiation (PAR) usually fails to reach the lower layer of the epilimnion, while PAR in clear water lakes may penetrate the epilimnion and even parts of the hypolimnion. Thus, the boundary of the trophogenic-tropholytic zones is a function of light penetration. At the compensation depth, photosynthesis is matched by respiration.
2. A body of colder, heavier, stagnant water mostly undisturbed by wind action, called the hypolimnion, which lies at the bottom.
3. An intermediate layer of water called the metalimnion or thermocline (see Fig. 2.7) where there is a marked drop of temperature with increasing depth. The thermocline is usually considered as being limited to a zone where the temperature drops at least by 1 °C per 1 m depth. The metalimnion is a region of high heterotrophic activity as the organic matter produced photosynthetically in the epilimnion is decomposed in the metalimnion.

Lakes can act either as sources or as sinks for CO_2. Being strongly supersaturated in CO_2, Arctic lakes are sources to the overlying atmosphere. The source of CO_2 supersaturation in the Arctic lakes is transport of tundra organic matter to surface waters. Those regions that lack the vast soil carbon storage of the Arctic may behave differently: thus several temperate and boreal lakes are net *sinks* for atmospheric CO_2 (Cole et al. 1994). Cole et al. have shown that several boreal, temperate, and tropical lakes are typically supersaturated with CO_2 and thus are net *sources* to the atmosphere.

In a large number of these lakes, the mean partial pressure of CO_2 averaged 1036 µbar, about three times the value in the overlying atmosphere, indicating that lakes are sources rather than sinks of atmospheric CO_2.

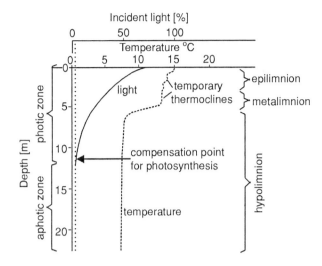

Fig. 2.7. Structure of a stratified temperate lake. The temporary thermoclines are formed by heating on calm days; they disappear each night by convective cooling or by wind

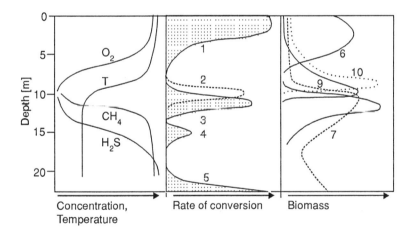

Fig. 2.8. An idealized vertical profile of a freshwater lake in the temperate zone, showing approximate concentrations, rates of conversion, and biomass. T temperature in °C, O_2, H_2S and CH_4 in mg l^{-1}, conversion rates in μg l^{-1} day^{-1}: *1* CO_2 fixation in the light, i.e., oxygenic photosynthesis, *2* CO_2 fixation in the dark, *3* CO_2 fixation in the light (anoxygenic photosynthesis), *4* and *5*, sulfate reduction, *6* to *10* biomasses in mg l^{-1}, *6* algae and cyanobacteria, *7* total bacterial mass, *8* phototrophic bacteria, *9* protozoa, *10* cladocera and copepods

On a global scale, the potential efflux of CO_2 from lakes (about 0.14×10^{15} g of carbon per year) is about half as large as riverine transport of organic plus inorganic carbon to the ocean. Lakes thus act as a small but potentially important conduit for carbon from terrestrial sources to the atmospheric sink (Cole et al. 1994).

2.4
UV-B Effects

The phenomena of acidification, climate warming, and the thinning of the ozone layer are usually considered separately. But Schindler et al. (1996) have reported that acidification and climate warming increase the exposure of aquatic organisms to deleterious UV-B radiation much more than ozone depletion. Between 1971 and 1990, they observed several lakes in north-western Ontario. During this period, the UV-B radiation increased by 10% as a result of depletion of the ozone layer.

The depth to which UV-B radiation can penetrate depends exponentially on the amount of dissolved organic carbon (DOC) in the water. Most boreal lakes have DOC concentrations of several mg l^{-1}, and this limits the ultraviolet penetration to a few centimeters. Thus, DOC effectively protects against exposure to UV-B radiation. This is where climate warming and acidification enter the picture. Both interact to lower the concentration of DOC. During the two decade period of observations, Schindler et al. noted an average climate warming of 1.6 °C. Along with the increase in air temperature, precipitation decreased by 25%, while evapotranspiration increased by 35%. The consequences were reduced inflows from the catchment areas and lesser amounts of DOC entering the lakes. Processes within the lakes also contributed to reducing the DOC in the water, causing concentrations to fall by 15–20%. This in turn enabled the UV-B radiation to penetrate 20–60% deeper. It may be noted that the 10% increase in the intensity of the radiation during the period only led to a 2% increase in the UV-B penetration in a lake whose water had only a relatively small amount of DOC.

In lakes that were deliberately acidified for experiments, the decline of DOC was much greater than in reference lakes. Similar observations have often been made in lakes acidified by rain both in Europe and North America. In the most extreme cases, the volume of water exposed to UV-B radiation was eight times larger than before the decline. Of nearly 700 000 lakes in eastern Canada, Schindler et al. (1996) found that the DOC concentrations in about 140 000 were low enough for UV-B penetration to be of concern, and that the highest concern must be for clear, shallow lakes, streams, and ponds, where even modest declines of DOC may eliminate the small regions that are deep enough to provide refuges from damaging UV-B radiation.

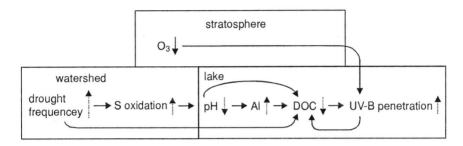

Fig. 2.9. Linkages among drought, acidification, DOC concentration and UV-B penetration depth in lakes. *Solid arrows* link processes; *dotted arrows* indicate changes in frequency, intensity or concentrations of constituents or processes. DOC concentration can be affected by acidification, aluminum increase and hydrological change (drought), also by changes in DOC decomposition rate induced by changes in UV-B penetration depth. In some lakes the chief mechanism for DOC decline can be drought-induced acidification. (After Yan et al. 1996)

Alpine lakes are also at risk, since they may receive incident UV-B that is more than 50% higher than that at sea level. High altitude species of trout have been shown to suffer sunburn, increased fungal infections, and higher mortality rates at environmentally realistic exposures of UV-B.

Climate warming, acid rain, and depletion of stratospheric ozone. interact in highly complex ways (Schindler et al. 1996). Huge amounts of carbon dioxide and sulfur dioxide have been released into the atmosphere through the combustion of fossil fuels and the smelting of metal-sulfide ores. These releases can alter ecosystems through global warming, more frequent droughts, acid rain, and sulfur dioxide toxicity to plants.

Yan et al. (1996) worked on Swan Lake, near Sudbury, Ontario, where the smelting of copper and nickel ores has harmed the landscape. Swan Lake received severe atmospheric deposition of sulfur from smelter fumes, and substantial amounts have been stored in sediments. During a drought the water level drops and the lake's surface area shrinks, leading to oxidation of the sulfur compounds contained in the uppermost layer of exposed near-shore sediments to sulfuric acid. Later the acid washes into the lake and lowers its pH greatly which in turn decreases DOC (see Fig. 2.9).

Rapidly flushing lakes with wetlands in their catchment areas are those most at risk, particularly in regions where acid deposition is severe. However, wetlands to some extent mitigate UV-B penetration by exporting dissolved organic carbon to lakes (Gorham 1996).

The level of UV-B irradiation of thousands of lakes has already increased because of stratospheric ozone depletion. Further increases, especially in the shorter, more photochemically active UV-B wavelengths, can be expected over the coming few decades. As global emissions of greenhouse gases will also continue to increase at least for some time, the detrimental interaction of global climate change, acid deposition and increased UV-B irradiance is likely to persist in some regions.

2.5
Seiching

Water in lakes slops back and forth like water in a bathtub. The surface of a lake very often does not remain still and flat. Besides the waves which break the surface, the entire lake surface oscillates. The vertical movement caused by these oscillations is sometimes substantial. This phenomenon is known as seiching. It begins when part of the lake surface is set-up (forced above its equilibrium level), usually by wind friction or a barometric pressure gradient, and then released to return to its equilibrium level. More unusual causes include heavy rain, snow or hail over a part of the lake and seismic movement of the lake bed. The seiche period is the time taken to complete one oscillation. It is a function of the lake size, geometry and bathymetry (depth contours).

2.5.1
Importance of Seiches

An accurate equilibrium lake level is needed for many hydrological calculations. In the past this was achieved by taking averages of long time intervals to eliminate the effect of surface oscillations but, in recent years, many lake water-level sites have been equipped with telemetry units which give immediate access to the data via radio connections to the recorders. Using the information as soon as it is recorded means that averaging over long time intervals is not possible. Therefore, knowledge of the surface oscillations becomes more important in the calculations (Carter and McKerchar 1995).

A good example is the calculation of lake inflows, needed for the management of, for example, hydroelectricity generation schemes. Often a lake's inflow cannot be measured directly because there are multiple sources, including seepage. The inflow must therefore be calculated from measurements of the outflow plus an estimate of the change in the lake volume (using lake area and lake level change). Errors in the lake level measurements because of seiching introduce a significant difference to the final inflow estimate (Carter and McKerchar 1995).

2.5.2
Circulation Patterns

On the basis of their major water circulation patterns, most lakes belong to one or the other of the following broad types:

1. *Dimictic*: have two seasonal periods of circulation or overturns; in spring after ice-break and in autumn after the decline of the epilimnion temperature and when the homothermic water mass becomes mixed. This pattern is characteristic of temperate lakes.
2. *Warm monimictic*: free water circulation in winter when the surface water sinks down. Usually lakes of temperate latitudes, Mediterranean and subtropical climates. The temperature does not drop below 4 °C.
3. *Oligomictic*: thermally stable tropical lakes with very slow or rare mixing.
4. *Polymictic*: irregular, fairly continuous mixing periods depending on the lake morphometry and climate. Usually equatorial and high attitude lakes without marked temperature and density gradients.
5. *Meromictic*: lacking complete circulation; no mixing of surface water with bottom water layers.

2.6
Regional Lake Quality Patterns

Understanding of regional lake quality patterns is important for lake restoration as it places specific lake conditions into perspective, constitutes a basis for establishing lake quality goals, identifies lakes which are most likely to benefit from restoration, and acts as a framework for assessing restoration success.

Restoration is a misnomer as it is frequently envisioned to mean pristine. A more accurate practical definition reflects a recreation of acceptable environmental conditions. As a rule these conditions reflect a local or regional perspective relative to uses for which a degraded lake was once suitable before its degradation (Bjork 1994). Inspired by the phosphorus limitation phenomenon in most freshwater lakes (see Dillon and Rigler 1974; Vollenweider 1975), restoration has usually focused primarily on reducing this nutrient in the lake to a level that gives more acceptable phytoplankton species composition and biomass. Prediction of the effectiveness of lake restoration improves if regional lake quality is taken into account in the evaluation of potential success.

The crucial issue is that the intensity of increasing anthropogenic usage and exploitation greatly impairs the natural basic functions of the lakes, namely the provision of clean water, a suitable habitat for aquatic flora and fauna and an intact countryside. Apart from the basic conflict between economy and ecology, conflicting interests also arise between differing anthropogenic uses as, for example, between expanding urbanization and tourism, and this can lead to self-induced restrictions on utilization in the leisure and tourism sector, for which clean water is at the same time both a basis and a commodity.

The principal aim for further development of the lakes should therefore be to ensure the provision of a well-balanced relationship between utilization requirements and settlement of conflicts so that the ecological sustainability of the lakes is ensured on a long term and continuing basis. This may be achieved firstly by adopting technical measures to reduce the existing impairment of the lakes' efficient functioning to an "ecologically acceptable" level, in particular relative to contamination of the water with harmful substances, and the prevention of the destruction of areas which are ecologically sensitive, and secondly by taking recourse to interdisciplinary concepts which can regulate and coordinate the uses to such an extent that serious impairment of individual functions is avoided.

2.6.1
Ecoregions

One approach for characterizing regional lake quality patterns is to integrate existing surrogate landscape level information pertaining to surface water quality in a manner such as Omernik's ecoregions (1987; 1995). Omernik's original ecoregions are based on regional similarities in a combination of spatial characteristics that influence aquatic resource condition, including soils, geology, land surface form, climate, potential natural vegetation and land use. These ecoregions, together with other data on lake properties make a basis for delimiting current conditions and expectations for lake quality and estimating the success of restoration treatments.

2.7
Inland Lakes

Many inland lakes have no natural outlets to oceans. Most of these are in arid or semiarid areas. Among them the Caspian, Aral and Dead Seas are lakes that have suffered strong environmental damage. The Caspian Sea suffered a 3 m fall in water level from

1932 to 1945 and all its ports had to be rebuilt to adjust to the lower water level. In 1977, the water level started to rise and has risen more than two meters in 15 years. The lake economy must once again adjust accordingly with changes to cities and other settlements, railways and roads.

In the Aral Sea the water level started falling in the 1960s as a result of inflowing water being extracted for agricultural purposes in upstream areas. The 10 m drop was catastrophic and the Aral Sea shrank to half its original size. The salt concentration in the water rose three times, to nearly the same concentration as sea water. The former lake bottom is now a salty desert with salt being blown away by the wind. This has led to an unacceptable quality of drinking water as well as salt enriched air. The Dead Sea has seen a drop in water volume due to water extraction from the River Jordan by surrounding countries, often involving political conflicts.

Supplying water to those lacking access to potable supplies, and protecting drinking water systems from unnecessary waste and contamination are two chief concerns related to the use of freshwater resources. The former problem usually requires laying out and installing a new supply system. The latter requires that an existing distribution system be improved and updated.

There are two main areas where distribution systems suffer reduced efficiency. The first is physical water loss, or water that does not reach the users, causing waste of a precious resource, and an increase in the water company's costs. This may be in the form of visible and invisible water leakage, reservoir overflow, defective control equipment and/or improper operational controls.

Another cause of water loss is inefficiencies in the system's measuring system, i.e. water may be reaching consumers but the users are not being properly billed. This may result from defective flow meters, illegal connections. insufficient water supply and/or users without flow meters or flow limiters. Water quality can be weakened in several ways, such as pollution of the raw water sources, contamination in the water distribution system and/or contamination in the user's home.

Field activities can include discovering leaks in the system and illegal connections, inspecting flow meter calibration, inspection of the water treatment system to detect quality problems, inspection of the distribution system for waste due to human or automation errors, auditing the billing process and preparation and training of technicians.

In Saudi Arabia, sea water is being converted into freshwater on a large scale using solar power. Powerful lenses intensify sunlight by 40 times to produce steam which is used to run coolers which freeze sea water. When sea water freezes, the salt collects on the outside of ice crystals and can be separated by washing the crystals in freshwater. The crystals are then melted into freshwater. This method uses wave, wind and solar energy, all abundant renewable energies, to convert sea water into potable water. This plant produces millions of gallons of water annually, enough to meet the domestic needs of 12 600 people daily.

Windmills could be used to lift sea water into stainless steel boilers and freezers, and a row of these units could bring large tracts of land under freshwater irrigation, promoting food production, employment and income, in addition to giving people water for drinking and bathing.

Direct applications are also available. Man-made sea water canals, such as the canal cutting across Dubai or the Buckingham Canal running parallel to the Indian east coast can be used for navigation, desalination, saline agriculture and for promoting sea food production.

When sea water is made available in open-cut canals, local residents can easily perform desalination by employing simple evaporation-condensation techniques. Canals also carry water for bathing cattle and domestic uses. Today, 40% of potable water used worldwide is used in the toilets of urban centers. With earthenware and rigid PVC pipes, sea water can be used for flushing toilets.

New crops are being found that can grow on salty soil irrigated with sea water. Hundreds of plants and trees have been identified in a USAID book on *Saline Agriculture*. Cotton has so adapted to salt water in Israel that yields have been increased by 20%. Some oil plants and medicinal plants are now being grown near the highly salty Dead Sea and in Eilat on the Red Sea.

2.7.1
Nutrients

Nitrogen, phosphorus and several other essential elements are needed by plants for their growth. In natural ecosystems one or more of the essential nutrients are sometime present in limiting amounts. Man tries to increase yields of crops by supplying them with the limiting nutrients (which are commonly nitrogen and/or phosphorus). However, some of the added fertilizers containing these nutrients are not absorbed by crops or otherwise leach off into water bodies where they stimulate the growth of nuisance algae. Thus the same nutrients are beneficial to us on land but become a liability in rivers, lakes and streams. In general, soils should be rich in nutrients but lakes and rivers should be deficient in nutrients. Overfertilization in rivers and lakes produces overgrowth of algae and macrophytes and can cause oxygen depletion, thus affecting the fishes there which often die of asphyxiation. Of course, sometimes in developing countries, excessive growths of macrophytes in ponds are deliberately desired as the plant is a valuable commercial/agricultural product (e.g., lotus or *Trapa bispinosa*). In developed countries, lakes are better valued for recreational use and therefore in those countries it is preferable to have as little fertilization in lakes as possible.

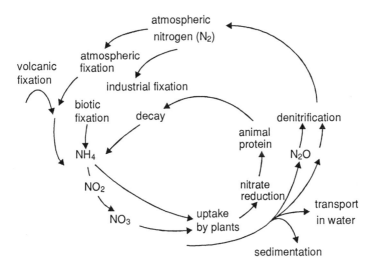

Fig. 2.10. Some major transformations of nitrogen in the biosphere and its transport in water

Eutrophic lakes have large inputs of nitrogen and phosphorus which cause explosive growth of algal populations, mostly confined to surface layers of water. When the algae die or the organic matter released by them decomposes, the oxygen of the water is used up and the deeper layers of the lake become anoxic, leading to asphyxiation of fishes.

Most of the nitrogen comes to the water (Fig. 2.10) through atmospheric fallout, leakage from farmland and forest soils, and sewage effluent. That which is deposited from the air consists almost equally of nitrogen oxides (from the combustion of fuel in road vehicles and stationary plants) and ammonia (evaporated from agriculture).

To this direct deposition must be added the airborne nitrogen that first falls on land and eventually becomes transported into the sea via the rivers. While the forest soil still seems to be capturing most of the nitrogen that is deposited on it, its critical limits for nitrogen saturation are being exceeded in most parts of central and Northern Europe. Gradually leakage will start, and consequently nitrogen will sooner or later reach the sea. Figure 2.11 illustrates nitrogen transformations near the sediment-water surface. Sewage is very rich in nitrogen and phosphorus and is a powerful cause of eutrophication. Most aquatic plants are P-limited. Thus, the addition of even small amounts of P stimulates their growth.

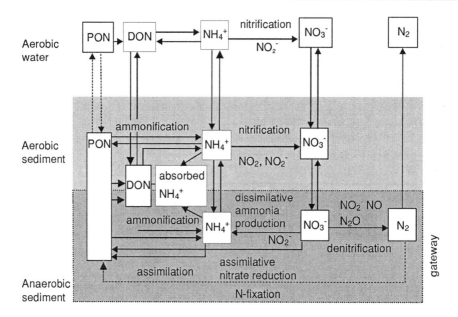

Fig. 2.11. Nitrogen transformations near the sediment-water interface. (After Santschi et al. 1990)

Sewage (including detergents) is the chief source of phosphorus inputs in water, and hence is the chief cause of cultural eutrophication (man-made eutrophication). Whenever the discharge of sewage into a lake has been stopped for long periods, the eutrophic lake has reverted back to a clean or oligotrophic state. Whereas phosphorus is the most important cause or source of cultural eutrophication, excessive use of inorganic nitrogenous fertilizers (not organic manure) is no less hazardous, especially from the viewpoint of human health. Even fairly low levels of nitrates in water can cause methemoglobinemia (and possibly also cancer). Nitrous oxides released into air can potentially thin out the protective ozone layer around the upper atmosphere, thereby permitting some more of the harmful ultraviolet radiation to penetrate to the Earth's surface.

When there is plenty of organic nitrogen or humus (organic compounds of carbon, nitrogen and other elements) in the soil, the nitrogen cycle tends to be quite closed and little if any nitrate is leached off into water. Additions of inorganic fertilizer interrupt the nitrogen cycle and much nitrate leaches off.

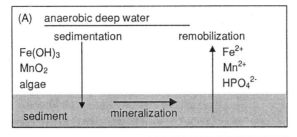

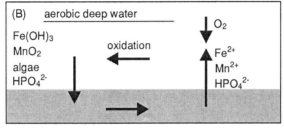

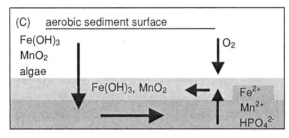

Fig. 2.12. The coupling between oxygen concentration in the deep water and phosphorus release from the sediments. **A** If oxygen is completely lacking in the deep water in summer, the deep water becomes enriched with reduced substances such as dissolved iron(II) and manganese(II) ions. Iron hydroxide, $Fe(OH)_3$, and manganese dioxide, MnO_2, onto which phosphorus can be adsorbed to a great extent. The degree of phosphorus retention by the sediments is low. **B** If the deep water contains oxygen, either naturally or as a result of artificial aeration, a redox cycle develops. Reduced iron and manganese ions are oxidized in the deep water. This causes a local increase in the rate of sedimentation of $Fe(OH)_3$ and MnO_2. These minerals are rapidly reduced again at the sediment surface. Phosphorus undergoes transport along with these substances, but the amount of phosphorus retained in the sediments remains low. **C** Not until oxygen penetrates into the sediment can an oxic layer develop, in which oxidized iron and manganese minerals become enriched. This gives rise to new phosphorus adsorption capacity. (After Wehrli et al. 1993)

Enrichment of phosphorus, reduced iron(II) and reduced manganese(II) occur simultaneously in the deep waters of eutrophic lakes during summer stagnation as soon as the oxygen is totally depleted (Mortimer 1941; Wehrli et al. 1993). Microorganisms at the sediment surface use nitrate, sulfate, iron(III) oxide and manganese(III) oxide as oxidizing agents for the oxidation of degradable biomass when oxygen is lacking. If and when these oxidizing agents are used up, fermentation and methanogenic bacteria take over, eventually releasing gaseous methane. Since amorphous iron hydroxide, $Fe(OH)_3$, shows a very high phosphate adsorption capacity, the reductive dissolution of iron(III) also mobilizes phosphorus (Fig. 2.12A). If oxygen is present in the deep water, dissolved iron(II) and manganese(II) are oxidized to solid iron hydroxide and manganese oxide, MnO_2, at the sediment surface. These minerals then become enriched in the transition zone between sediment and water, and dramatically increase the phosphorus retention capacity of the sediment (Fig. 2.12C). With little mixing above the lake bottom, the connection between oxygen and phosphorus loosens; if the sediment surface becomes anaerobic, dissolved iron(II) and manganese(II) begin to be exported unhindered into the deep water. Since these elements do not undergo oxidation until they have reached the bulk water, their concentration in the vicinity of the sediment surface remains low. Under these conditions, an efficient final storage of the algal nutrient phosphate in the sediments does not occur (Fig. 2.12B). Instead, an intense redox cycle develops in the deep water, resulting in the reduction of particulate iron oxide and manganese oxide in the upper layers of the sediment. The dissolved ions Fe^{2+} and Mn^{2+} are transported into the aerobic deep water, where they are oxidized back to particulate $Fe(OH)_3$ and MnO_2. By analogy with an engine transmission, with regard to phosphorus, such a cycle (Fig. 2.12B) corresponds to the gears being in neutral, since, although the phosphorus is retained within the cycle, phosphorus retention in the sediments is hardly affected (Wehrli et al. 1993). Average values of total concentrations of metals and ranges of concentrations of ligands (complexing sites) in natural freshwaters are shown in Fig. 2.14.

Figure 2.13 shows the transformations of Fe(II,III) at an oxic-anoxic boundary in a water column or sediment. Peaks in the concentrations of solid Fe(III) (hydr.) oxides and of dissolved Fe (II) coincide with the depth of maximum Fe (III) and Fe (II) production. Good dissolution of Fe (III) (hydr.) oxides results from the combination of ligands and Fe (II) produced in the underlying anoxic layer or zone.

Figure 2.15 illustrates how the residual metal ion concentrations in oceans and lakes are regulated by the adsorption of metal ions to settling particles and by the tendency of ligands to form soluble complexes. Jorgensen and Vollenweider (1989) developed a conceptual model illustrating the connection between the external load and compartmental routes inside a lake (Fig. 2.16).

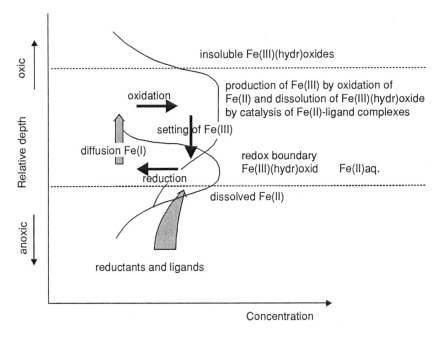

Fig. 2.13. Transformations of iron at an oxic-anoxic boundary in a water column or sediment. (After Davison 1985)

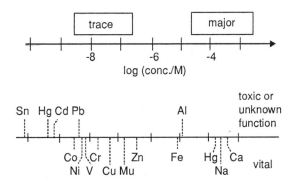

Fig. 2.14. Mean values of metal concentrations and ranges of ligands in natural freshwaters. (After Buffle and Altmann 1990)

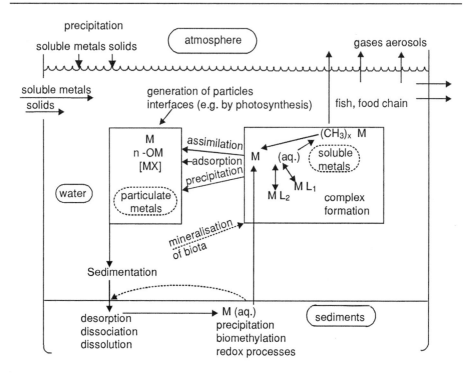

Fig. 2.15. Regulation of residual metal ion concentrations in oceans and lakes by the adsorption of metal ions in the settling particles and by the tendency of ligands to form soluble complexes. (After Morgan and Stumm 1990)

Many surface waters possess a fairly strong buffering capacity. Figure 2.17 illustrates the bicarbonate buffering system in soils and surface waters. Losses in soil fertility caused by raising crops should be restored by supplying organic manure or excreta, not by overuse of inorganic fertilizers. Manure and wastes from cattle, poultry and pig farms etc. should not be allowed to go into waterways but should be taken to agricultural fields where their addition to soil greatly improves soil fertility and crop yields.

Mercury and other toxic metals have attracted concern because of their accumulation in the aquatic food chain. The toxicity of mercury has been known since the Minamata disaster in 1953. Methylmercury is not tightly bound to sediments. It is fairly water-soluble and assimilable by biota. Methylmercury causes severe neurological disorders and even death in humans. Several instances of serious illness and death resulting from eating contaminated fish, shellfish or Hg-treated cereal grains have been reported not only from Japan but also from Iraq, Sweden, Canada and Ghana.

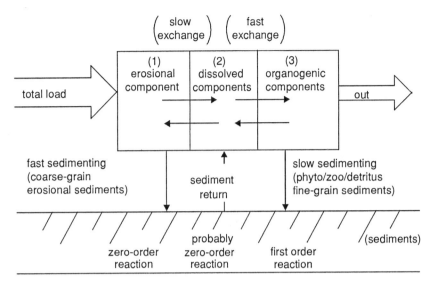

Fig. 2.16. Conceptual model illustrating the connection between the external load and major processes and compartmental routes inside a lake. (After Jorgensen and Vollenweider 1989)

Lead is a source of brain damage, mental deficiency and nephritis. Drinking water can be contaminated from lead pipes or from storage in lead-glazed ceramics and improperly baked earthenware containers. Untreated wastes rich in chromium can poison rivers, lakes, and drinking water, and several such episodes are known from Japan and other countries. Taiwan, Argentina, Chile and Japan are known to have suffered from outbreaks of arsenic poisoning resulting from consumption of milk powder that had been preserved with arsenic contaminated sodium phosphate. Some sources of toxic metals in water bodies are shown in Table 2.4.

The eutrophication concept has been confused since the early 1920s because it was based on such instantaneous parameters as oxygen tension, nutrient concentration and phytoplankton density. This kind of empirical and static concept did not support any functional relationship between the different state variables nor the prediction of the dynamics of the system response to perturbations. Later, the concept of **rates**, i.e., the velocity of exchange between various mass and energy compartments, was introduced and it then became possible to relate concentrations to supply and consumption as well as phytoplankton density to primary production.

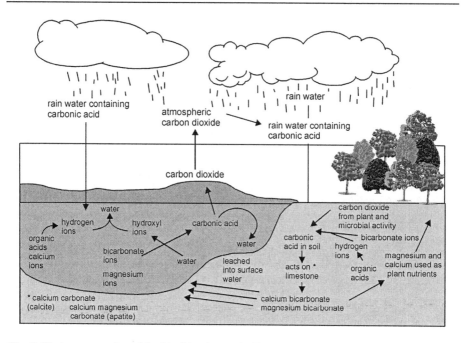

Fig. 2.17. A conceptual model of the bicarbonate buffering system in soil and surface water

Table 2.4. Toxic metals in water bodies

Metal	Sources	
	Natural minerals	Man-made
As	$FeAsS$, As_2S_2, AsO_2,	Herbicides, fertilizers,
	$FeAs_2$, As_4S_4	Detergent presoaks
Cd	$CdCO_3$, CdS, CdO	Electroplating, photography, dyes
Cr	$Fe_2Cr_2O_4$	Metal plating, dyes, ink
Pb	PbO, PbS, $PbCO_3$, $PbSO_4$	Ammunition, motor fuel
Hg	Hg_2Cl_2, HgS, HgO	Chlorine manufacture, electronics, pesticides, fungicides
Se	SeO_3, SeO_4	Copper smelting
	(traces in sulfide ores)	
Ag	Ag_2O, AgCl, AgS, AgF	Electroplating, photography, food and beverage processing

The classical trophic state indices are connected to external parameters by the rate of primary production. Qualitatively, eutrophication can be simply defined as an increase in the primary productivity of a lake, and researchers have used primary production per unit of lake surface and time as a quantitative measure of its trophic state (see Imboden and Gachter 1979). These workers have chosen annual primary production as a trophic state index and have shown that this index is greatly influenced by such physical processes as the intensity of vertical mixing. Of the nutrients entering the lake, a major proportion becomes incorporated into algae which release the nutrients back into water, either when still alive or after their decay. In fact, certain blue-green algae can serve to indicate eutrophic conditions. *Oscillatoria rubescens, Aphanizomenon flosaquae* and *Microcystis aeruginosa* have often been regarded as indicators of lake eutrophication. Biologically, oligotrophic lakes are poor in plankton, but support many diverse species of Chlorophyta, Chrysophyta and Bacillariophyta (*Tabellaria, Cyclotella*), and these algae are distributed to great depths. Algal blooms are formed very rarely. In contrast, eutrophic waters are rich in plankton, have larger numbers of individuals but these belong to few species, and water blooms occur frequently. The common algae of such waters belong to Cyanophyta and Bacillariophyta (*Melosira, Fragilaria, Asterionella*). Chemically, oligotrophic lakes are rich in dissolved oxygen and poor in nitrogen and phosphorus whereas eutrophic lakes are poor in dissolved oxygen and rich in nitrogen and phosphorus.

2.8
Trophic Status and Hypolimnetic Oxygen Concentration

Lakes are important sources of drinking water. As "natural" habitats, they not only produce fish but also guarantee the survival and natural reproduction of all other organisms present in the unpolluted water. Algal blooms drastically lower the oxygen concentration in the hypolimnion. For a minimal acceptable water quality, the dissolved oxygen should never fall below 4 g O_2 m^{-3} anywhere in the lake at any time. The general trophic structure in a typical take is illustrated in Fig. 2.18.

In order to grow, algae require various nutrients. According to the limitation law, the growth of algae is limited by the nutrient in shortest supply compared with demand. In lakes, the production of algae is generally limited by the concentration of available phosphorus compounds (Fig. 2.19). In oligotrophic lakes, primary production increases in proportion to the P concentration. In eutrophic lakes, it reaches a plateau at about 460 g C m^{-2} yr^{-1}. In mesotrophic lakes, primary production ranges between 150 and 200 g C m^{-2} yr^{-1} (Gachter and Wuest 1993). On average, production rates of this magnitude are attained at phosphorus concentrations approaching 20 to 30 mg m^{-3}.

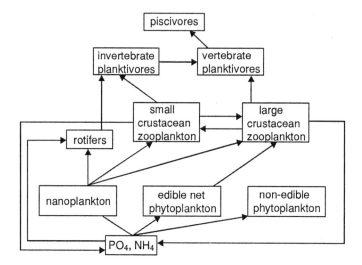

Fig. 2.18. A conceptual model of trophic structure in a typical lake. (After Carpenter et al. 1985)

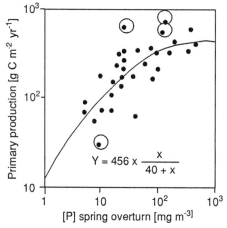

Fig. 2.19. Annual primary production as a function of total phosphorus concentration at spring turnover. (After Fricker 1980)

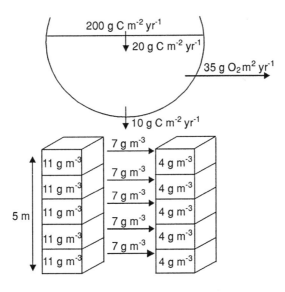

Fig. 2.20. Schematic illustration of primary production and mineralization in a mesotrophic lake. (After Gachter and Wuest 1993)

With respect to oxygen, conditions are considered to be unfavorable if the natural O_2 supply is insufficient to prevent the hypolimnetic oxygen concentration from falling below 4 g O_2 m^{-3} at a primary production rate of 200 g C m^{-2} yr^{-1} (Fig. 2.20). Oxygen is not consumed in the water column alone. Sediments consume even more. Thus, the water quality goal for oxygen may not be attained even in deep mesotrophic lakes if the ratio of sediment surface to water volume is large in the layers close to the lake bottom. Eutrophication not only increases the demand of the sediments for oxygen, but may also inhibit its resupply.

According to Gachter and Wuest (1993), in Swiss lakes, reductions in phosphorus loading have resulted in decreased phosphorus concentrations and hence in a reversal of eutrophication. However, the dependence of primary production, chlorophyll concentration or phytoplankton biomass on phosphorus concentration varies from lake to lake, and also from year to year within a single lake.

Based on sediment records, some lakes experienced anoxic conditions long before sewer systems were invented and fertilizers spread on agricultural land. Such lakes will most probably remain partially anoxic even after the introduction of optimal wastewater treatment and increased care in the use of manure in agriculture. If all the above criteria are satisfied except for the oxygen criterion, however, there is no reason to aerate these lakes artificially (Gachter and Wuest 1993).

Artificial mixing during winter is a relatively inexpensive way of increasing the oxygen supply from the atmosphere to those lakes in which eutrophication has caused an accumulation of salts in the hypolimnion sufficient to impede natural mixing. As a beneficial side-effect, artificial mixing increases nutrient concentrations at the surface, and hence the export of nutrients from the lake. An artificially increased oxygen concentration in the hypolimnion extends the habitat of those organisms that depend on oxygen and lowers the accumulation of phosphorus in the hypolimnion. However, it neither lowers the trophic status of lakes significantly nor guarantees that the eggs of sensitive fish will successfully hatch. Artificial mixing and oxygenation of the hypolimnion should therefore only be considered if:

1. The phosphorus loading of a lake has been lowered as far as possible and the phosphorus concentration has reached a new steady state,
2. The sediment record indicates that the lake was oxic at lower rates of phosphorus loading prevailing in the past, and
3. Artificial mixing and oxygenation are reasonable tools to use in overcoming remaining deficiencies in water quality.

2.9
Acidification

In lakes, acidification sometimes results in oligotrophy because of reduced input of phosphorus from the drainage area and also because algal utilization of phosphorus in acidified lakes is impaired, yielding lower biomass than could be expected from ambient phosphorus concentrations (see Table 2.5). But this hypothesis has been questioned by some workers (see Olsson and Pettersson 1993). According to these studies, experimental lake acidification, generally, did not result in significantly decreased phosphorus concentrations or phytoplankton production. Phosphorus concentrations are naturally low in runoff from areas sensitive to acidification, which makes it difficult to quantify the potential importance of acidification in low phosphorus leaching. According to physiological indicators of phosphorus deficiency, the phytoplankton community of acidified lakes may be adapted to an oligotrophic environment. At high aluminum concentrations, often found in acidified lakes, the algal utilization of phosphorus may be impaired (Olsson and Pettersson 1993).

Table 2.5. Some processes and mechanisms potentially leading to oligotrophication of acidified lakes. (After Olsson and Pettersson 1993)

Type of Process	Mechanism
Processes in the drainage area	Effective P fixation to Al complexes in podzol soil or decreased degradation of organic material, increased trapping of P in the drainage area
Processes in the pelagic zone	Precipitation of P with Al compounds, increased phosphorus sedimentation, decreased availability of phosphorus, dominating zooplankton (calcanoid copepods) produce refractory fecal pellets, high levels of aluminum prevent enzymatic recycling of organic phosphates, decreased intracellular phosphorus turnover, dominating zooplankton is characterized by low phosphorus turnover rate
Processes in the littoral zone	Development of *Sphagnum* mats accumulates P and prevents sediment-water exchange, *Sphagnum* can rapidly trap and assimilate P, *Odonata* produce large refractory fecal pellets, decreased fragmentation of detritus by macroinvertebrates
Processes in the profundal zone	Maximum adsorption at pH 3.8 to 6.3, increased binding capacity of phosphorus to sediments

2.9.1
Acidification in Streams

Streams and rivers are more susceptible to human influences than lakes. Running waters are sensitive indicators of how whole watersheds become anthropogenically affected by, say, acidification. Changed biotic patterns usually involve increased occurrence of those green algae that indicate an increase in nutrients (nitrogen), reduced species richness of invertebrates (especially mayflies, crustaceans, gastropods), a general shift in proportion from invertebrate grazers towards shredders and decreasing populations of fish, salmonids, roach, burbot, minnow (Herrmann et al. 1993). The mechanisms for the changes in individual, population and community levels include elevated hydrogen, aluminum and cadmium concentrations that affect ion balance and respiration in fish and invertebrates, but also various behavior patterns (avoidance reaction, downstream movement, choice of spawning site), and developmental stages (molt and emergence of insects, hatching and growth of early fish stages). Aluminum ameliorates low pH temporarily but does not biomagnify along food chains, and neither predatory insects nor flycatchers accumulate aluminum. It seems less likely that cadmium is a serious threat to invertebrates in "normal" concentrations at low pH. Iron precipitation can affect feeding ability and respiration of mayfly nymphs. That humic substances may mitigate metals still seems uncertain for fish and invertebrates (Herrmann et al. 1993). Generally, most changes in the biotic patterns of streams seem to be related to abiotic impact routes. In most cases of sublethal acidification stress, growth, development and reproduction of the organisms are retarded.

Unlike in lakes, very few food-web mediated changes have been confirmed for running waters, especially in relation to acidification. This indicates that mainly abiotic factors are responsible for the occurrence and dynamics of lotic invertebrates. This may be because the physical environment is more unstable and, of course, man-induced impacts like acidification add to the list of stress factors. The reduced effectiveness of biotic interactions such as predation, and the rapid dispersal of stream-living organisms, make it difficult to demonstrate such impacts (Malmqvist 1993).

In lakes, after moderate acidification, the growth rate of fish often increases because of reduced competition, whereas fish surviving in severely acidified water tend to show reduced growth. The same may also happen in stream populations, unless the invertebrates serving as food are reduced too much.

At falling pH value, the competitively superior amphipod *Gammarus pulex* may retreat, which allows the more acid-resistant, but competitively inferior isopod *Asellus aquaticus* to expand (Hargeby 1990). The food-web structure proves useful for assessing the status of a stream locality. Scrapers (mostly mayflies) are positively correlated with pH, but negatively correlated with the water color. Altered algal species composition can affect food quality for macro-invertebrates.

The increasing proportion of shredding caddis larvae in acidified streams appears to be due either to the elevated amounts of coarse detritus that remain undegraded over a longer period in acid streams, or the absence of fish predators.

2.9.2
Restoration of Acidified Waters

Liming is the method of choice to counteract acidification. Raised abundance or numbers of species of benthic invertebrates, and a general increase in abundance of several fish species can follow liming. The realized colonization of plants and animals depends on several abiotic and biotic factors such as stability, predictability and quality of bottom substrate and water chemistry, the number of species available for recruitment, the distance to the nearest location with potential colonizers, their dispersal biology, and the composition of the new invertebrate community (Henrikson and Brodin 1993; Herrmann et al. 1993).

2.10
Large Lakes

The proper functioning of the ecosystems of many large lakes has been greatly endangered by increasing environmental stresses. The total area of lakes on Earth amounts to approximately 2.5×10^6 km^2 or 1.8% of the continental area, containing 1.2×10^5 km^3 of water. The 253 largest lakes (larger than 500 km^2) contain an estimated 78% of the world's unfrozen surface freshwater and represent an essential global life support system. Large lake ecosystems have already degraded in many regions. Sustainable conservation requires proper control of human activities in the catchments of lakes.

2.10.1
Distinction Between Large and Small Lakes

Three main criteria used to distinguish large from small lakes are:

1. Only large lakes have a true pelagic zone that is physically distinct from the littoral zone.
2. The mixing depth of the epilimnion which is a function of mean wind velocity and frequency, and of wind fetch in general is greater than in small lakes.
3. Only in large lakes can differences in physical forcing over the water surface allow the formation of areas whose physical structure differs from other areas of the lakes. Physically different structures can cause patchiness, i.e., horizontal differentiation of biological processes (Tilzcr and Bossard 1992).

Large lakes tend to be deeper and have smaller watersheds than small lakes. They have longer water retention times than small lakes. Many receive proportionately less sediment and/or nutrient inputs and, consequently, are clearer and more oligotrophic than small lakes. The area-to-volume ratio is less in large than in small lakes. The freewater zone is the main site for the bulk of metabolic processes which drive the biogeochemical cycle. In a system that has a long water retention time, any substance entering the lake remains there for long periods. Larger lakes therefore usually respond more slowly to external inflow, and also have longer recovery times to external influxes of any undesirable substances. The time elapsing before a particle reaches the lake bottom is longer in a deep lake, thus providing more opportunity for chemical interaction with the surrounding water.

During the growing season nutrient regeneration is more intense in oligotrophic than in eutrophic systems. Because large lakes in general are more oligotrophic than smaller lakes, regenerated production usually comprises a greater proportion of total production than in small lakes. Furthermore, the food web structure usually, but not always, tends to be controlled by bottom-up rather than top-down mechanisms (Tilzer and Bossard 1992).

2.11
Biogeochemistry

Biogeochemical processes in lakes exist as a balance between the interactions of the lake with its catchment, the atmosphere and in-lake processes. This balance can shift dramatically with basin size, latitude and climate. Our understanding of lake ecosystems has been influenced by a strong reliance on small, mid-latitude temperate lakes, whose main feature is their pronounced seasonality. Much less information about high-latitude and low-latitude lake systems is available. Large lakes are excellent sites to assess ecosystem variation at different latitudes and in different vegetational regions. They respond to their environment, but are also products of their environment. Some important physical determinants for lake response are latitude, topography, structure and composition of soils and bedrock, groundwater-aquifer relations, and geothermal factors. These characteristics react with climate and organisms in basin-specific fashions. The ratios of drainage basin area to lake surface area and to lake volume interact with water budgets in fundamental ways (Tilzer and Bossard 1992).

2.11.1
Biological Interactions

Many large lakes contain large numbers of endemic species. Although the biological components of any ecosystem can be characterized by simplistic descriptions (e.g., taxonomic lists, diversity indices, etc.), such components are best described by the complex interactions among organisms. These interactions determine the structure and efficiency of the food web and how it responds to external and internal perturbations. Food web dynamics are highly sensitive to physical and chemical processes and are indicators for stresses caused by human activities.

2.11.2
Land-Water Interactions

Lakes depend on water balances maintained by hydrological inputs from their drainage basins, direct precipitation on surface waters and losses through evaporation and outflow. The expected consequences of future global changes in human population and climate have highlighted the need for accurate hydrological models owing to increasing demand for freshwater.

Drainage basins furnish particulate matter which affects the geochemical characteristics of lakes and contributes to the transport, transformation, and removal of constituents from water.

2.11.3
Physical Processes

The hydrodynamics of large deep lakes, and especially the renewal of deep waters, is important both because it increases biological productivity through replenishment of nutrients in surface waters and because it modulates the responses of the lake ecosystem to anthropogenic and climatic influences.

In large, deep, temperate, freshwater lakes the vertical stability of the deep water column is orders of magnitude smaller than in tropical lakes. Moreover, the temperature of maximum density and the pressure dependence severely constrain deep water renewal during spring and fall. Large portions of the water in deep lakes are not directly affected by wind-mixing, and are virtually independent of processes in the littoral zone (Tilzer and Bossard 1992).

2.11.4
Biogeochemistry and Nutrient Relations

Large lakes are well suited for the study of the responses of aquatic ecosystems to internal physical, chemical or biological perturbations. Biogeochemistry deals with the transport and transformation of chemicals or elements from their arrival in the ecosystem to their ultimate export, burial or gaseous release. These fluxes constrain the extent of ecosystem productivity. Basic physical characteristics of lakes and their catchments aid in the determination of water retention times and inference of whether or not the dominant chemical inputs to a lake are via chemical weathering and runoff or via direct atmospheric input. Fluxes of trace chemicals and key nutrients are under biological control to varying degrees between lakes, depending on latitude, topography, geology and climate. Fluxes under tight biological control are sensitive to alterations in food web structure. Possibly internal physical processes dominate the observed lake water chemistry of large, high-latitude lakes, whereas lake chemistry is increasingly controlled by biotic communities as latitude decreases. Large lakes respond to environmental forces on a regional scale (Fig. 2.21). Some of the key processes are those controlling primary biological production, vertical transport of nutrient elements and flux rates at benthic-pelagic and littoral-pelagic interfaces (Tilzer and Bossard 1992). Equatorial lakes having permanent pycnoclines become dominated by new inputs, diffusion across chemoclines and by high concentrations of dissolved organic carbon. Shallower temperate lakes that undergo annual holomixis are dominated by interactions with the sediments whereas very deep lakes that mix infrequently can mimic the ocean.

2.12
Food Webs

In temperate large lakes, two basic components of the planktonic food web prevail in oligotrophic situations: (1) The "microbial" food chain from auto- and heterotroph picoplankton to ciliates and rotifers, and (2) the "classic" food chain from nano- and microalgae to herbivores and further to predatory copepods and cladocerans to exclusively pelagic zooplankton-feeding fish as final consumers (Tilzer and Bossard 1992). The connection between these two different transfer paths of energy is rather weak in oligotrophic systems. In temperate lakes with higher nutrient levels and food concentrations, an additional third food chain from pico- and nanoplankton to filter-feeding consumers like *Daphnia* can link the "microbial" and the "classic" food chains closely, if filter-feeders graze on both pico- and nanoplanktic size classes.

Over 75% of the sinking flux of particulate organic matter may be utilized by free-living bacteria, which make the major food source of particle-eating metazoa and protozoa. This "microbial loop", which depends on the efficiency of liberation of dissolved organics from particulate matter, can be a major route for the flow of material and energy in pelagic food webs. In many respects the bacteria associated with particles appear to be more versatile than free-living bacteria in their metabolic activities.

The microbial food web (see Fig. 2.22) appears to be more important for processing the primary production and cycling of nutrients in large lakes than in small lakes. The relative contributions of free-living bacteria, phototrophic picoplankton, metazoa and protozoa to carbon flux and nutrient dynamics is greater in oligotrophic than in eutrophic ecosystems (Tilzer and Bossard 1992). In some organically rich systems, pelagic

ciliates are major bacterivores. In freshwater, phagotrophic phytoflagellates (mixo-
trophs) consume the bacteria.

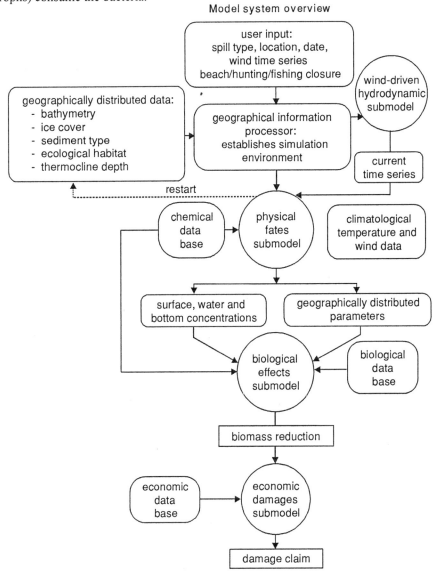

Fig. 2.21. An overview of the natural resource damage assessment model for a large lake environment

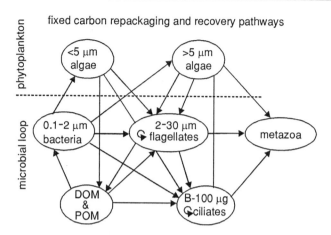

Fig. 2.22. Trophic interactions within the microbial food web, separated here into phytoplankton and "microbial loop" (i.e. bacteria and protozoa) components. Nonliving pools of dissolved and particulate organic matter (DOM and POM) are also included. Note the many direct links between heterotrophic and autotrophic microbes, such as ingestion of phytoplankton by flagellates and ciliates and ingestion of bacteria by mixotrophic phytoplankton. The curved arrows in the flagellate and ciliate compartments indicate further predator-prey interactions within these broad classes of organisms. In this model, production of phytoplankton smaller than 5 μm is accessible to metazoa only after being repackaged into larger protozoan cells

Both flagellates and ciliates usually graze on larger bacteria. Since the larger bacteria are those most actively growing and dividing, bacterivorous protozoa are selectively grazing bacterial production instead of randomly cropping the standing stock of suspended bacteria. The phytoplankton communities of large lakes experience changes due to alterations in mixing regimes and external nutrient loading. Their species composition responds to the nutrient regime and sedimentation rates. These effects may propagate community-wide through trophic relationships. Proportional rates of utilization of the major nutrient elements (P, Si, N) during bloom periods are expected to be useful indicators of the new production of organic matter available to secondary production (Tilzer and Bossard 1992).

According to Tilzer and Bossard, modifications of the planktonic system along large-scale gradients of latitude are chiefly characterized by controlling physical factors. In lakes of comparable size and trophy, the productivity is expected to decrease from low to high latitudes, mainly because of shorter growing seasons, exhibiting larger effects in eutrophic than in oligotrophic lakes.

From the temperate region to the subtropical arid climatic zone there occurs a gradient from freshwater lakes to salt lakes and a decreasing diversity of plankton from low to high salinity. Latitudinal gradients of physical limnological characteristics within the temperate zone, such as mixing depth, thermal regime and stability of stratification, are expected to change plankton communities. There is need to check these hypotheses for large lakes at different latitudes.

The size relationships between prey and their predators are characteristic features of any food web structure. In nutrient poor lakes, long food chains with many steps and low transfer efficiency are dominant. As nutrient levels increase, there is a threshold concentration above which there is a change from predatory to filtering consumers. The food web structure in large, deep lakes is usually simpler than in small lakes. Changes in planktivore abundance and feeding behavior alter zooplankton communities, their grazing rates and water clarity in some lakes. Such "top-down" changes in food web structure can exert significant effects on the efficiency of energy transfer through pelagic food webs. In large lakes much of total productivity is associated with the pelagic zone. There is need to find out how the zooplankton and phytoplankton communities and apparent water quality of the pelagic zones of large lakes may respond to changes in fish stocks.

2.13
Modeling of Lake Ecosystems

Though a vast amount of data concerning lake ecosystems have been analyzed, the models evolved are not 100% foolproof, as their complexity increases geometrically with the increasing number of parameters. Despite these limitations, however, these models do have some limited predictive value, especially for anticipating eutrophication and thermal pollution effects in lakes not yet studied. Overall productivity relationships can be predicted from biomass models. Table 2.6 lists the primary criteria to assess the eutrophication status of a water body and table 2.7 lists some physical, chemical and biological parameters and their responses to increased eutrophication.

Table 2.6. Primary criteria for assessing the eutrophication status of a water body (Ryding and Rast 1989)

Parameter	Unit[a]
Morphometric conditions	
Lake surface area	km^2
Lake volume (average conditions)[b]	$10^6\,m^3$
Mean and maximum depth	M
Location of inflows and outflows	-
Hydrodynamic conditions	
Volume of total inflow (including groundwater) and outflow for different months	$m^3\,day^{-1}$
Theoretical mean residence time of the water (renewal time, retention time)	yr
Thermal stratification (vertical profiles along longitudinal axis, including the deepest points)	
Flowthrough conditions (surface overflow or deep release, and possibility of bypass flow)	
In-lake nutrient conditions	
Dissolved reactive phosphorus, total dissolved phosphorus and total phosphorus	$\mu g\,P\,l^{-1}$
Nitrate nitrogen, nitrite nitrogen, ammonia nitrogen, and total nitrogen	$mg\,N\,l^{-1}$
Silicate (if diatoms constitute a large proportion of phytoplankton population)	$mg\,SiO_2\,l^{-1}$
In-lake eutrophication response parameters	
Chlorophyll a, phaeophytin a	$mg\,l^{-1}$
Transparency (Secchi depth)	m
Hypolimnetic oxygen depletion rate (during period of thermal stratification)	$g\,O_2\,day^{-1}$
Primary production[c]	$g\,C\,m^{-3}\,day^{-1}$ or $g\,C\,m^{-2}\,day^{-1}$
Diurnal variation in dissolved oxygen[c]	$mg\,l^{-1}$
Dissolved and suspended solids[c]	$mg\,l^{-1}$
Major taxonomic groups and dominant species of phytoplankton, zooplankton and bottom fauna[c]	-
Extent of attached algal and macrophyte growth in littoral zone[c]	-

[a]The terminology and units proposed by the International Organization of Standardization is recommended for expressing the parameters.

[b]A bathymetric map and hypsographic curve is necessary in many cases.

[c]Can provide additional information on the trophic conditions of a water body, recommended if resources are adequate or if special situations require more detailed information.

Table 2.7. Trophic criteria and their responses to increased eutrophication[a]

Physical	Chemical	Biological[b]
Transparency (D)	Nutrient concentrations (I)	Algal bloom frequency (I)
Suspended solids (I)	Chlorophyll *a* (I)	Algal species diversity (D)
	Electrical conductance (I)	Phytoplankton biomass (I)
	Dissolved solids (I)	Littoral vegetation (I)[c]
	Hypolimnetic oxygen deficit (I)	Zooplankton (I)
	Epilimnetic oxygen supersaturation (I)	Fish (I)[d]
		Bottom fauna (I)[e]
		Bottom fauna diversity (D)
		Primary production (I)

[a](I) signifies the value of the parameter which generally increases with the degree of eutrophication, (D) signifies the value which generally decreases with the degree of eutrophication.
[b]The biological criteria have important qualitative (e.g. species) changes as well as quantitative (e.g. biomass) changes, as the degree of eutrophication increases.
[c]Aquatic plants in the shallow, near shore area may decrease in the presence of a high density of phytoplankton.
[d]Fish may be decreased in numbers and species in bottom waters (hypolimnion) beyond a certain degree of eutrophication, as a result of hypolimnetic oxygen depletion.
[e]Bottom fauna may be decreased in numbers and species in high concentrations of hydrogen sulfide (H_2S), methane (CH_4) or carbon dioxide (CO_2), or low concentrations of oxygen (O_2) in hypolimtic waters.

2.14
Lake Morphometry

The depth, size, and basin shape of a lake are important factors determining its potential for fish production. Most lakes are elliptical sinusoids, elliptical cones, elliptical paraboloids, or ellipsoids in shape. The ratio between their mean depths and maximum depths varies between about 0.3 to 0.7. Generally, lakes with lower maximum depths have more U-shaped basins than do deeper lakes. Further, in lakes of different morphometry in similar climate regions, both surface and deep water temperatures are fairly different. In general, small shallow lakes are warmer than large, deep lakes. Mean Secchi disc depth strongly depends on lake morphometry, but maximum Secchi depth is also influenced by variables concerning latitude. Duration of stratification and depth of thermocline depend on both morphometry and latitude. Frequent observations of hyperbolic declines of production with depth have been made, and these drops seem to be caused by decreasing nutrient concentrations with increasing depth of lake.

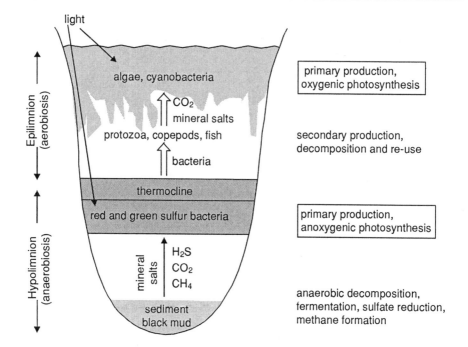

Fig. 2.23. Vertical section through a eutrophic lake, to illustrate an aquatic ecosystem

2.15
Eutrophication

Eutrophication of lakes and reservoirs is a most pervasive water quality problem worldwide. Eutrophication refers to the excessive nutrient enrichment of water which results in a series of adverse symptomatic changes, including nuisance production of algae and other aquatic plants, deterioration of water quality, taste and odor problems and fish kills. These problems interfere with human use of water resources. Despite efforts at many levels to control the causes of eutrophication, water quality has continued to deteriorate in many streams, lakes, reservoirs and coastal areas. Therefore, development of a sound eutrophication management strategy is essential for a satisfactory solution to this significant water pollution problem. Figure 2.23 shows a vertical section through a typical eutrophic lake.

The effects of eutrophication are considered negative in many places around the world, and often reflect human perceptions of good versus bad water quality. Excessive algae and aquatic plant growth are highly visible and usually interfere with the uses and aesthetic quality of a water body. One of the consequences is the production of taste and odor problems in drinking water drawn from a lake or reservoir, even though the water may be treated and filtered prior to use. The water treatment process itself can become more expensive and time-consuming for eutrophic waters. The water transparency of a lake or reservoir may be greatly reduced.

Eutrophication is commonly accompanied by a concomitant change in the species composition of the aquatic biota. These changes may be either directly due to the nutrients, or more commonly, may be caused by some underlying homoeostatic factors, e.g., change in grazing behavior of animals. Such changes in homoeostatic relations can in certain cases delay or prevent the required desirable changes or improvements in a lake even after the nutrient load has been diminished significantly. Another similar factor that can cause delayed effects is the deposition of nutrients in bottom sediments. Such nutrients may continue to diffuse into and enrich the water long after further inputs of nutrients into the lake have been checked.

With increasing eutrophication, the diversity of the phytoplankton community of a lake decreases and the lake finally becomes dominated by cyanobacteria. Some striking examples of cyanobacteria dominated lakes are the hypertrophic Wolderwijd and Veluwemeer lakes in the Netherlands. In these lakes, *Oscillatoria agardhii* is the dominant species. The natural populations of this species are successively limited by phosphorus, light, and combined nitrogen. One interesting observation is that even at times when the growth of *O. agardhii* is N-limited, this species is not succeeded by any nitrogen-fixing algae. These lakes are much less favorable to nitrogen fixers even when N-limitation prevails, and the critical factor determining as to which algal species will dominate is the trophic state of the lake, especially with reference to organic matter and phosphorus.

The phenomenon of hypertrophy involves such a high enrichment of a freshwater system with nutrients that a pronounced increase of biomass and a strong decrease in the number of species results. In those lakes which are only slightly eutrophic the maximum biomass of phytoplankton is controlled by the amounts of nutrients; species having a high affinity for the limiting factor and showing a marked light tolerance tend to be selected. At somewhat greater eutrophication levels, the selection pressure of the limiting factor is somewhat lower, and a more diverse phytoplankton community develops whose biomass is higher. With continuing increase in eutrophication level, the light availability in the epilimnion is adversely affected and as a consequence those algal species having a low light requirement (i.e., cyanobacteria) preferentially survive. Most planktonic cyanobacteria have gas vacuoles which enable them to regulate their buoyancy in response to the available light conditions. These vacuoles enable the cyanobacteria to maintain themselves in water layers with optimum light conditions.

Deep lakes are generally more sensitive to hypereutrophication than shallow lakes (Mur 1980). The biomass concentration in deep lakes is low and the cyanobacteria species are all gas vacuolate forms that grow in distinct strata. In contrast, in shallow lakes stratification does not occur and the light intensity becomes low only when the numbers of algae become very large. Non gas-vacuolate species tend to dominate here, and good examples are species of *Oscillatoria* (*Microcystis, Aphanizomenon* and *Anabaena* exemplify gas vacuolate forms of deep lakes).

The crops of zooplankton in most hypertrophic lakes are also quite high. These microanimals feed on phytoplankton. The populations of zooplankton are regulated by such fishes as roach and bream. Any marked increases in the populations of planktivorous fishes lead to steep declines in zooplankton and corresponding rises in the phytoplankton biomass. According to Mur (1980), a selective removal of planktivorous fishes exerts a positive effect on the trophic state of a water body. This may be caused not only by predation of fish on zooplankton but also by the predation on zoobenthos that leads to a mobilization of phosphorus from the bottom sediments. The control of those fish populations that feed on plankton and benthos seems to constitute a promising method of lowering the trophic level of a lake.

Although planktonic algae constitute a very significant factor in lake eutrophication, the effects of nutrient enrichment on higher aquatic plants are by no means insignificant, especially in shallow waters. It has long been realized that primary production in aquatic habitats is under the influence of both solar energy and nutrients. The latitude has also been found to influence productivity (Brylinsky and Mann 1973). A regression of plankton production on latitude brought forth the interesting conclusion that whereas lakes at high latitudes had low productivity, those situated at low latitudes exhibited a wide range of levels of productivity. This work was done on Canadian lakes. This conclusion is, however, not accepted by some other Canadian researchers, e.g., Schindler and Fee (1974). These latter workers are of the opinion that the observed correlation between geographical latitude and plankton production is due only in part to the effect of solar radiation. The major factor determining phytoplankton productivity seems to be phosphorus content rather than latitude.

Village waters are generally not affected by point sources of pollution but are mainly subject to non-point sources. Eutrophication of rural ponds is a good example of non-point pollution. Management of the effects of nutrient increases depends chiefly on the Vollenweider phosphorus-loading concept. Schindler (1980 has discussed the long-term background and a number of eutrophication and lake research programs in North America and Canada. Golterman (1980) pointed out that although much is known about nutrient loading and lake productivity, there is a danger of oversimplification due to limitations of available models in describing the complexities of the phosphorus cycle. It has been felt that such key processes as recycling of nutrients and sediment formation or release should be included to represent the dynamic nature of ecosystems.

Kajak (1980) related the roles of phosphorus loading and dynamics in lake water to phosphorus nonpoint loads, agricultural impacts, epilimnetic recirculation, the role of biota in the phosphorus cycle, and phytoplankton impacts.

In the American Lake Mendota, Fallon and Brock (1980) studied growth, primary production and sedimentation over two annual phytoplankton cycles to observe the succession of cyanobacteria following stratification of the lake. He interpreted the observed declines in standing crop in terms of epilimnetic decomposition and sedimentation.

Considerable attention is now being given to stop or reverse the eutrophication process with a view to ensuring an adequate supply of clean water for drinking and other purposes, and also with a view to preventing the lakes from premature death. The following procedures have been recommended by various limnologists to slow down the eutrophication process:

1. Limiting the amount of nutrients entering the lake,
2. Reduction in the amounts of nutrients solubilized in water through microbial decomposition of bottom sediments; this can often be achieved by the bottom-sealing technique of Sylvester and Seabloom (1965), i.e., artificially planting an inert layer which covers bottom sediments,
3. Harvesting and removal of algal blooms and mechanical removal of higher plants; this can reduce the amount of nutrients recycled into the water upon decay of algae and higher plants,
4. Removal of dissolved nutrients from water chemically or physically,
5. Encouraging the setting up of natural food-webs (e.g., daphnids and fishes), which can remove the algae, and subsequently harvesting the fish, and
6. Controlling the growth and multiplication of algae and higher plants through application of appropriate doses of copper sulfate and sodium arsenate, respectively.

Of all the above methods of controlling eutrophication, that involving limitation of nutrient input is certainly the best and the most reliable. Although various kinds of major and minor nutrients are essential for plant growth, in natural habitats it is mostly nitrogen and phosphorus which are critical. However, certain cyanobacteria can even fix the atmospheric nitrogen, and hence their growth does not become limited by the deficiency of nitrogenous compounds in lake water, and in such cases phosphorus becomes the most critical element limiting algal growth. Most instances of man-made eutrophication have been traced directly to a superabundance of nitrogenous and phosphatic compounds in water.

There are also serious ecological consequences related to cultural (or human-induced) eutrophication. As algal populations die and sink to the bottom of a water body, their decay by bacteria can reduce oxygen concentrations in bottom waters to levels which are too low to sustain fish life, resulting in fish kills. Such oxygen-deficient conditions can also result in excessive levels of iron and manganese in the water, which interferes with drinking water treatment. Negative potential health effects, especially in tropical regions, are related to such parasitic diseases as schistosomiasis, onchocerchiasis and malaria; these become aggravated by cultural eutrophication. To control the problem, we need the following:

1. Quantitative tools for assessing the state of eutrophication of lakes and reservoirs,
2. A framework for developing cost-effective eutrophication management strategies,
3. A basis upon which strategies can be designed for each specific case according to the physical, social, institutional, regulatory and economic characteristics of the local area or region, and
4. Specific technical guidance and case studies regarding the effective management of eutrophication.

2.15.1
Eutrophication: Hysteresis and Remediation

Before World War II and for a few decades afterwards, most Dutch lakes were clear and, being mostly shallow, had lots of macrophytes, which provided shelter and hatching areas for fish of prey that controlled planktivorous fish that grazed on zooplankton. Zooplankton was well developed and as a consequence, in combination with low nutrient levels, phytoplankton levels remained low. In the 1960s, detergents were introduced with a high content of polyphosphates. Extension of sewerage and wastewater treatment systems caused a gradual loading of the lakes, later greatly enhanced by intensive cattle breeding (Lijklema 1995). The manifestation of negative symptoms accompanying eutrophication was delayed: growth of periphyton, declining macrophyte stands, increasing phytoplankton and cyanobacteria, dominance of bream and other fish species feeding on zooplankton and grubbing in the sediments, reduced transparency, increasing oxygen fluctuations and decreasing biodiversity. According to Lijklema (1995), all these dormant effects were held up during the slow and gradual saturation of agricultural soils and sediments with phosphate. To some extent the lakes resisted the adverse impacts, but when around 1965 certain thresholds were exceeded, a rapid deterioration ensued. As usual, society delayed taking remedial and preventive actions. However, in time phosphate was banned in detergents (during 1980–90). Implementation of tertiary treatment was also delayed. In many lakes little or no improvement has been observed thus far. Although, in several systems, the external P loading has been reduced in many lakes, perpetual eutrophication symptoms are observed due to resilience and hysteresis effects which are now working in undesired directions. In some cases, biomanipulation has been successful, but not in others.

There are few, if any, natural aquatic systems left anywhere in the world that have not been impacted by human activities. The main reason for this is related to chemical changes in the atmosphere, in particular the long-range transport of air borne contaminants to even the remotest parts of the globe, such as the high Arctic. However, chemical changes to some lakes probably have little significant impact on natural biodiversity. This contrasts with other lakes so strongly degraded by human activities that they are totally unlike their original natural state. The lower Great Lakes fall into this latter category.

There are five primary ways in which biodiversity may be anthropogenically affected: (1) direct exploitation or removal of species, (3) physical alteration, (3) species invasions or introductions, (4) the addition of nutrients and (5) persistent toxic contaminants. The stressors most associated with some large lakes in recent years have been toxic, persistent, bioaccumulating chlorinated organic chemicals and, to some extent, heavy metals. Some of these chemicals have affected several of the top predators (both fish and birds).

In Lake Erie overfishing appears to have been the major factor in the reduction and/or extirpation of sturgeon (*Acipenser fluvescens*), lake trout (*Salvelinus namaycush*), burbot (*Lota lota*), whitefish (*Coregonus clupeaformis*) and blue pike (*Stizostedion vitreum glaucum*) populations.

2.15.2
Physical Modifications

Direct physical modifications of the lentic and lotic habitat, as well as changes to the terrestrial compartment of the watershed (e.g. deforestation and dam building) often have substantial impacts on the ecosystem. This combination of severe changes to the lotic environment undoubtedly resulted in the extinction of the Atlantic salmon (*Salmo salar*) in some Canadian lakes. Wetlands and marsh areas serve as breeding and growth areas for a variety of wildlife communities including warm and cool water fish, birds, waterfowl and amphibians. Since the mid 1700s, 70% of the wetlands in the Great Lakes basin have been lost, primarily due to agricultural drainage, and more than 20 000 acres of wetland are now being lost annually (Allan and Zarull 1995).

No doubt the most immediate impacts of navigational work, shoreline protection, harbor and marine construction, lake-filling and dredging tend to be very localized, but all these local effects can collectively reduce the littoral zone of some large lakes to the extent that they do exert a major impact on the recruitment and survival of many species of fish.

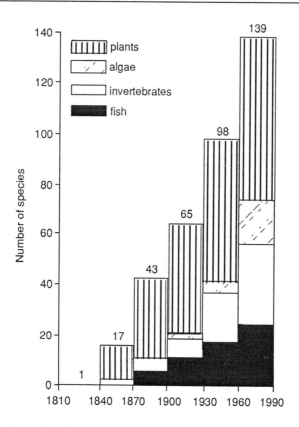

Fig. 2.24. Introduction of exotic species to the Great Lakes. (Adapted from Mills et al. 1993)

Since 1800, over 130 non-indigenous aquatic organisms have become established in the Great Lakes; (Fig. 2.24) most of these are plants, about 25 fishes, 24 algae, 14 mollusks and other invertebrates, and 17 disease pathogens. Several of these have become established through deliberate release or through official stocking programs. Others have found their way into the lakes through accidental releases, either from hatcheries or ships. Still others have entered as opportunist invaders via the canals and locks of the St. Lawrence seaway. There is evidence that at least 10% of the successful invading species have had substantial impacts on the Great Lakes (Mills et al. 1993).

Nutrient loadings to the lower lakes increased significantly after their basins were deforested, and this was augmented by the loss of wetlands and the intensification of farming. But it was really the widespread application of artificial fertilizers and the discharge of untreated or partially treated sewage that caused the most dramatic and visible impacts of nutrients on the ecosystem. Cultural eutrophication led to an increase in the total biomass of the system, accompanied by pelagic and benthic species changes, generally the replacement of more sensitive species with pollution-tolerant ones. While the total number of species declined, the number of trophic levels also decreased, resulting in less complex food webs.

The peak inputs of synthetic organic chemicals and heavy metals to the lower Great Lakes occurred in the 1960s and 1970s, followed by declines in the 1980s. In the Great Lakes basin some 10 wildlife species (top predators) have suffered physiological impacts, population declines, or reproductive effects related to persistent toxic substances since the 1960s (Table 2.8) Happily, over the last decade, there have been recoveries in reproductive success and increases in populations for the most affected bird species, for example the herring gulls (*Larus argentatus*) and cormorant (*Phalocrocorax auritus*) suggesting a cause and effect relationship between persistent toxic contaminants and avian biodiversity. Restoration, protection and conservation of lake ecosystems and their natural biodiversity are complex issues. Several factors act simultaneously and synergistically (Fig. 2.25) and can complicate restoration and protection actions in many near shore areas, as well as the rest of a large lake.

Environmental consciousness is also increasing. In the original sense, eutrophication refers to the natural aging process of a lake. A lake receives inflows of water from its surrounding drainage basin, along with materials carried in the water from the land surface (e.g., following a rain storm or from irrigation drainage). Materials associated with rain, snow and wind-blown substances, as well as groundwater inflows also enter a lake. The water quality and biological communities in a lake reflect the cumulative impacts of all the water and material inflows into the lake.

Table 2.8. Species of fish and wildlife affected by contaminants in the Great Lakes. (After Allan and Zarull 1995)

Species	Effects on reproduction	Population decrease	Mortality
Mink	x	x	x
Otter	x		
Cormorant	x	x	
Night heron	x	x	
Bald eagle	x	x	
Herring gull	x	x	
Terns		x	x
Snapping turtle		x	
Lake trout		x	

x means known effects (recorded)

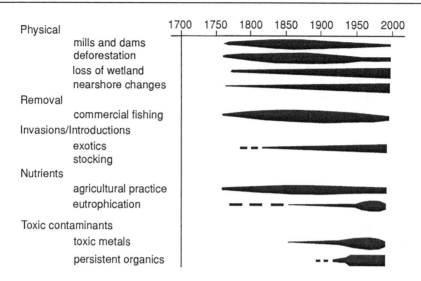

Fig. 2.25. Historical evolution of stresses affecting Lake Ontario. (Adapted from Sly 1991)

Over thousands of years, a lake slowly fills up with soil and other materials carried by inflowing waters, and eventually becomes a marsh and, ultimately, a terrestrial system. Lakes undergoing such natural eutrophication generally have good water quality and support a fairly diverse biological community. The growth of algae and other aquatic plants in a lake usually remains minimal and generally in balance with the input of plant nutrients. However, human settlement in the drainage basin, the associated clearing of forests and developmental activities of man change the natural eutrophication process in a dramatic way. The runoff of most materials from the land surface to the water body is greatly accelerated. An increased input of plant nutrients (mainly phosphorus and nitrogen), stimulates growths of algae and aquatic plants which, in turn, stimulate the growth of fish and other organisms in the aquatic food chain. It is often possible to treat a lake or reservoir undergoing cultural eutrophication in such a way that it will again experience an "aging" rate more typical of natural eutrophication i.e., to restore it (see Fig. 2.26).

During the last 3 decades, the term "eutrophication" has been used more and more to mean this artificial and undesirable addition of plant nutrients to water bodies. In some situations, this view can be misleading, since what is an undesirable addition to one water body may be harmless, or even beneficial, in another. Nevertheless, eutrophication is commonly known as the state of a water body manifested by an intense proliferation of algae and higher plants. Their accumulations produce adverse changes in the water quality and interfere significantly with human uses of the water resource.

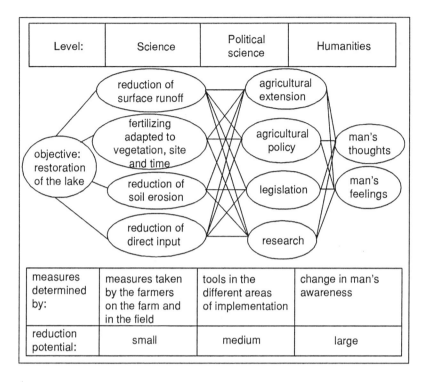

Level:	Science	Political science	Humanities

(diagram)

measures determined by:	measures taken by the farmers on the farm and in the field	tools in the different areas of implementation	change in man's awareness
reduction potential:	small	medium	large

Fig. 2.26. Some measures to restore a lake

The above general description of the eutrophication process applies both to natural lakes and reservoirs (man-made impoundments). Reservoirs are water bodies which have been created artificially by the construction of a dam across a flowing river or stream. The important factors to be considered in selecting eutrophication control measures for these two types of water bodies are usually sufficiently similar that the terms "lake" and "reservoir" can be used interchangeably. Figure 2.27 outlines the sequence of decisions to be made in the development of a suitable eutrophication management and control program. The approach is sufficiently general to be applied to the assessment and control of other environmental problems as well.

The eutrophication process involves complex interactions between several natural and anthropogenic factors (Fig. 2.27). In general there is much evidence linking increased eutrophication to the excessive input of plant nutrients from point and non-point sources in the drainage basin. Consequently, nutrient loading concepts are frequently used in the assessment of nutrient control measures.

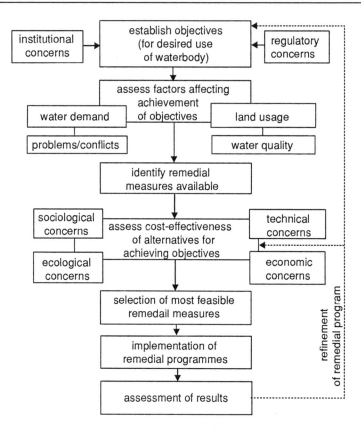

Fig. 2.27. Sequence of decisions to be made in the development and implementation of eutrophication control programs

The main objective of traditional water pollution control efforts was to clean up raw wastewater and gross industrial wastes, which are potential sources of pathogens and toxic materials. Ever since treatment of such effluents became more common, especially in industrialized nations, environmental impacts of other types of pollution in the drainage basin have tended to assume greater importance (Fig. 2.28). For example, pollution from urban and rural runoff is being seriously considered in the development of effective water pollution control programs. Accordingly, there is a definite, continuing need to develop an integrated view of land, atmosphere and water interactions in the drainage basin, as they relate to the assessment and treatment of cultural eutrophication.

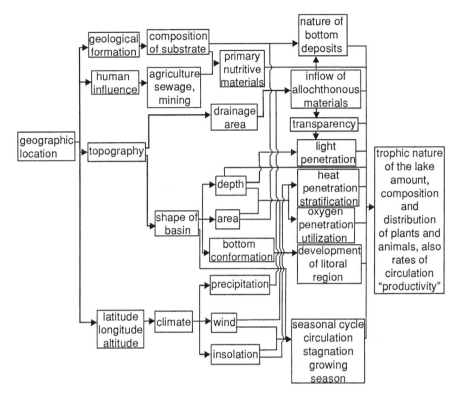

Fig. 2.28. Interrelationships of factors affecting the metabolism of lakes and reservoirs. (Stewart and Rohlich 1967)

In industrialized countries, the need for an integrated concept of land, atmosphere and water interactions in the drainage basin has become necessary because of increasingly serious water quality problems. In arid and semi-arid regions, socio-economic development can result in rapidly increasing water demands and the availability of water can become a significant constraint to development. This is very often due to the combined effect of large populations and climatically related water demands for increasing agricultural production.

Information required for the assessment and control of lake or reservoir eutrophication normally includes such items as the depth, volume and water flushing rate, the in-lake concentrations of nutrients and algae, the occurrence of nuisance growth of algae and other aquatic plants, the occurrence of oxygen depleted bottom waters in the lake and related fish kills, the annual nutrient loading to the lake and the population and land use characteristics of the drainage basin

Although it is well known that nitrogen and phosphorus are perhaps two of the main factors determining eutrophication, there is some indication that in some cases carbon, rather than phosphorus and nitrogen, may be the chief nutrient limiting the production of algal blooms (King 1970), and that free CO_2 concentrations may regulate the structure of algal communities. In cyanophyta-bacteria associations, bacteria degrade organic matter, producing CO_2 which is utilized by the cyanobacteria in their photosynthesis. This photosynthesis generates oxygen which is, in turn, used by bacteria to oxidize or decompose organic matter. During this efficient symbiotic system, nitrogen, phosphorus and other nutrients are cycled between the algae and bacteria, and amongst the two and the environment. It is also known by means of CO_2 enrichment studies that carbon may stimulate primary productivity when other factors are not limiting.

Nitrogen is, together with phosphorus, the most production-stimulating nutrient and catalyzes the decomposition of organic material in sediments. With increasing knowledge about phosphorus leading to upgrading of sewage-treatment plants to three-stage treatment for high performance removal of phosphorus, the composition of nutrient loading on aquatic ecosystems has changed during recent decades. Although phosphorus loading has decreased, there has been an increase in nitrogen loading, resulting chiefly from precipitation of the airborne nitrogen originating from combustion in transportation and industrial processes, the intensive use of nitrogen fertilizers and changes in agricultural and silvicultural practices (Jansson et al. 1994). Larger quantities of nitrogen are being supplied to and cycled in lakes, streams and marine environments, and the N:P ratio in aquatic systems has tended to increase. In some countries increased nitrogen transport in waterways has resulted in high nitrogen loading in those coastal areas where lakes that are efficient nitrogen traps are uncommon. The effects of increased nutrient loading to some marine environments have been observed since the 1960s, with the eutrophication of the Baltic Sea. In the 1980s eutrophication became a severe problem in some coastal areas. The symptoms included algal blooms, expansion of reduced bottom areas with sulfide production and killing of fish and crustacean populations. This eutrophication is largely an effect of increased nitrogen supply since nitrogen happens to be the key limiting factor in these marine areas. To improve this situation several Northern European countries have decided to halve nitrogen discharge, by reducing discharge from sewage-treatment plants and industrial sources, reducing leakage from agriculture and forestry, and reducing atmospheric deposition on land and in marine zones. One strategy has been to determine whether the nitrogen load entering marine areas might be reduced by complementary control of the nitrogen transport in streams. Since wetlands are known to retain nitrogen, restoration and establishment of wetlands have been proposed as effective measures (Fleischer et al. 1991).

Jansson et al. (1994) have studied the nitrogen retention capacity in diverse wetlands in agricultural areas in southern Sweden in order to determine whether wetlands could be used to reduce nitrogen export from farmland to the coastal marine ecosystems. Nitrogen retention mechanisms were also investigated in lakes with increased nitrogen loading. Their findings demonstrated that nitrogen removal in wetlands depends mainly on denitrification. Sedimentation can also be quantitatively important in lakes. The water retention time is the most critical single factor for removal of nitrogen. Thus, lakes remove more nitrogen than small wetlands even though the specific nitrogen retention is generally considerably higher in wetlands. According to Jansson et al. ponds are the most suitable type of wetland for nitrogen removal.

The ponds used for nitrate removal should have a volume and an area that is as large as possible in order to increase the retention time of water in the pond and to obtain a large sediment area for denitrification (Fleischer et al. 1994). Also, macrophytes should be established in the pond. Macrophyte production should be large enough to produce organic substrates for denitrifying bacteria, but should be restricted to ensure that incoming water can be uniformly distributed in the pond (Jansson et al. 1994). The larger the area converted to wetland the greater will be the nitrogen removal. However, since nitrate removal is proportional to loading, wetlands located downstream of other wetlands will probably remove less and less nitrogen per wetland area. Also it is desirable not to locate wetlands upstream of lakes with high nitrate-retention capacity, unless reduction of nutrient loading on the lake is the goal.

If the nitrogen-removal capacity happens to be low, wetlands still contribute to a more diversified landscape and act as important refuges for animals and plants. Restoration of wetlands in modern farming landscapes therefore needs to be encouraged in those cases where it does not conflict with, e.g., land use for other purposes.

2.15.3
Reversibility of Anthropogenic Eutrophication

Man-induced eutrophication has several manifestations of which the most prominent is the nuisance growth of planktonic algal blooms and the luxuriant growth of benthic algae and rooted macrophytes in shallow near-shore areas. Although the growth of algae and plants is influenced by several environmental and physiological factors, the greatest single factor influencing productivity is the nutrients, especially phosphorus. Hence, the control of eutrophication is generally directed toward preventing the input of nutrients into the lake. A phosphorus control program is the most effective way of combating lake eutrophication. Such a program involves the reduction or removal of phosphorus from detergents, a nutrient control program at waste water treatment plants and processing or storage of urban runoff and agricultural wastes. There are two broad approaches to lake recovery, viz., (1) methods to limit fertility and/or sedimentation in lakes, and (2) procedures to manage the consequences of lake aging (Ryding 1981). The former approach tackles the root cause of the problem whereas the latter approach attempts to treat the superficial symptoms.

Several physical, chemical and biological factors affect water quality in a lake. Some of these factors are light, wind, nitrogen, phosphorus, and chlorophyll content, the growth limiting effect of nutrients, and phosphorus release from sediments. Different combinations and degrees of these and other related factors generate different results and experiences in attempts at rehabilitating different lakes. Even though the relationship between nutrient loading and chlorophyll standing crop has been recognized for a long time, it remains impossible to predict precisely how any given lake will respond to an artificial reduction in nutrient inputs. In some cases, sewage diversion and improved wastewater treatment operations have successfully reversed eutrophication (Forsberg et al. 1978). In these lakes the major source of nutrient input was domestic wastewater and, when its input was reduced, the lakes began to recover. However, similar attempts in Lake Leman and Lake Sillen have failed to reverse cultural eutrophication so far. Many shallow, heavily polluted lakes have likewise not responded to advanced wastewater treatment and reduced phosphorus input efforts mainly because of recycling of phosphorus from bottom sediments.

On the basis of his work on several Swedish lakes, Ryding (1981) recommended the adoption of algal assay results into lake management schemes for assessing the effect of excess plant nutrients in polluted lakes; plotting the algal growth potential in terms of chlorophyll against lake water chlorophyll can give some indication of the extent of inter-relationship between these parameters in lake recovery studies (Ryding 1981). In those lakes where no signs of decrease in the algal standing crop are manifested after the adoption of remedial measures, it may be worthwhile to determine the C:N:P ratio of the algal particulate matter.

2.16
Eutrophication Control

Eutrophication control programs can be planned either toward treating the basic causes or the symptoms. In some cases, a combination of the two proves useful. Alternatively, programs can focus on treating primarily point sources or nonpoint sources of nutrients. Examples include limiting nutrient inputs from municipal wastewater treatment plants and controlling runoff from farms and urban areas, respectively. Further, the program can be either structural or non-structural in form (e.g. building a wastewater treatment plant versus changing agricultural fertilizer application practices).

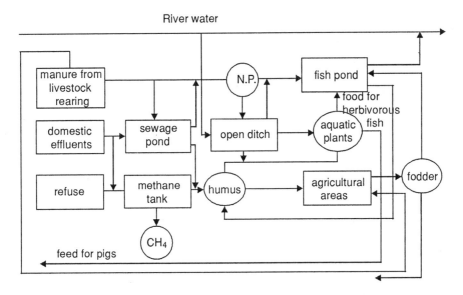

Fig. 2.29. Schematic representation of different possibilities for the integrated management of nutrient-rich water bodies in tropical settings

Table 2.9. Water quality problems treatable by in-lake restoration measures

Control measure	Water quality problem				
	Odors	Fish kills	Reduced commercial fishing	Poor drinking water quality	Excessive algal blooms
Dredging		X	x		x
Hypolimnetic aeration	x	X		x	
Nutrient inactivation			x		x
Altered circulation		X			x
Algicides					x
Biomanipulation					x
Dilution/flushing	x	X	x	x	x
Removal of hypolimnetic water	x	X	x		x
Lake drawdown	x				x
Harvesting					
Covering sediments	x	X			x

Attempts should be made to treat the underlying and most readily controllable causes of eutrophication rather than merely to alleviate the symptoms. Usually, this means reduction or elimination of the excessive nutrient inputs that stimulate excessive growths of aquatic plants. This approach is usually the most effective long term strategy.

The alternative strategy is to treat the specific symptoms of eutrophication. This is the logical and perhaps the only option if the costs of treating the basic cause (excessive nutrient inputs) are too high, or if additional treatment is necessary in a given case. In some cases, several "in-lake" treatment options may offer some temporary relief from eutrophication.

2.16.1
In-Lake Control Measures

Table 2.9 lists some treatment measures that can be applied directly in a lake or reservoir to alleviate the symptoms of eutrophication. Examples of in-lake methods include the harvesting of aquatic plants, the use of algicides, in-lake nutrient inactivation or neutralization, artificial oxygenation of bottom waters, dredging or covering of bottom sediments, increasing the water flushing or circulation rates, and "biomanipulation" (Cooke et al. 1993; Ryding and Rast 1989). These measures offer an effective means of combating, at least temporarily, the adverse impacts of eutrophication (Fig. 2.29).

2.16.2
Technology of Lake Restoration

It is usually believed that oxygen fixes phosphate, whereas lack of oxygen results in phosphate release. However, when a lake is treated with oxygen by any technical means, phosphate is released from the sediment even under aerobic conditions. The presence of dissolved oxygen does not prevent the release of phosphorus from the sediments; this release, however, is mediated by biochemical instead of purely chemical processes. Changes in nutrient cycles initiated for lake restoration should be oriented toward specific water quality goals.

2.17
Use of Ecological Indicators for Lake Assessment

An increasing number of environmental managers are now including ecological considerations in their management strategy. The environmental manager searches for ecological indicators that can assess the health of an ecosystem. Ecosystems being highly complex, it is quite difficult to find suitable ecological indicators to give reliable information on ecosystem health.

According to Costanza (1992), the concept of ecosystem health involves the following: (1) homeostasis, (2) absence of disease, (3) diversity or complexity, (4) stability or resilience, (5) vigor or scope for growth and (6) balance between system components. All or most of these elements must be considered simultaneously. He proposed an overall system health index, $HI = V*O*R$, where V is system vigor, O is the system organization index and R is the resilience index. Costanza's proposal was further refined by Jorgensen (1995) who introduced the concepts of exergy, structural exergy, and ecological buffer capacities, as ecological indicators. These three concepts combined the elements given above and the three components in Costanza's index.

2.17.1
Exergy, Structural Exergy and Buffer Capacity

Exergy means the amount of work a system can perform when brought to thermodynamic equilibrium with its environment. The exergy measures directly the distance between the present state of the considered ecosystem and the thermodynamic equilibrium. It expresses the organization of the ecosystem by the living components. Organic matter is a fuel for heterotrophic organisms and can perform work (averaging 18.9 kJ g^{-1}, mineral oil has 40–42 kJ g^{-1}), and biomass therefore contains exergy. Besides their contribution to exergy, biological components (but not detritus) also contain information in the genes or the amino acid sequence, which determines the organization of living matter.

It is not possible, because of the enormous complexity, to determine the exergy of an ecosystem as it is too vast a task to determine the concentrations of all components. It is, however, possible to determine the exergy of the components that are most influenced by a given change, for instance in the context of model calculations (Jorgensen 1995).

Exergy used as a measure for ecosystem health covers points (1), (2), some of (3), and (4) and (5) of the points given in the introductory definitions (by Costanza 1992, see above). Therefore, there is a need for supplementary ecosystem health indicators in addition to exergy. To overcome these shortcomings, the concept of structural exergy, Exst, has been introduced by Jorgensen (1995). Exst is exergy that is calculated relative to the total biomass. Structural exergy becomes non-dependent on the nutrient level or the amount of resources. It measures the ability of the system to utilize the available resources. Exst provides better coverage of points 3 and 6 above. Resilience is usually defined as the ability of the ecosystem to return "to normal" after perturbations, but this concept cannot be strictly applied in a real ecosystem because an ecosystem is a soft system that will never return to exactly the same point again. No doubt it will attempt to maintain its functions on the highest possible level, but never with exactly the same biological and chemical components in the same concentrations. Some concept of buffering is needed here.

Forcing functions are those external variables that drive the system, for example, discharge of waste, precipitation, wind and solar radiation, while state variables are the internal variables that determine the system, for instance in a lake the concentrations of soluble phosphorus and of zooplankton. This concept has to be viewed multidimensionally with all combinations of state variables and forcing functions. Even for one type of change there are many buffer capacities corresponding to each of the state variables (Jorgensen 1995).

2.17.2
Practical Assessment of Ecosystem Health

Jorgensen (1995) has proposed the following procedure for practical assessment of ecosystem health:

1. Assess the most important mass flows and mass balance involved,
2. Draw a conceptual diagram of the ecosystem, containing the components of importance for the mass flows defined above,
3. Develop a dynamic model (for inadequate data, a steady state model is to be applied),
4. Calculate exergy, structural exergy and relevant buffer capacities by the use of the model using the mathematical equations given in Jorgensen (1995) (not included here). If the model is dynamic, it reveals the seasonal changes in exergy, structural exergy and buffer capacities. If observations of the state variables are available one can calculate the annual average values of the exergy and structural exergy without the use of a dynamic model. Buffer capacities can only be calculated by the application of a dynamic model, and
5. Assess the ecosystem health: high exergy, structural exergy and buffer capacities imply good ecosystem health. If the exergy and structural exergy are high, but one of the focal buffer capacities is low, the remedy is to improve the structure of the

ecosystem to assure a higher focal buffer capacity. If the exergy is high, but the structural exergy and some focal buffer capacities are low, the system is, most likely, eutrophic and therefore a reduction of nutrient loading is warranted.

2.18
Role of Industry in Lake Management

A widely-held perception in the public mind is that industry has been the main culprit associated with the deteriorating environmental quality of rivers and lakes. Although historically the rapid industrialization after the industrial revolution has contributed much to water pollution, there has also been much progress in reducing emissions in OECD countries after World War II. The situation for many developing countries and in the countries formerly under communist regimes is, however, not good in this context. The role of industry today in preserving lakes and rivers must focus on three challenges:

1. Avoiding the emissions of pollutants into water,
2. Access to freshwater, for example for cooling and production processes, and
3. Corporate involvement in conservation efforts.

Since the Rio Earth Summit in 1992, the general public has realized that the business sector is very necessary to fund and implement solutions to the global environmental challenges. Industry can achieve results if its leaders regard the environment as a strategic priority, as industry has: (1) the technological capacity to find solutions, (2) the managerial skills to govern and change complex systems involving technology and (3) the financial resources and the market access to implement new solutions (Willums 1995). According to Willums, the new management principle of "ecoefficient leadership" may be a useful way for industry to contribute towards reducing the environmental impact on lakes and rivers.

2.19
Wetlands

Wetland is defined as land that has the water table at, near, or above the land's surface or which is saturated for a long enough period to promote wetland or aquatic processes as indicated by hydric soils, hydrophytic vegetation, and various kinds of biological activity that are adapted to the wet environment.

Wetlands are a much misunderstood resource. The application of terms such as swamp or wasteland point to a lack of understanding of the value of wetland environments. In fact, wetlands are among the richest of environments which provide many benefits to society. An environment without wetlands is incomplete and may be unable to support the functions upon which we depend for livelihood and life support.

Quite often, wetlands are considered as wasteland. Because of this erroneous thinking wetlands have frequently been altered or lost simply because their value was not appreciated. Wetlands in fact play very important roles, e.g. wildlife production, flood protection, nature study, aquifer recharge, toxic buffering, and recreation, in our total environment. An environment without wetlands is a potential threat to our well-being. Wetlands are often as diverse as rain forests and are central to the life cycles of many plants and animals, some of them endangered. They provide a habitat as well as spawning grounds for an extraordinary variety of creatures and nesting areas for migratory birds. High-latitude northern wetlands and peatlands help moderate climatic change by serving as a sink for carbon dioxide.

Wetlands are sources of rich harvests of wild rice, fur bearing animals, fish and shellfish. Wetlands limit the damaging effects of waves, convey and store floodwaters, trap sediment and reduce pollution, thereby functioning as nature's kidneys (Kusler et al. 1994).

All wetlands are shallow-water systems or areas where water is at or near the surface for some time. They have plants adapted to flooding, and hydric soils, which, when flooded, develop colors and odors that distinguish them from upland soils. Wetlands occur in diverse topographical settings: in flat, tidally inundated but protected areas, such as salt marshes and mangrove swamps; next to freshwater rivers, streams and lakes and their floodplains; and they form in surface depressions almost anywhere (Williams 1991; Mitsch and Gosselink 1993). Such wetlands comprise freshwater marshes, potholes, meadows, and vernal pools where vegetation is not woody, as well as swamps where it is. Some wetlands flourish on slopes and at the base of slopes, supplied by springs and as bogs and fens fed by precipitation and groundwater (Kusler et al. 1994).

One central, common feature of all wetlands is their fluctuating water levels. The extent of the fluctuation often differs from site to site. Because water levels rise and fall continuously, some portions of wetlands at times resemble true aquatic systems, at times terrestrial systems and at times intermediate systems. Such shifts explain the immense biodiversity of wetlands. Alterations in their water levels give rise to a series of ecological niches that can support terrestrial, partially aquatic and fully aquatic plants and animals. Vertical gradients caused by differing depths of water and saturation create further environmental variation. Wetlands essentially contain species from both aquatic and terrestrial domains (Mitsch and Gosselink 1993; Kusler et al. 1994).

Even a temporary niche can be crucial to the nesting, spawning, breeding or feeding patterns of a particular species. Short-legged birds such as greenbacked herons and limpkins feed along shallow-water shorelines. Longer-legged species (e.g., egrets and great blue herons) feed in deeper water. Shifts in water levels trigger breeding by ducks in prairie potholes (Kusler et al. 1994).

2.19.1
Importance

Wetlands can have a wide range of functions, which include provision of products, services and life support. Wetlands serve as incubators for aquatic life and shelter higher ground from tides, waves and flooding. These highly complex and varied eco-systems are now endangered by the growing demands for real estate property, construction sites and cropland. There is an urgent need for a meaningful policy that reconciles society's entrepreneurial efforts with its need to conserve intact wetlands. Some wetlands are of international, national, provincial or regional significance according to their biological, hydrological, social/cultural and/or economic production functions.

The value of wetland functions may or may not be quantifiable. For example, it may be possible to describe the number of shorebird and waterfowl that frequent a wetland (Finlayson and Moser 1991). It may also be possible to measure the economic benefits associated with these birds, whether they accrue locally (for example through hunting or viewing) or far away at the other end of a migration flyway. Wetlands are also key elements of the life-support system, having ecological benefits which present different challenges to evaluators (Bond et al. 1992).

Value to society comes from use value, for either consumptive uses (e.g. hunting, rice harvest) or non-consumptive uses (e.g., recreation or water purification). In some countries wetlands have suffered degradation.

Land use decisions affecting wetlands have so far tended to be based primarily on the direct benefits predicted for the proposed development. Economic worth is, of course, important but other costs or impacts of such activity should also not be overlooked.

Sustainable development means development which builds on the strengths of the environment and does not waste environmental resources. It also implies strengthened planning procedures to anticipate and prevent negative environmental impacts. Wetlands are natural systems worthy of careful evaluation for their biological, hydrological and socio-cultural values. Application of the concept of sustainable development to wetlands will require careful consideration of the full range of values derived from wetland environments with a view to making optimal long-term use of environmental resources (Finlayson and Moser 1991). Some wetlands are significant because of their uniqueness; others are important due to cumulative losses of typical wetlands which reduce the overall number of wetlands approaching threshold limits for specific functions in some areas. Wetlands may have fresh, brackish or saline waters. They may be permanent, seasonal or temporary. Depending on wetland location, class and function, their ecological values can include sustenance for enormous numbers of waterfowl, sources of fish production, storage and slow release of large quantities of water, erosion protection and recreational properties.

With increasing competition for land in urban areas, changes to agricultural production techniques and increased demand for hydro-electric power, wetlands have continued to be impacted through dyking, filling, drainage, flooding and other forms of conversion. Such use has caused the number and extent of wetlands to decrease substantially. Sediment and water chemistry characteristics and the nutrient status are fundamental to the study of wetland functional values (Table 2.10; Whitaker et al. 1995). Among important wetland sediment attributes is the capability to accumulate organic matter, hold water and store nutrients. These abiotic features of the substrate are closely interdependent on the hydrological conditions of wetlands. Water flow is the principal forcing function that drives the physicochemical environment, the nutrient level and the biotic responses in wetlands. The interaction of flowing water with vegetation creates a spatial heterogeneity and some physical diversity in wetlands. Consequently, hydrological conditions can directly modify both physical and chemical properties. Hydrodynamic energy gradients should correspond to gradients in ecosystem characteristics including nutrients and dissolved oxygen availability, organic deposition, water storage and depth, soil salinity and pH (Mitsch and Gosselink 1986).

Ammonium trapping in the interstitial water and on the cation exchange complex imparts added value to wetland sediments. The regulation of the nutrient inflow from surrounding terrestrial systems is particularly notable in freshwater riverine marshes. The open characteristic of these wetlands results in a continual subsidy and withdrawal of nutrients. The river waters flowing through the marsh act as a nutrient medium with unidirectional motion.

2.19.2
Classes

The five wetland classes are bog, fen, swamp, marsh and shallow open water. Variables such as hydrology, fauna, vegetation, soil, local climate, landscape setting and existence influence their development. Ecological classification is useful to conceptualize wetlands but the classes are by no means mutually exclusive; frequently wetlands combine several complex units. For instance, marshes are often associated with shallow open waters.

Table 2.10. Methods used for the physical and chemical analysis of the substrate compartments of a wetland

Variable	Method	Observation	Overlaying water	Interstitial water	Sediment
Dissolved oxygen	Titrimetric Winkler	In situ incubation	x		
pH	Potentiometric	pH meter, lab.	x	x	
Electrical con-ductivity	Electrometric	Conductometer, lab.	x	x	
Organic matter	Loss on ignition	At 600 °C, for 2 h			x
Moisture content	Gravimetric	At 105 °C, for 24 h			x
Dissolved ammonium	Indophenol blue	Spectrophotometer	x	x	
nitrate	Cadmium	Spectrophotometer	x	x	
Nitrite	Reduction				
Exchangeable ammonium	Equilibrium extraction	KCl (1 M) for 2 h			

1. **Bogs** are peat-covered wetlands in which the vegetation shows the effects of a high water table and a general lack of nutrients. The surface waters of bogs are strongly acidic. They usually have *Sphagnum* (moss) and heath shrub vegetation both with and without trees,

2. **Fens** are peatlands characterized by a high water table, with slow internal drainage by seepage down low gradients. They may have low to moderate nutrient content and may contain shrubs, trees or neither. Like bogs, most fens occur away from agricultural or urban areas,

3. In **swamps**, standing or slow-flowing water occurs seasonally or persists for long periods, leaving the subsurface continuously waterlogged. The water table may seasonally drop below the rooting zone of vegetation, creating aerated conditions at the surface (Bond et al. 1992). Swamps are nutrient-rich, productive sites. Vegetation may consist of dense coniferous or deciduous forest, or tall shrub thickets,

4. **Marshes** are periodically or permanently inundated by standing or slow-flowing water and are rich in nutrients. Marshes are mainly wet, mineral soil areas. They are subject to a gravitational water table, but water remains within the rooting zone of plants for most of the growing season. There is a relatively high oxygen saturation (Bond et al. 1992). They have a characteristic emergent vegetation of reeds, rushes, cattails and sedges, and

5. **Shallow open waters** include ponds and waters along some river, coast and lake-shore areas. They are usually relatively small bodies of standing or flowing water commonly representing a transitional stage between lakes and marshes. The surface waters appear open, generally free of emergent vegetation. The depth of water is usually less than two meters in summer (see Finlayson and Moser 1991).

2.19.3
Functions

Table 2.11. Translating wetland functions into benefits of value to society. (After Bond et al. 1992)

Function/capabilities	Services and experiences	Examples of benefits or products to society derived from or supported by wetlands
Life support		
Regulation/ absorption	Climate regulation, toxics absorption, stabilization of biosphere processes, water storage, cleansing	Flood control, contaminant reduction, clean water, erosion control
Ecosystem health	Nutrient cycling, food web support, habitat, biomass storage, genetic and biological diversity	Environmental quality, maintenance of ecosystem integrity, risk reduction (and related option values)
Social/Cultural		
Science/information	Specimens for research, zoos, botanical gardens, representative or unique ecosystems	Enhanced understanding of nature
Aesthetic/ Recreational	Viewing, photography, bird watching, hiking, swimming	Direct economic benefits to users' personal enjoyment and relaxation, benefits to tourist industry, local economy
Production		
Subsistence production	Natural production of birds, fish, plants	Food, fiber, self-reliance for communities
Commercial production	Production of foods (e.g. fish, crops), fiber (e.g. wood, straw), soil supplement (e.g. peat)	Products for sale, jobs, income

These are the capabilities of wetland environments to provide goods and services including basic life-support systems. Such functions may directly or indirectly provide benefits to society.

A given wetland can, for instance, support water storage, habitat for many species, scenery, fish habitat, toxic buffering and flood control. Based on these functions, many benefits can be derived from the wetland: e.g. clean drinking water, a place to swim, photography, duck hunting, reduction of flood damage, reduced drought risk on adjacent fields or harvest of *Trapa bispinosa*.

Wetland functions provide many benefits to society. These benefits do have some value: food, risk reduction, jobs, lifestyle, life support for humans and other species. Wetland functions may or may not provide benefits that are readily measurable. Some benefits from a wetland may have no measurable immediate value to society.

Wetlands regulate and maintain essential ecological processes and life-support systems. Some of these functions are the following (see Table 2.11):

1. Climate regulation,
2. Watershed protection and water catchment,
3. Erosion prevention and soil protection,
4. Storage and recycling of human waste,
5. Storage and recycling of energy, and
6. Toxics absorption.

Wetlands act as useful "environmental filters" in agricultural and urban areas where runoff carries with it an excess of nutrients and often toxic chemicals. Through wetland vegetation life cycles, these chemicals are frequently removed from the water. This "cleansing" benefits environment and society by reducing water contamination in downstream and groundwater areas. Wetlands can be used as sites for secondary sewage or storm water treatment. In some regions, wetlands influence the micro-climate and groundwater by stimulating local precipitation and replenishing groundwater supplies.

Wetland marshes, swamps, and shallow water areas are often highly productive or "fertile" ecosystems. Wetlands support a complex web of energy transfers and associated flora and fauna. For instance, marsh and swamp habitats can produce up to four times the net primary nutrient production of lakes. However, nutrient-poor wetlands, such as bogs and some fens, are biologically more simple, having limited biological diversity (Bond et al. 1992).

Wetlands support a variety of mammals and a large number of birds of prey, songbirds and shorebirds. The biological functions, including diversity of habitat, are some of the most significant elements of the social tourism and cultural value of wetlands.

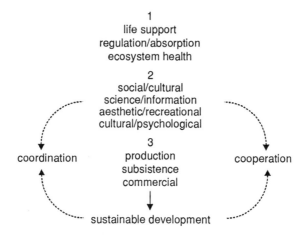

Fig. 2.30. Relationship of wetland functions to societal values

2.19.4
Subsistence Production and Commercial Production

Wetland production functions incorporate a complex variety of aspects, and fall into two general categories: subsistence and commercial. Some production functions of wetlands have a bearing on the following areas/aspects:

1. Industry,
2. Water supply,
3. Food,
4. Building, construction and manufacturing materials,
5. Fuel and energy,
6. Minerals, and
7. Medicinal resources.

Figure 2.30 illustrates the relationship of wetland functions to societal values. Wetland functions are varied and diverse, depending upon wetland class, location, and size. According to Bond et al. (1992) any evaluation of wetland functions must take into account the following parameters:

1. Regional and inter-regional linkages of these functions,
2. The associated social/cultural and production functions of biological and hydrological/biogeochemical natural system attributes, and
3. The monetary and non-monetary value of such functions and relationships.

References

Allan RJ, Zarull MA (1995) Impacts on biodiversity and species changes in the lower Great Lakes. Lakes and Reservoirs: Research and Management 1:157–162

Ambuhl H, Buhrer H (March 1993) The lake as an ecosystem. EAWAG News 34:2–6

Arditi R, Ginzberg LR (1989) Coupling in predator-prey dynamics: ratio-dependence. J Theor Biology 139:311–326

Bjork, S (1994) Overview. In: Eisettora M (ed) Restoration of lake ecosystems: a holistic approach. IWRB Pub No 32, pp 1–5

Brooks JL, Dodson SI (1965) Predation, body size, and composition of plankton. Science 105:28–35

Brylinsky M, Mann KH (1973) An analysis of factors governing productivity in lakes and reservoirs. Limnol Oceanogr 18:1–14

Buffle J, Altmann RS (1990) Interpretation of metal complexation by heterogeneous complexants. In: Stumm W (ed) Aquatic surface chemistry. Wiley-Interscience, New York, pp 351–383

Charles DF (1991) Acidic deposition and aquatic ecosystems: regional case studies. Springer, New York

Carpenter SR, Kitchell JF, Hodgson JR (1985) Cascading trophic interactions and lake productivity. Bioscience 35:634–39

Carter G, McKerchar A (1995) Hydrology and hydraulics. Water and Atmosphere (NIWA, New Zealand) 3:23–24

Cole JJ, Caraco NF, Kling GW, Kratz TK (1994) Carbon dioxide supersaturation in the surface waters of lakes. Science 265:1568–1570

Cooke GD, Welch EB, Peterson SA, Newroth PR (1993) Restoration and management of lakes and reservoirs. 2nd ed. Lewis Publishers, Boca Raton

Costanza R (1992) Ecosystem health. Columbia University Press, New York

Davison W (1985) Conceptual models for transport at a redox boundary. In: Stumm W (ed) Chemical processes in lakes. Wiley Interscience, New York, pp 31–53

Dillon PJ, Rigler FH (1974) The phosphorus-chlorophyll relationship in lakes. Limnol Oceanogr 19:767–773

Cairns J, Lanza GR, Parker BC (1972) Pollution related structural and functional changes in aquatic communities with emphasis on freshwater algae and protozoa. Proc Acad Nat Sci Phila 124:79–127

Fallon RD, Brock TD (1980) Planktonic blue-green algae: production, sedimentation, and decomposition in Lake Mendota, Wisconsin. Limnol Oecanogr 25:72–80

Finlayson M, Moser M (1991) Wetlands. Internat Water Fowl and Res Bureau, Slimbridge, UK

Fleischer S, Stibe L, Leonardson L (1991) Restoration of wetlands as a means of reducing nitrogen transport to coastal waters. Ambio 20:271–272

Fleischer S, Gustafson A, Joelsson A, Pansar J, Stibe L (1994) Nitrogen removal in created ponds. Ambio 23: 349–57

Forsberg C, Claesson A, Ryding SO, Forsberg A (1978) Research on recovery of polluted lakes. I Verh Internat Verein Limnol 20:825–32

Fretwell SD (1977) The regulation of plant communities by food chains exploiting them. Perspectives in Biology and Medicine 20:169–185

Fricker H (1980) OECD eutrophication programme. Regional Project "Alpine Lakes" Berne: Bundesamt fur Umwelt, Wald und Landschaft (BUWAL)

Gachter R, Wuest A (1993) Effect of artificial aeration on trophic status and hypolimnetic oxygen concentration in lakes. EAWAG News 34:25–30

Golterman H (1980) Quantifying the eutrophication process: difficulties caused by sediments. Progr Water Technol 12:63–70

Gorham E (1996) Lakes under a three-pronged attack. Nature 381:109–110

Gulati RD (1990) Zooplankton structure in the Loosdrecht lakes in relation to trophic status and recent restoration measures. Hydrobiologia 1991:173–88

Hansson LA (1992) The role of food chain composition and nutrient availability in shaping algal biomass development. Ecology 73:41–47

Hargeby A (1990) Effect of pH, humic substances and animal interactions on survival of *Asellus aquaticus* (L) and *Gammarus pulex* (L.) Oecologia 82:348–354

Henrikson L, Brodin Y (1993) How to save acidified waters – a synthesis of the Swedish liming programme. Springer, Berlin

Herrmann J, Degerman E, Gerhardt A, Johansson C, Lingdell PE, Muniz IP (1993) Acid stress effects on stream biology. Ambio 22:298–307

Hunter MD, Matson PA (1992) (eds) Special feature: the relative contributions of top-down and bottom-up forces in population and community ecology. Ecology 73:723–65

Imboden DM (1990) Mixing and transport in lakes: mechanisms and ecological relevance. In: Tilzer M, Serruya C (eds) Large lakes: ecological structure and functions. Springer, Berlin

Imboden DM, Gachter R (1979) The impact of physical processes on the tophic state of a lake. In: Ravera O (ed) Biological aspects of freshwater pollution. Pergamon Press, Oxford, pp 93–110

Imboden DM, Sigg L, Schwarzenbach RP (1993) The distribution of substances in lakes: interactions between physical and chemical processes. EAWAG News 34:7–11

Jansson M, Andersson R, Berggren H, Leonardson L (1994) Wetlands and lakes as nitrogen traps. Ambio 23:320–25

Jorgensen SE (1995) The application of ecological indicators to assess the ecological conditions of a lake. Lakes and Reservoirs: Research and Management 1:177–182

Jorgensen SE, Vollenweider RA (1989) (eds) Guidelines of lake management. Vol I Principles of lake management. Internat Lake Environ Committee Foundation, Shiga, Japan, pp 13–17

Kajak Z (1980) Influence of phosphorus loads and of some limnological processes on the purity of lake water. Hydrobiol 72:43–50

King DL (1970) The role of carbon in eutrophication. Jour Water Poll Contr Fed 42:2035–51

Kusler JA, Mitsch WJ, Larson JS (1994) Wetlands. Sci Amer pp 64–70

Lijklema L (1995) Perspectives in environmental restoration and nature conservation in lakes. Lakes and Reservoirs: Research and Management 1:151–155

Malmqvist B (1993) Interaction in stream leaf packs: effects of a stonefly predator on detritivores and organic matter processing. Oikos 66:454–462

Matveev V (1995) The dynamics and relativel strength of bottom-up vs top-down impacts in a community of subtropical lake plankton. Oikos 73:104–108

Mazumder A (1994) Patterns of algal biomass in dominant odd – vs. even-link lake ecosystems. Ecology 75:1141–49

Mazumder A, Taylor WD, Lean DRS, McQueen DJ (1992) Partitioning and fluxes of phosphorus: mechanisms regulating the size-distribution and biomass of plankton. Archiv für Hydrobiologie, Ergebnisse der Limnologie 35:121–143

McQueen DJ, Post JR (1988) Cascading trophic interactions: uncoupling at zooplankton-phytoplankton link. Hydrobiologia 159:277–96

McQueen DJ, Post JR, Mills EL (1986) Trophic relationships in freshwater pelagic ecosystems. Can J Fish Aquat Sci 43:1571–1581

Mills EL, Leach JH, Garlton JT, Secor CL (1993) Exotic species in the Great Lakes: history of biotic crises and anthropogenic introductions. J Great Lakes Res 19:1–54

Mitsch WJ, Gosselink JG (1986, 1993) Wetlands. Van Nostrand Reinhold, New York

Morgan JJ, Stumm W (1990) Chemical processes in the environment: relevance of chemical speciation. In: Merian E (ed) Metals and their compounds in the environment. VCH Verlagsgesellschaft, Weinheim, pp 67–103

Mortimer C (1941) The exchange of dissolved substances between mud and water in lakes. J Ecol 29:280–329

Mur LR (1980) Concluding remarks. In: Barica J, Mur, LR (eds) Developments in hydrobiology. W Junk, The Hague, pp 313–33

OECD (1982) Eutrophication of water. Monitoring, assessment and control. Final Report. OECD Cooperative Program on Monitoring of Inland Waters (Eutrophication Control). Environment Directorate, OECD, Paris

Olsson H, Pettersson A (1993) Oligotrophication of acidified lakes – a review of hypotheses. Ambio 22:312–317

Ryding SO (1981) Reversibility of man-made eutrophication. Experiences of a lake recovery study in Sweden. Int Revue Ges Hydrobiol 66:449–503

Ryding SO, Rast W (1989) The control of eutrophication of lakes and reservoirs. MAB Digest, 1. UNESCO, Paris and Parthenon Publishing Co, Carnforth

Ryding SO (1981) Mass-balance studies to, within, and from drainage basins. A new approach for water monitoring programmes (In Swedish). Nordic Cooperative Organization for Applied Research (NORDFORSK) Publication 1983. NORDFORSK, Helsinki, pp 149–162

Sakamoto M (1966) Primary production by phytoplankton community in some Japanese lakes and its dependence on depth. Arch Hydrobiol 62:1–28

Santschi P, Hohner P, Benoit G, Buchhol T, Brink M (1990) Chemical processes at the sediment-water interface. Marine Chem 30:269–315

Sarnelle O (1992) Nutrient enrichment and grazer effects on phytoplankton in lakes. Ecology 73:551–560

Schindler DW (1980) Evolution of the experimental lakes project. Canad J Fish Aquat Sci 37:313–318

Schindler DW, Curtis PJ, Parker BR, Stainton MP (1996) Consequences of climate warming and lake acidifiction for UV-B penetration in North American boreal lakes. Nature 379:705–708

Schindler DW, Fee EJ (1974) Primary production in freshwater. Proc Ist Internat Congr Ecol The Hague, pp 155–58

Seko I (1994) Survey on the state of world lakes. INSIGHT Newsletter of the UNEP Internat Environ Technol Center (IETC), Osaka/Shiga (Japan) Fall 1994, pp 1–2

Shapiro J, Wright DI (1984) Lake restoration by biomanipulation: Round Lake, Minnesota, the first two years. Freshwater Biol 14:371–383

Slobodkin LB (1992) A summary of the special feature and comments on its theoretical context and importance. Ecology 73:1564–66

Sly PG (1991) The effects of land use and cultural development on the Lake Ontario ecosystem since 1750. Hydrobiologia 213:175

Stewart KM, Rohlich GA (1967) Eutrophication, a review. Publ No 34. State Water Quality Control Board, Sacramento, Calif

Sylvester RO, Seabloom RW (1965) Quality of impounded water as influenced by site preparation. Dept of Civil Engineering Univ, Washington, Seattle

Tilzer MM, Bossard P (1992) Large lakes and their sustainable development. Aquatic Science 54:91–103

Vollenweider RA (1975) Input-output models with special reference to the phosphorus loading concept in limnology. Schweiz Z Hydrol 37:53–84

Walters CJ, Park RA, Koonce JF (1980) Dynamic models of lake ecosystems. In: LeCren, ED, Lowe-McConnell, RH (eds) The functioning of freshwater ecosystems. Cambridge Univ Press, Cambridge, pp 455–79

Wehrli B, Ventling A, Muller R (1993) Biogeochemical processes at the sediment surface. EAWAG News 34:17–20

Whitaker VA, Matvienko B, Tundisi JG (1995) Spatial heterogeneity of physical and chemical conditions in a tropical reservoir wetland. Lakes and Reservoirs: Research and Management 1:169–175

Williams M (1991) Wetlands: a threatened landscape. Basil Blackwell, Oxford

Willums J-O (1995) The management of lakes and the role of industry. Lakes and Reservoirs: Research and Management 1:187–189

Yan ND, Keller W, Scully NM, Lean DRS, Dillon PJ (1996) Increased UV-B penetration in a lake owing to drought-induced acidification. Nature 381:141–143

3 Marine Environment

3.1
Introduction

All life on Earth came from the sea. The oceans determine the climate. They supply a vital source of food for half the population. The marine environment is highly fragile because of the wide variety of foreign objects that are being intentionally or unintentionally introduced into it. The sea deceptively appears to us to be uniform; exchanges between the water and the atmosphere and the effects of atmospheric particles linked to climatic processes in fact introduce modifications to the ocean mass, which is also very varied. Close to the coast, the diversity is even more marked, because of the variety of substances entering the sea via rivers. No matter how much the composition of the sea may seem to be uniform, the many micro-environments it is made up of are in fact localized and short-lived. The marine environment in some areas in now a receptacle for untreated sewage (Fig. 3.1), oil and industrial wastes, fertilizers and pesticides. Table 3.1 lists the sources of marine pollution around the world.

The major threats for the marine environment come from the land. The most important sources of pollution are nutrients, products linked with microbial activity and pesticides used in agriculture. Pollution by hydrocarbons and from the dumping of toxic products is, surprisingly, usually less important. Industrial waste, which can sometimes be locally strong, is less important than other substances which reach the sea from several different sources. Sea-level rise and increased desertification pose serious threats to the more vulnerable developing countries. As climate zones shift farmers will be forced to adapt. The countries and populations which face the greatest threats from anthropogenic climate change are those that are least able to bear its costs.

As the total emissions of developing countries begin to exceed those of the developed world, the rapidly developing countries will have to think seriously about how to control their emissions. They will need to improve supplyside energy generation, expand and modernize the use of renewables and increase energy efficiency. The wealthier countries will have to take up a stronger role in supporting these efforts.

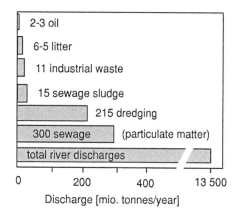

Fig. 3.1. Discharges in the marine environment. (UNEP 1993)

The coastline and the nearby continental shelf suffer most from the effects of riverine inputs into the sea. The waters from rivers pouring into open seas are dispersed by ocean currents thus reducing their contaminating effects. In contrast, the possible toxicity of sediments can be reinforced when they are trapped by phenomena inherent to specific coastal systems such as deltas or coastal lagoons, whose fragility is much higher. In closed or semiclosed seas, exchanges with the continental shelf occur largely by drainage from the land, because the dispersal effect as a result of marine currents is weak. In these environments life conditions are often suddenly and irreversibly altered.

Pollutants from natural sources, and those dumped in the marine environment intentionally, inflict strong damage on the oceanic animal and plant species. In fishing grounds, the damage done by pollution of the sea is often serious, and it also applies to the overexploitation of the marine fauna and flora in coastal waters and on the continental shelf. The eradication of marine mammals, the destruction of the sea bed by trawling, the devastation of the fauna of the coral reefs, overdevelopment of the coast, destruction of fish breeding areas and the dumping of toxic wastes are some examples of the exploitation and violation of the seas today.

The sea should be regarded as a living system. About three-fourths of marine contamination comes from the mainland (Fig. 3.2). All the water circulating in the world ends up in the sea, and airborne pollutants are also eventually deposited on the surface of the water. The sea is the Earth's temperature regulator. If it were not for the sea's capacity for absorbing heat, the average temperature of the planet would be 1 or 2 °C higher. The sea is also one of the main carbon dioxide regulators, and it is estimated that it stores 20 times as much carbon as the world's forests put together.

Table 3.1. Sources of marine pollution around the world. (From GESAMP 1995)

Water discharge or other process or activity potentially causing contamination	1[a]	2	3	4	5	6	7	8	9	10	11	12	13	14	15
Sewage	x	x	x	x	x	x	x	x	x	x	x	x	x	x	
Petroleum hydrocarbons (maritime transport)	x	x	x	x	x	x	x	x	x	x	x	x			
Petroleum hydrocarbons (exploration and exploitation)		x		x	x			x		x	x	x	x	x	
Petrochemical industry		x		x	x					x	x	x			
Mining			x				x			x			x	x	
Radioactive wastes	x	x	x				x		x	x		x			
Food and beverage processing	x	x	x			x				x	x	x	x	x	x
Metal industries		x	x		x				x	x		x			x
Chemical industries	x	x	x						x	x					
Pulp and paper manufacture	x				x					x			x	x	x
Agricultural runoff (pesticides and fertilizer)			x		x		x	x		x					x
Siltation from agriculture and coastal development						x	x	x			x				
Sea-salt extraction							x				x				
Thermal effluents						x	x		x	x	x	x			
Dumping of sewage sludge and dredge spoils		x							x	x					

[a]1 Baltic Sea, 2 North Sea, 3 Mediterranean Sea, 4 Persian Gulf, 5 West African areas, 6 South African areas, 7 Indian Ocean region, 8 South-East Asian region, 9 Japanese coastal waters, 10 North American areas, 11 Caribbean Sea, 12 South-West Atlantic region, 13 South-East Pacific region, 14 Australian areas, 15 New Zealand costal waters

Coastal waters, which provide 90% of the living harvest of the seas, are exposed to multiple stresses. Most wastes from land end up in the sea (Fig. 3.1) and remain trapped near the shore, poisoning marine life. Coral reefs – the tropical forests of the oceans, home up to a third of the world's fish species – are being destroyed by pollution and overexploitation; so is another vital nursery for fisheries and wildlife-the mangroves (see Table 3.2). With them goes not only a source of food, but also a defense against coastal erosion.

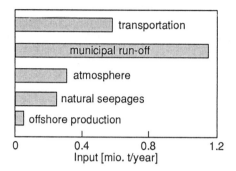

Fig. 3.2. Various sources of petroleum input in the marine environment

Table 3.2. Some examples of the different types of impacts that can affect mangroves

Type	Spatial extent	Magnitude of impact	Duration	Frequency	Reversibility
Biological impact					
Invasion and establishment of populations of exotic species	May eventually cover whole region	Absolute or relative abundance	Long term	Continuous new state	Often irreversible
Chemical impact					
Non-point source, chronic poisoning	May cover local scales or whole regions	Biological effects (e.g. fish kills)	Long term	Continuous	Slowly reversible
Accidental, acute poisoning	Local	Biological effects (e.g. fish kills)	Short term	Low frequency	Reversible for populations and individuals
Physical impact					
Structural change due to hydrologic change	Local	Magnitude of change in discharge and distribution of discharge over time	Long term	Variable, depending on type of regulation	Often reversible

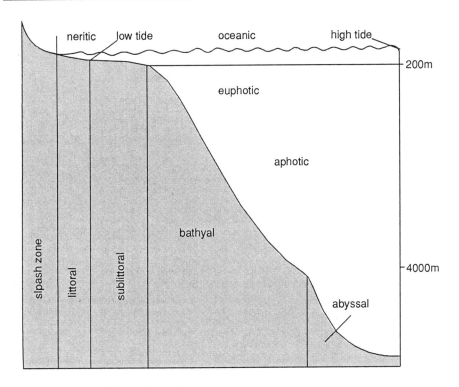

Fig. 3.3. Divisions of the marine environment

Shallow and virtually land-locked, the Persian Gulf is overloaded by wastes from the cities and industries around its shores. Two thirds of the world's sea-borne oil exports cross its waters. On the other side of the Arabian peninsula, the countries of the Red Sea and Gulf of Aden have agreed on an action plan to pre-empt the problems which rapid development may bring. The West and Central African plan is concerned with land-based pollution and stresses the fight against coastal erosion, a serious concern to countries which are already losing vast tracts of land to the inland desert. The Philippines, Thailand, Malaysia, Indonesia and Singapore are emphasizing the protection of coral reefs and fisheries in their action plan.

Figure 3.3 illustrates the divisions of the marine environment. The oceanic province consists of those waters which are deeper than 200 m. This includes the euphotic or epipelagic zone, and photosynthesis occurs in its upper portion (up to 200 m). The aphotic zone is at depths greater than 200 m and is non-photosynthetic. Figure 3.4 shows the processes involved in air-sea exchange of contaminants.

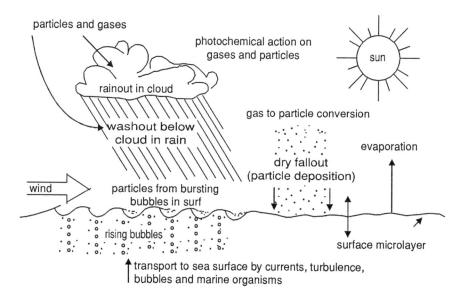

Fig. 3.4. Processes in air-sea exchange of contaminants. (After Waldichuk 1990)

The Indian Ocean ranges from the East Coast of Africa to Western Australia and, unlike the Pacific and Atlantic Oceans, is bordered to the North by the Asian continent. It covers an area of approximately 73 600 000 km². It encompasses 38 countries including the island states. The Indian Ocean shows great diversity, extending from open ocean to semi-enclosed seas, from huge to minor islands and archipelagos, from narrow to broad watersheds, from the most arid to the most humid coasts. Despite such diversity, the Indian Ocean should be considered as one large ecosystem because of the physical, chemical and biological interactions between the various water bodies. However, for effective conservation, cooperation and rational management it is divided into several sub-ecosystems.

The coastal zone is the interface between land and the sea and its boundaries depend on several ecological and anthropogenic (political, administrative and legal) issues. In a broader context the coastal zone may include the coastal hinterland, the lowlands, the coastal waters and the deep sea up to the exclusive economic zone (EEZ) because activities that seem remote do affect and interact with this zone. The coastal area is characterized by a variety of forms: rocky shores, sandy beaches, estuaries, lagoons, intertidal flats, wetlands, and islands. These forms provide habitats for specific biological communities including intertidal communities, mangroves, seagrasses, coral reefs, and the deep/open sea communities. These different forms and habitats are closely interlinked and act as a unified system. They house numerous resources which support a significant proportion of the mostly impoverished coastal populations.

Man has settled in the coastal zone in search of the many amenities that it provides. Several decades ago, the presence of man caused no harm to the coastal environment. Fishing and other forms of use of the marine and coastal resource caused local environmental changes but did not threaten sustainability. Today, technological advancement, coupled with the growing resource needs for the ever increasing population, has led to the use of numerous resources from the marine and coastal zone. We are now extracting hydrocarbons, mining coral reefs, cutting mangroves, and promoting coastal tourism, marine transportation and exploitation of fisheries. In addition we have beach erosion and a sea level rise. It is becoming clear that without proper management, these uses would result in conflicts and endanger the resources.

Recent years have seen an alarming increase in the numbers of coastal inhabitants worldwide. By the year 2000 it is estimated that about 70% of the world's population may live in close proximity to the coasts. The pollution in coastal zones caused by human activities is threatening some fisheries which make up the primary source of food in coastal countries.

The marine environment – including the oceans and all seas and adjacent coastal areas – forms an integrated whole that is an essential component of the global life-support system and a positive asset that presents opportunities for sustainable development.

As the major outcome of UNCED, organized in Rio de Janeiro, Brazil, in June 1992, Agenda 21 outlines the main issues, objectives and action required to achieve, or at least move towards, sustainable development as we approach the next century.

Oceans and coastal seas were long considered as inexhaustible sinks where waste and other pollutants could safely be dumped without any significant adverse effects. It is now realized that environmentally irresponsible actions and activities can cause potential public health hazards through exposure to contaminated food and water. In recent years, another threat to the living marine resources has been identified, the one resulting from global environmental change. Global warming may raise the temperature of the oceans, and have catastrophic effects on, inter alia, the sensitive corals. The depletion of the atmospheric ozone layer is causing more ultraviolet radiation to reach the Earth's surface, affecting the plankton in the oceans.

Generally, two different levels of global environmental change can be identified: global environmental change embracing those changes that are truly global and affect the Earth as a whole (e.g. the growing hole in the ozone layer), and those environmental changes that occur locally which relate to global phenomena through their cumulative effects (e.g. deforestation and soil erosion).

The governance of ocean space requires international cooperation and mechanisms that supersede national jurisdiction. It was recognized at UNCED that the implementa-

tion of strategies and activities related to marine and coastal areas and seas would require effective institutional arrangements at national, regional and global levels.
Agenda 21 identifies the difficulties in the management of high sea fisheries, including
the adoption, monitoring and enforcement of effective conservation measures. Fishing
fleets operating in international waters are utilizing inappropriate and indiscriminate
fishing methods resulting in over-harvesting of living marine resources. Similarly,
monitoring and preventing marine pollution on the high seas from sea-based sources,
including dumping of hazardous wastes, is quite difficult.

3.2
The Intertidal Zone

A combination of habitat type and adult population level can affect settlement by juveniles of some shellfish species. This has important implications for the management of
some intertidal zones in the coastal strips which lie between high tide and low tide.
Some factors important in determining the structure of intertidal communities are
summarized below.

Important biological factors that act to structure sand flat communities are: species
interactions, recruitment settlement of juvenile organisms, food resources and disturbance caused by, for example, burrowing activity.

One of the most important physical factors involved in the distribution and abundance of organisms is sediment grain size. Its significance comes from its effect on
water retention and its suitability for burrowing. Whereas fine sand tends to hold water
in its interstitial spaces after the tide has retreated, coarse sand and gravel permit quick
water drainage. Therefore, fine sand gives more protection against desiccation than
coarse sand, making the latter less hospitable. Fine sand and mud are also more suitable for burrowing than coarser sediments (Wilkinson 1995).

Environmental effects on the intertidal zone caused by human activities fall broadly
into four categories: sedimentation, nutrient input, contamination, and habitat modification. Some of these impacts are mere amplifications of natural environmental conditions. One such case is increased sedimentation. Increases in both suspended sediments
and sea floor sedimentation have direct adverse effects on plants and on suspension-
feeding organisms such as bivalve mollusks and polychaete worms. Suspension-
feeding bivalves often have a strong influence on energy flows and plankton population dynamics of coastal ecosystems, so any effect on these organisms can potentially
affect the whole harbor ecosystem. Suspended sediments also alter water clarity. Decreased light penetration weakens photosynthesis and, therefore, the productivity of a
system. Sand flats occur in those areas of the intertidal zone which are exposed to

wave activity as the tide comes in. Mud flats are generally restricted to intertidal areas completely protected from open ocean wave activity.

Some of the best-developed mud flats occur in partly enclosed bays, lagoons, harbors and estuaries where there is a source of fine-grained sediment particles. These areas are more stable than beaches and sand flats and are more conducive to the development of permanent burrows (Wilkinson 1995). The fine particle size coupled with the flat angle of repose of muddy sediments means that the water does not drain away from the substrate. This, together with poor interchange of the interstitial water with the seawater above and a high internal bacterial population, often results in complete depletion of oxygen in the sediments below the first few centimeters of the surface. This is one of the most important characteristics of a mud flat which distinguishes it from a sand flat.

To survive while buried in the substrate, organisms must either be adapted to live under anaerobic conditions or must have some way of bringing the overlying surface water with its oxygen supply down to them. The latter is often the most common adaptation (Wilkinson 1995).

For soft-bottom marine habitats, interactions between established adults and settling larvae determine population and community dynamics. These interactions can include: ingestion of the larvae by the adults, interference by increased sediment mixing resulting from the digging of burrows, modification of local flow patterns by either physical structures (such as bivalve shells protruding above the sediment surface) or feeding currents, the provision of refuges from predation, and chemical cues (Wilkinson 1995). Thus, the adults may either help or hinder settling juveniles.

Hydrodynamic processes not only influence sediment grain size but also affect the strength of biological interactions. Usually, in areas with greater wind-wave exposure the strength of biotic interactions tends to be weaker than in calmer areas.

3.3
Hydrothermal Vents

High-temperature hydrothermal vents, called black smokers, at oceanic ridges are important agents of mass and fluid transfer between the oceanic crust and the ocean. Much of the iron, copper, zinc and other metals mobilized from the oceanic crust during fluid-rock interaction are discharged into the ocean as black smoke but some metals are precipitated as sulfide deposits at the vent site and within underlying volcanic rocks. These sea-floor mineral deposits are the modern equivalents of the economically valuable sulfide deposits preserved as parts of ophiolites on land. Huge sea-floor deposits have been found that are commensurate in size with their ancient counterparts.

The discovery, in the 1970s, of some sulfide-forming hydrothermal vents confirmed the submarine origin of volcanic-associated massive sulfide (VMS) deposits preserved in phiolite and island-arc terranes of many orogenic belts. VMS deposits consist of a concordant lens-shaped body of massive sulfide (60% or more of sulfide minerals) overlying a discordant pipe- or funnel-shaped zone of altered volcanic substrate known as the stockwork (Koski 1995).

During the formation of VMS deposits on the ocean floor, recrystallization and replacement processes continually modify the active deposits, and it is unlikely that any of the early mineralization survives. Once formed, the sulfide deposits are degraded by mass wasting and oxidation. Over time, deposits are further modified by burial, metamorphism, deformation and, sometimes, subaerial weathering (Koski 1995).

On bare rock ridges, hydrothermal fluids with high temperatures and salinities and low pH carry much greater loads of dissolved metals, including copper and zinc, than does sea water, and are potent ore solutions. Among the diverse sea-floor hydrothermal deposits, current focus has been on the most pristine and accessible constructions – the black smoker chimneys growing on the tops of mounds. Chimneys are essentially minideposits, formed by precipitation of sulfides and sulfates by the steep temperature and chemical gradients existing within the zone where hydrothermal fluids and seawater mix.

The mounds contain extensive breccia deposits, indicating that both high and low temperature chimneys atop the mound may have repeatedly grown and collapsed to produce sulfide fragments that, along with fragments of the underlying volcanics, are gradually cemented and replaced during subsequent reactions with hydrothermal fluids percolating through the porous infrastructure. During replacement, more soluble metals e.g. zinc, are remobilized from chimney fragments to the outer reaches of the mound contributing to large-scale metal zonation within the deposit (Koski 1995). This cycle of reworking and refining is repeated many times as the mounds enlarge.

3.4
Marine and Coastal Pollution: Regional Sustainability

The methods and techniques required for research on and management of the land-sea interactions are distinctive because of short-term geomorphological and ecological dynamics and the responses of coastal people to coastal changes (Uitto 1993). Coastal zones are areas of population concentrations, major urban centers, as well as recreation and tourism. They are also used for various competing economic activities, such as mariculture and fishing, waste disposal, transportation and energy production. All this human activity places heavy pressures on coastal zones and marine areas beyond.

The coastal areas of Southeast Asia have been particularly exposed to these pressures, as seen from the threatened status of some fisheries, as well as the more frequent occurrence of toxic algal blooms and red tide. Monitoring of coastal pollution is a high priority for the international community.

The world's first major incidence of marine pollution with serious health effects to the local population residing in affected areas was the case of Minamata, in Japan. The Minamata disease was identified in 1956 when patients who had consumed sea food caught in Minamata Bay suffered from severe neurological disorders frequently leading to death. In 1959, a study group from Kumamoto University identified organic mercury as the cause of the disease, but this was officially recognized only in 1968. The mercury had entered the sea through the discharge of waste water by the Chisso Chemical plant. Chisso Ltd. began waste treatment in 1959, but sporadic discharge and spills continued until 1968. The Minamata episode has affected the entire community for about 4 decades.

There is now a strong need for protection of the world's marine and coastal environments. Much remains to be done to understand and manage complex marine ecosystems. It is necessary to devise national, regional and global mechanisms for governance of oceans and coastal areas for sustainable development. Effective mechanisms are required for the establishment, enforcement and monitoring of shipping regulations, fisheries and international agreements, as well as international harmonization of policies and actions, together with the continuous monitoring of marine pollution and improved understanding of its effects, sources and socio-politico-economic causes.

3.5
Living Resources

The living creatures of the sea have long supplied mankind with food, oils and useful materials. Overfishing and pollution are threatening these resources. Fish provide nearly a quarter of the animal protein consumed by people worldwide. The global catch has recently been rising by 7% a year and at least 25 major fisheries are depleted. By the year 2000, annual fish supplies may fall short of demand by about 10 million tonnes. About 160 species of mammals live in the world's oceans. They include some of the most intelligent, best loved and most exploited creatures on Earth. Several species of great whale are endangered, as are all species of sea cow and some species of seals, dolphins and otters. The humpback whale is endangered by hunting, disruption of its breeding grounds and human competition for food supplies. Its numbers have also fallen to as low as 4000. For centuries, indigenous communities have hunted marine mammals on a sustainable basis. Advanced hunting methods are now taking their toll. Hundreds of thousands of marine mammals are also killed accidentally, when they become entangled in fishing nets. The use of drift nets, sometimes 60 km long, is causing particular concern. Marine mammals are also threatened by habitat destruction, pollution and competition over food supplies.

Many marine animals liberate large numbers of larvae or gametes into the water column where they are subject to various sources of mortality, including starvation, predation, physiological stress and advection from adult habitats. The sheer numbers of larvae led to the belief that these mortality factors overwhelm larvae except during certain years when planktonic conditions happen to be particularly good (Mullin 1993). Planktonic prey is patchy; fish larvae that locate and remain in favorable patches tend to survive well.

Tiny fish larvae commonly disperse widely for a long time, and there is a need to study larval mortality on a variety of spatial and temporal scales. One way is to study individuals on a small scale under experimentally controlled conditions and then sum the outcomes of these small-scale interactions to infer how fish recruitment on a broad scale may be affected by environmental variation. Another strategy is to work on large spatial and long temporal scales which make it possible to study the effects of environmental change on fish recruitment directly in the complex natural ecosystems. Unlike terrestrial ecosystems, marine ecosystems consist of living creatures which are either microscopic or else so scattered as to be hardly visible.

Overexploitation has caused devastation in important fishing grounds. The fishing techniques have become more and more efficient, boats are safer and equipped with sonar, telecommunications, satellite positioning, longer and more resistant nets, refrigeration tanks, fish-processing rooms and air support. As a result, fish catches since 1989 have fallen alarmingly. Fishing is the harvesting of natural production; therefore, catches must not exceed the natural population's capacity for renovation. The capture of one species can have devastating effects for others. One example: in the last 40 years, more than 7 million dolphins have died in the Pacific alone, caught in tuna nets. There is need to practice fishing on a sustainable basis.

Certain marine mollusks and crustaceans are priced delicacies, being luxury products brought to the consumer by the large fishing fleets without our having any idea of the damage this causes to their habitats. Fish trawling and drift nets capture everything in their way indiscriminately; sometimes up to 70% of the catch has to be thrown away (see Barbault 1995). In some places, for each kg of shrimp caught, 30 kg of commercially useless but environmentally very valuable fish are thrown away. These practices destroy the marine habitats. As fish provides 40% of the animal protein for 60% of the inhabitants of the less-developed countries, aquaculture could provide a way of producing fish without damaging natural populations. Unfortunately, aquaculture seriously degrades the coastline and especially coastal wetlands.

3.6
Marine Biodiversity

The term "biodiversity" encompasses the variety of all living organisms found on Earth. It includes diversity within species, between species and of ecosystems. There is ample justification for conserving biodiversity on ethical, bio-ecological and economic grounds. The current concern for its conservation is not due to its loss per se; the very nature of biodiversity makes a certain amount of loss inevitable. It is the unprecedented rate at which this loss is now occurring that is causing concern. Unless stringent measures are taken internationally, irreversible damage is bound to occur.

Strangely, though biodiversity encompasses all living organisms, people generally tend to think of biodiversity in terms of terrestrial living organisms. Nearly 70% of the Earth's surface is covered by oceans and it is extremely important that marine biodiversity be at least as well preserved as the terrestrial. Paradoxical as it may seem, it is a fact that scientists today have a better idea of the surface of the dark side of the moon than the depths of the oceans. Nevertheless, our understanding of marine biodiversity has increased in recent years and though not as well documented as the terrestrial, marine biodiversity is probably far greater. For instance, at least 43 of the more than 70 phyla of all life forms are found in the oceans, whereas only 28 are found on land. At the species level, it may be safely assumed that at least half of the Earth's living species are to be found in the diverse marine and coastal habitats, ranging from coral reefs, mangroves, sea grasses, rocky or sandy beaches down to the soft sediments of the deepest ocean floors and through the water column in between.

Of course, conservation of marine biodiversity does pose problems, some similar to those concerning terrestrial biodiversity, others specific to the marine environment which is extremely fragile, due to the large variety of elements that make it up and the foreign objects that can be introduced into it. Exchanges between the water and the atmosphere, effects of atmospheric particles linked to climatic change, the variety of substances flowing into the sea through rivers, and direct human intervention in the various forms of marine exploitation, are but a few examples of the causes and agents of changes in the marine environment.

The principal threats for the marine environment come from the land. Over 70% of marine contamination comes from the mainland (dumping of waste products, pesticides, hydrocarbons or toxic products). Even air pollution originating on land is eventually deposited on the water surface. Pollutants from natural sources as well as those dumped into the marine environment greatly damage the marine biota.

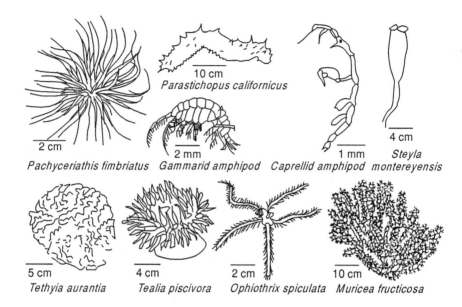

Fig. 3.5. Invertebrate filter, suspension and detritus feeders

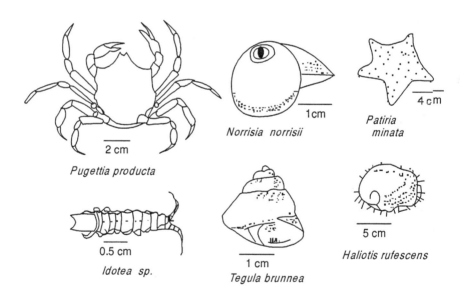

Fig. 3.6. Some common invertebrate grazers

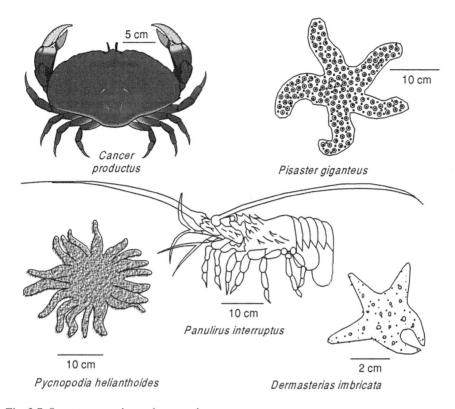

Fig. 3.7. Some common invertebrate predators

The damage can differ in magnitude, depending on the type of environment and the state of natural evolution of the contaminated system. The harmful effects of this contamination tend to be mitigated in the high seas, but they are magnified in specific marine environments such as closed or semi-closed seas, deltas, estuaries and lagoons.

Ever increasing demands are being made on marine resources. Human impact is largely responsible for the loss of over 50% of the world's mangroves and is threatening coral reefs. Fishing techniques are increasingly sophisticated, leading to the over-exploitation of marine resources with devastating impact on important fishing grounds. Ten of the world's 17 fishing grounds are already facing exhaustion. Overall fish catches since 1990 have fallen alarmingly. Figures 3.5 to 3.7 illustrate the biodiversity commonly observed in marine kelp forests. These forests, particularly common in Pacific and Atlantic oceans, are dominated by giant benthic brown algae (Phaeophyta) such as *Laminaria, Postelsia, Macrocystis* and *Nereocystis.*

3.6.1
Biodiversity in Coral Reefs

Continuing human development accompanied by the virtually unsatisfiable greed for materials, resources and higher and higher standards of living and enjoyment result in many unwelcome byproducts, including deforestation, chronic pollution and rapid climate change, all of which accelerate habitat destruction across the globe. It is a complex and difficult task to assess the long-term impacts of these new levels of habitat destruction, but such study is essential until such time that these disturbing trends reverse. In this context, a study of habitat fragmentation is particularly relevant, for it attempts to model the dynamics of the species extinction process. The model makes unusual non-intuitive predictions, and demonstrates how extinction may be delayed, often occurring many years after initial habitat destruction. Thus today's environmental fragmentation and destruction, for which we must accept responsibility, can build up an extinction debt that will have to be faced by future generations.

Another model describes coral reefs where species are understood to interact in a hierarchical web of competitive interactions. This general model permits an examination of the trade-offs between a set of species in terms of their colonization potentials, mortality rates, and competitive abilities. One main virtue of the model is that it is spatial by construction and can be set up to study the effects of patch removal and increasing habitat fragmentation, which is invaluable for the study of sessile coral and forest communities where the competition between species is largely for space, and where the effects of habitat destruction on community structure are non-trivial. Field data from Israeli coral reefs suggest that a large class of corals have both poor recruitment abilities and low abundances – characteristics that make these corals highly sensitive to disturbance. Stone's model found coral reefs to be extremely fragile, and predicts a relatively large number of species extinctions with only mild habitat destruction. The unusual effects of indirect species interactions were also analyzed, and the equilibrium distributions of species abundances were determined analytically for a wide range of parameter regimes.

Reef ecosystems are well known for their high species diversity: their impending widespread devastation is much less well known. Some 10% of reefs have already been completely destroyed, another 30% will surely be destroyed within the next 20 years while another 30% will be lost within 20–40 years (Wilkinson 1995). Surveys made both before (1969) and after (1973) major oil spills, showed coral communities to be particularly fragile and susceptible to habitat destruction as caused by chronic oil pollution. Only about 50% of species remained by the time of the 1973 census and there have been few, if any, signs of improvement over the intervening years.

The oceans contain about fifty times as much carbon dioxide as the atmosphere. Small changes in the ocean carbon cycle can therefore result in large atmospheric consequences. Primary producers or the phytoplankton form the base of the biological food web. All other life in the open ocean depends on new organic matter being produced by these tiny plants. The interactions of phytoplankton with their environment are highly complex. Unlike land plants, which are largely static, phytoplankton inhabit the surface layer of the ocean that is often undergoing vigorous mixing. They may be transported to depths beyond the limit of light penetration. This means that phytoplankton cannot always rely on a steady supply of light. For understanding how the amount (biomass) of phytoplankton varies in space and time, it is necessary to study:

1. The interactions between phytoplankton growth in relation to light, nutrients and mixing,
2. The rates at which small planktonic animals feed on phytoplankton, and
3. The rates at which they sink out of the surface water layers to support biological communities in the depths of the ocean (Fig. 3.8).

In some subtropical waters, total primary production and underlying phytoplankton biomass can be moderately high and support a large biomass of zooplankton (over 3000 mg C m^{-2}), despite phytoplankton nutrients being quite low.

3.7
Oceanic Productivity and Water Color

Much useful information can be gathered about ocean waters by taking small samples (one part in 10^{15}) of their volume for laboratory analysis on a typical voyage of a research ship. This is how most of our present knowledge of the chemistry and microbiology of the oceans was obtained. Samples of a few 100 l are taken to characterize, say, 100 km x 100 km of ocean with an average depth of 5 km (Davies-Colley and Kirk 1995). Moored sensors can record physical and chemical characteristics over time at a fixed point in the ocean. Increasingly, information about the oceans is now being obtained by remote sensing techniques which can greatly improve our coverage of these huge areas. Airborne or satellite-borne instruments produce instant images of the distribution of certain features of the ocean surface water over larger areas than could conceivably be characterized by large fleets of research ships. Sea surface temperature imaging from satellite sensors of the thermal infrared has become routine now. Imaging of visible light ("optical imagery") within a selected range of wavelengths is also quite valuable. Studies of the marine phytoplankton, bacterioplankton and other organisms at the base of marine food webs will continue to depend ultimately on taking water samples as in traditional oceanography. But this conventional sampling may be complemented by images of the ocean surface water temperature and color.

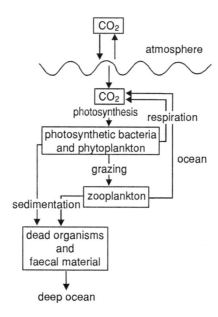

Fig. 3.8. The flow of carbon from the atmosphere to the deep ocean via a simplified planktonic food web

Remote sensing can monitor only the surface layer (Piazena and Häder 1995; 1997). Sufficient light for algal photosynthesis penetrates down to about 100 m. Since most of that depth is usually fairly well mixed, useful information on the ecologically crucial sunlit zone can be obtained by sensors imaging sunlight backscattered upwards from phytoplankton and associated particles in the surface waters (Davies-Colley and Kirk 1995). Thus, quantitative information may not be of high accuracy unless the vertical distribution of phytoplankton is known. Use of optical imagery for remote sensing of water properties greatly depends on a good understanding of how the optical properties of water relate to the phytoplankton and other light-absorbing and light-scattering constituents (Kirk 1994).

Measurements can be taken of light moving upwards in the water column as a result of light "scattering" by phytoplankton and other small particles in the water. Although this upwelling light typically amounts to only 3% of that moving downwards, it does determine the hue of the sea as viewed from above the surface by satellite radiometers. Phytoplankton absorbs blue and red light. The ubiquitous humic material known as "yellow substance" (gelbstoff in German) absorbs mainly blue light, so that sea waters containing these materials look greener than the open ocean. This explains why coastal sea waters containing appreciable yellow substance and phytoplankton are green in color, in contrast to the typically blue-violet color of open ocean waters (Davies-Colley and Kirk 1995). The ratio of upward directed to downward directed light at each wavelength is known as "reflectance". It is this variation of reflectance across the spectrum that determines the color of sea water. Figure 3.9 shows the reflectance spectra of ocean water modeled for different concentrations of chlorophyll *a*. Increasing phytoplankton biomass, as indicated by chlorophyll *a*, progressively decreases the ratio of reflectance in the blue (at the 443 nm absorption peak of chlorophyll *a*) to that in the green (at 550 nm where phytoplankton absorption is minimal). Efforts to estimate chlorophyll *a* from ocean color imaging have generally used the ratio of signals near 443 and 550 nm.

Primary production is usually a more ecologically relevant quantity than chlorophyll *a* concentration. Productivity is an estimate of the rate of synthesis of organic material rather than of the amount of potential photosynthesizing matter present. Some workers have estimated chlorophyll *a* from optical imagery and have then deduced production rates from chlorophyll *a* and other information. Efforts have also been made to relate carbon-14 fixation measurements directly to ocean color. Berthelot and Deschamps (1994) found a remarkably good correlation of the 443 nm to 550 nm reflectance ratio with carbon-14 fixation rates over a wide range of ocean waters, from eutrophic areas of nutrient upwelling with rapid production, to oligotrophic areas poor in nutrients and phytoplankton.

Compared with the water surface reflectance spectra shown in Fig. 3.9, the optical signals received by satellite or even airborne radiometers become degraded owing to light scattering by atmospheric gases and aerosols. Attempts are now underway to develop techniques for correcting this degradation. This is one of the areas where the linkage between water and atmospheric sciences is strong (Davies-Colley and Kirk 1995). Overfishing, dumping, silting, dredging and mining are some of the threats to oceans today. It is a formidable task to translate concern about loss of species into productive biological study, to detect patterns in biological diversity and to relate these to ecosystem function. Currently, efforts are being made in the more tractable terrestrial ecosystems from which the marine environment differs significantly.

Species richness on land is known to increase greatly towards the equator, but such latitudinal clines in biodiversity turn out to be far less obvious in the sea. Deep-sea experts see clear evidence for increasing diversity towards the tropics, but evidence for such clines in shallow seas is taxon-specific as some taxa show the predicted increase at lower latitudes (mollusks, macrozooplankton, and the dinoflagellate *Ceratium*) but others (polychaetes and amphipods) do not (see Vincent and Clarke 1995). The situation is even more confused in some oceans with, for example, marked longitudinal clines deriving from the intense hot spots of biodiversity in the Indo-West Pacific (mollusks and fishes).

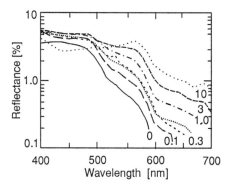

Fig. 3.9. Subsurface reflectance curves (*dashed lines*) modeled for phytoplankton-dominated waters with different chlorophyll *a* concentrations (curve labels in mg m) and constant yellow substance content. (After Davies-Colley and Kirk 1995)

Deep seas used to be considered species-poor but it has now become known that although individuals are spread thinly, diversity is high. While species-richness clines are confusing, deep-sea animals are known to be generally smaller than those in shallower seas, and body size relationships may be a key to understanding the structure and function of ecosystems.

Increasing spatial heterogeneity can result in greater diversity, even in the deep sea. The forces effecting such heterogeneity range from currents, riverine influences, and benthic storms to small-scale disturbance caused by grazers, burrowers and other bioturbators. Sessile organisms (e.g. oysters) can act as templates on which complex communities are based. The question of scale assumes significance in the marine environment where there are only about 10% as many species as on land but many more phyla.

3.8
Life in Oceans: Past and Present

Ecology interconnects the null hypothesis of geology (i.e., uniformitarianism) and the null hypothesis of organic evolution (viz., speciation by natural selection). Gradual changes in Earth processes drive corresponding shifts in species and communities that are quite gradual rather than abrupt (Overpeck et al. 1992; Jackson 1994). At times, changes in climate are abrupt and unstable, with marked effects on the subsequent climate and biota, and sudden environmental change sometimes underlies the near-universal occurrence of discontinuities in morphology and community composition which mark the age, stage and epoch boundaries of the geological time scale (Levinton 1988). Several palaeontological studies of patterns and rates of morphological change in the sea that allow distinction between punctuated and gradual speciation are known now (Gould and Eldredge 1993).

Most gastropod mollusks have a discontinuous evolution. Cases of punctuation and stasis outnumber gradual evolution by more than ten to one for the over 100 species studied (Stanley and Yang 1987; Jackson 1994). In general, species of marine invertebrates exist as discrete packages in space and time.

A large number of marine species appear abruptly in the fossil record, persisting unchanged for millions of years. Speciation and extinction commonly occur in pulses so that groups of species come and go as ecological units that dominate the seascape for millennia (Jackson 1994). According to Jackson (1994) there does not seem to exist any necessary correlation between the magnitude of environmental change and the subsequent ecological and evolutionary response.

Stasis – with inferred punctuational speciation – is a common feature of the fossil record, but it appears that on geological time scales and for physical variables, the more unstable the environment, the more the lineages are prone to stasis, and conversely, the more stable the environment, the more the lineages tend to evolve continuously (Sheldon 1993). Most of the fossil record comes from relatively dynamic, shallow marine settings, and we could gain the impression that, unless extreme, any forthcoming climatic and other physical changes are not likely to pose a particularly serious threat for most species.

Many long-term evolutionary trends in the sea may well have been due to interactions among species almost regardless of fluctuations in the physical environment, but synchronous turnover of faunas appears to have been more closely dependent on climatic changes and oceanographic conditions.

Careful consideration of both modern and paleo-oceanographic time series has indicated that marine populations and their environments have been far from constant over time. There has been much past variability of marine ecosystems. Low frequency temporal variability in the ocean may be the norm rather than the exception. Important time scales range from interannual to millennial, with the largest differences often occurring over the longest time spans. Very long time series are needed to discriminate statistically between "trend" and "fluctuation" or to validate statistical associations among variables. Conversely, interpretation of long series is often confounded by changes in spatial coverage and data collection methods. The biotic changes are large enough to be ecologically and socially significant, irrespective of whether they have been caused by anthropogenic or natural causes.

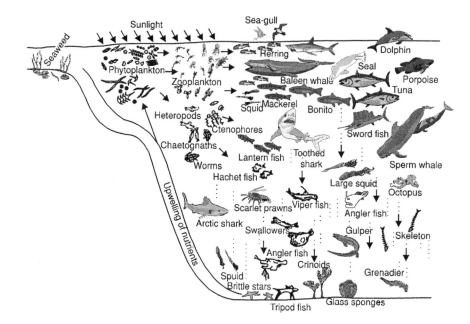

Fig. 3.10. Simplified sketch to illustrate life in the ocean (the organisms and depths are not drawn to the same scale)

The spatial extent of both biological and physical signals quite often extends between hemispheres and/or ocean basins. It has become known that biological "output" variables often co-vary strongly and for prolonged periods with variables that affect the physical dynamics of the ocean, pointing to close correlation between physical and biological processes. One disturbing aspect is that most of the available time series have been made from only a few variables. Biological data are dominated by fishery records, and reflect both environmental and harvest rate effects. Physical data are mostly coastal or atmospheric measurements. Quite often, the biological and physical measurements are not well matched and there is no information on key intermediate variables. Consequently cause-effect relations of observed statistical associations become difficult to delineate.

Most of the basic organic material that fuels and builds life in the sea is synthesized by oceanic phytoplankton within the surface layers (euphotic zone) of sea water where light is available. These microscopic plant cells are consumed by the herbivorous zooplankton (Figs. 3.10 to 3.12) and by some small fishes, which in turn are eaten by other carnivores.

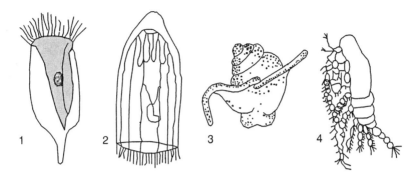

Fig. 3.11. Four representatives of marine filterfeeding zooplankton. *1* Tintinnid (*Parundella*), *2* Medusa (*Aglantha*), *3* Pteropod (*Limacina*), *4* Copepod (*Calanus*) (drawn to different scales)

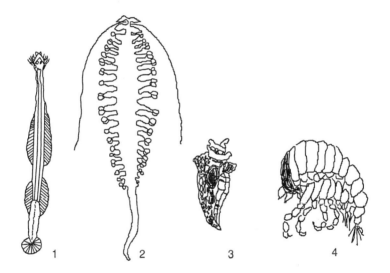

Fig. 3.12. Four representatives of marine raptorial feeding zooplankton. *1* Chaetognath (*Sagitta*), *2* Polychaete (*Tomopteris*), *3* Pteropod (*Clione*), *4* Amphipod (*Hyperia*) (all drawn to different scales)

The "snow" of the wastes and dead bodies of organisms occurring in the upper layers of water, shown in Fig. 3.10 by dots and short downward arrows, serves as the main source of food for the varied inhabitants of the lower oceanic depths. In the shallower zone, food is available from large rooted plants and drainage from the land.

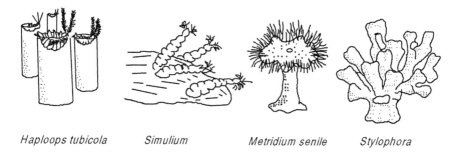

Haploops tubicola   Simulium   Metridium senile   Stylophora

Fig. 3.13. Some marine filter feeders

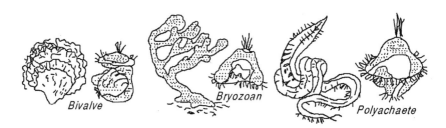

Fig. 3.14. Three marine benthic invertebrates having planktonic larvae

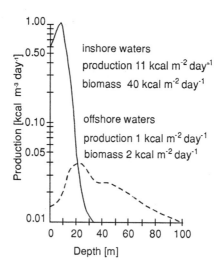

Fig. 3.15. The vertical distribution of primary productivity in the sea

The typical coastal upwelling (long arrows at left) fertilizes the surface waters with nutrients of organic matter, by the decomposer organisms such as bacteria etc., from near the bottom. This process keeps the surface waters continually supplied with nutrients, leading to the survival and growth of phytoplankton and zooplankton in the upper layers. Apart from the zooplankton, there are other filter-feeders (Fig. 3.13). Figure 3.14 shows three marine benthic invertebrates which produce larvae that are planktonic. Figure 3.15 shows the general pattern of vertical distribution of primary productivity in the sea. A generalized marine food web is illustrated in Fig. 3.16.

3.9
Phytoplankton

Knowledge of global-scale features of phytoplankton biomass and productivity is essential to the understanding of the global carbon cycle and the relation between the distribution of primary producers and higher trophic-level consumers on a planetary scale. Phytoplankton in the sea transform about 100 Gt of carbon (in the form of carbon dioxide) annually into fixed organic carbon (phytoplankton biomass; see Sullivan et al. 1993). In contrast to terrestrial ecosystems, phytoplankton represent a rather small standing crop but the biomass productivity is enormous: many of the organisms can double their biomass once per day. Therefore these algae constitute a major sink for atmospheric carbon dioxide in the sea as well as the major source of carbon and energy for the oceanic food web. Accurate assessments of large-scale phytoplankton distribution, abundance, productivity and sedimentation rates are difficult to obtain by conventional ship-based studies with their characteristic low spatial and temporal resolution.

According to Sullivan et al. (1993), local distributions of phytoplankton in the Southern Ocean are influenced by the following chemical or physical factors: (1) low silicic acid concentrations in the southeastern Pacific, (2) low light availability in areas of deep mixing, but a favorable light environment in stratified waters, and (3) iron availability downstream of continents over shallow areas, and near icebergs.

The southwest Atlantic (45° to 55°W) appears to be the region of most persistent biological activity and is also a major international fishery for ground and demersal fish and for the highest concentrations and largest areal extent of Antarctic krill, *Euphausia superba* (Dana), in the circumpolar ocean. The co-occurrence of dense phytoplankton blooms and high standing stocks of krill suggest that grazing activity and fecal pellet production in this region is probably a major determinant of the locus of intense fluxes of particulate matter to the ocean interior. The coherence of phytoplankton and zooplankton stocks points to a grazer-dominated or "top-down" regulation of production in localized areas. Silica-containing diatoms seem to dominate the downward flux of particulate matter ("oceanic snow") as seen by the rich siliceous oozes that characterize sediments in the region. Thus, phytoplankton distributions at the planetary scale are predictive of distributions of living marine resources and associated biogeochemical processes. Sullivan et al. (1993) have shown that satellite-derived pigment distributions constitute a useful guide and may also be used to find potential sites of enhanced flux of particulate matter to the deep ocean, the nature and extent of deposits of siliceous sediments, as well as likely locations of concentrations of living marine resources such as krill, fish, marine mammals and birds. The geographical extent and magnitude of persistent blooms makes it possible to evaluate the role of phytoplankton and biological productivity in ecological, biogeochemical and climatological investigations of the region (Sullivan et al. 1993).

Graneli and Haraldsson (1993) put forward the view that atmospheric deposition of acidifying substances to terrestrial ecosystems increases the leaching of trace metals from soils and thereby changes the balance between phytoplankton species in coastal waters. The hypothesis is based on the following arguments:

1. pH has a strong influence on the sorption of metals in soils and sediments and hence on the concentrations of metals in water,
2. Acidification increases the mobility of several trace metals in soils and in lake sediments,
3. Trace metals can both enhance and inhibit the growth of marine phytoplankton in a species specific manner, and
4. Several trace metals essential for phytoplankton growth are found in growth-limiting concentrations in seawater, whereas concentrations are high in acidified river water. It is also shown that the effects of cobalt on the growth of coastal phytoplankton is species-specific. Cobalt concentrations are usually inversely related to the pH of the water especially in acidified water.

The availability of the macronutrients nitrogen, phosphorus and, with respect to diatoms, silica is the major regulating factor for biomass accumulation and species composition of phytoplankton in both lakes and marine waters. Phytoplankton also require several elements in much smaller amounts than the macronutrients, e.g. Se, Co, Fe, Mn, Cu, and their requirements for refined and "clean" methods of sampling and analysis have shown that earlier measurements of trace-metal concentrations in natural waters were sometimes overestimated by several orders of magnitude (Bruland et al. 1991; Graneli and Haraldsson 1993). The fact that trace metal concentrations, especially in the open oceans, are extremely low, has generated much interest in the idea of trace-metal regulation of total phytoplankton biomass and/or species composition.

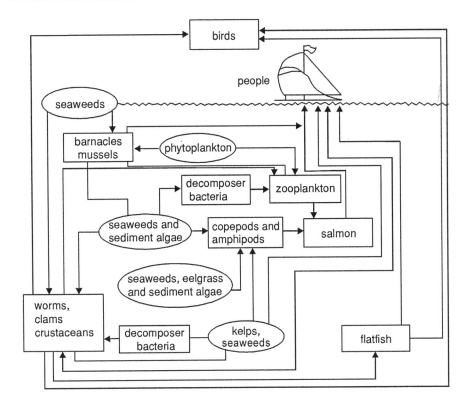

Fig. 3.16. A generalized marine food web. Ovals primary producers, rectangles consumers

In recent years, interest has focused mainly on iron, as several scientists have found evidence of iron limitation of phytoplankton, e.g. in the subarctic Pacific (Martin and Fitzwater 1988) and in Antarctic waters (Martin et al. 1990). However, Zn, Mn and Co can also be limiting. As in the case for other trace metals, cobalt concentration in oceanic waters is very low, usually 4–50 pM (pico mole = 10^{-12} M). More than two thirds of the cobalt flux to the ocean comes from rivers. Thus, if the trace metal is limiting for phytoplankton growth, increased riverine additions of the metal could cause increased phytoplankton production in coastal water, especially if concentrations in river water are markedly higher than for offshore waters. Cobalt is needed for the synthesis of cobalamin, or vitamin B_{12} and is therefore one of the most important trace metals for algal growth. Cobalt stimulates growth and nitrogen fixation of blue-green algae and can substitute for zinc, another limiting trace element in marine diatoms.

About one fifth of the world's ocean surface waters have sufficient light, nitrate, phosphate and silicate, and yet their standing stocks of phytoplankton are low. Several factors that might limit phytoplankton growth and biomass in these high-nitrate, low-chlorophyll (HNLC) areas have been debated (see Kolbar et al. 1994) but not resolved. Some researchers have suggested (see Martin et al. 1994) that increased phytoplankton production in these areas could remove significant amounts of carbon dioxide from the atmosphere.

In some of these areas, zooplankton grazing may possibly contribute to the maintenance of low chlorophyll levels. Strong turbulence at high latitudes may also mix phytoplankton below the critical depth, resulting in light-limitation of growth. Besides these factors, iron appears to have the potential to limit phytoplankton production in HNLC areas. Oceanic iron concentrations in surface waters are usually 10^{-12} M, too low to support high phytoplankton biomass or maximum growth rates. It appears that iron availability may regulate ocean production in HNLC areas. Martin et al. (1994) tested the idea that iron might limit phytoplankton growth in large regions of the ocean by enriching an area of 64 km^2 in the open equatorial Pacific Ocean with iron. This resulted in a doubling of plant biomass, a threefold increase in chlorophyll and a fourfold increase in plant production. Similar increases were found in a chlorophyll-rich plume downstream of the Galapagos Islands, which was naturally enriched in iron. These findings demonstrate a direct and firm biological response of the equatorial Pacific ecosystem to added iron. The response observed in the fertilization experiment was similar in magnitude and character to the increased production and chlorophyll found in the Galapagos plume. The observed presence of elevated concentrations of iron in the downstream plume is consistent with the hypothesis that the high chlorophyll concentrations are supported by iron, originating from the Galapagos platform. Martin et al. also noted increased grazing within the iron fertilized patch, in which microheterotrophic biomass increased by about 50% over the course of the patch experiment. This increase was rapid and appeared to level off quickly. The corresponding average increase in grazing pressure was approximately 4.5 μg C l^{-1} d^{-1}.

The biomass increase estimated from microscopic counts of cell numbers and volumes was about 15 μg Cl^{-1} (a doubling). Qualitative observations of filters and plankton tows suggested that the mesozooplankton also increased, presumably when vertically migrating plankton encountered the high biomass in the patch and remained there to feed (Martin et al. 1994).

This HNLC phenomenon has been ascribed to "top-down" grazing pressure by herbivores, which prevent the phytoplankton from fully utilizing the available nutrients. Using a sensitive fluorescence method, Kolber et al. (1994) followed changes in photochemical energy conversion efficiency of the natural phytoplankton community, both before and after artificial enrichment with iron, of a small area (7.5 x 7.5 km) of the equatorial Pacific Ocean. Their results suggest that iron limits phytoplankton photosynthesis in all size classes in this region by impairing intrinsic photochemical energy conversion, thereby supporting the hypothesis of physiological ("bottom up") limitation by this element (Kolber et al. 1994). Large-scale deposition of iron in the ocean might be an effective way to combat the rise of anthropogenic CO_2 in the atmosphere. As part of an experiment in the equatorial Pacific Ocean, Watson et al. (1994) observed the effect on dissolved CO_2 of enriching a small patch of water with iron. They noted significant depression of surface fugacities of CO_2 within 48 h of the iron release, which did not change systematically after that time. But the effect was only a small fraction (about 10%) of the CO_2 drawdown that would have occurred had the enrichment resulted in the complete utilization of all the available nitrate and phosphate. Thus, artificial fertilization of this ocean region did not cause a very large change in the surface CO_2 concentration, in contrast to the effect observed in laboratory-scale incubation experiments, where addition of similar concentrations of iron usually results in complete depletion of nutrients. Thus the transient nature of the iron fertilization does not support the idea that iron fertilization would significantly affect atmospheric CO_2 concentrations (Watson et al. 1994).

The residence time of bioavailable iron added to surface waters was very short and iron disappeared rapidly from the patch, being undetectable far from the presumed source downstream of the islands. Within a given parcel of water, both the patch and plume systems reflected a transient addition of iron rather than a sustained addition.

3.10
Clouds and Sea Surface Temperature

In the tropical western Pacific (TWP) Ocean, the clouds and the cloud-radiation feedback can only be understood in the context of air/sea interactions and the ocean mixed layer. There has been much interest in determining why sea surface temperature (SST) rarely exceeds 30 °C. Observations on this issue have usually been made using monthly cloud and SST data, with the focus being on intraseasonal and interannual time scales. For the unstable tropical atmosphere, use of monthly averaged data misses a key feedback between clouds and SST that occurs on the cloud-SST coupling time scale, estimated to be 3–6 days (see Webster et al. 1996). This is the time needed for a change in cloud properties, due to the change of ocean surface evaporation caused by SST variation, to feed back to the SST through its effect on the surface heat flux.

To clarify the SST it is necessary to distinguish between the "skin" SST, i.e. the radiometric temperature of the sea surface, and the "bulk" SST. The true bulk SST is defined to be the temperature within the upper few centimeters of the ocean surface. The bulk SST determined from buoy measurements is typically obtained at a depth of 0.5

m, while the bulk SST determined from ship measurements may be obtained from depths as large as 5 m. The radiative, latent and sensible heat exchange between the atmospheric and oceanic boundary layers depends on the actual skin temperature of the ocean, making this the critical SST for examining air/sea interactions. The skin temperature can significantly differ from the bulk water temperature in the tropics.

There exists a complex relationship between SST and deep convection. In general, deep convection occurs more frequently and with greater intensity as SSTs become higher. This is based on the assumption that the atmospheric stability is sufficiently reduced to allow the onset of moist convection. However, the amount and intensity of convection observed decreases with increasing SST because very warm SSTs may occur only under conditions of diminished convection. This suggests that convection decreases SSTs, probably in view of the enhancements to surface fluxes of heat and moisture out of the ocean surface because of the vertical overturning associated with deep convection (Hong and Raman 1996).

There are marked differences in daytime versus night-time net radiative cooling in clear versus cloudy areas of the tropical atmosphere. Daytime average cooling is approximately -0.7 °C per day, whereas night-time net tropospheric cooling rates are about -1.5 °C per day. The comparatively strong nocturnal cooling in clear areas gives rise to a diurnally varying vertical circulation and horizontal convergence cycle. Various manifestations of this cyclic process include the observed early morning heavy rainfall maxima over the tropical oceans. The radiatively driven day versus night circulation modulates the resulting diurnal cycle of intense convection which creates the highest, coldest cloudiness over maritime tropical areas and may well be a basic mechanism governing both small and large scale dynamics over the tropical environment. Table 3.3 shows the changes in surface heat flux components associated with a 1 °C change in SST for average conditions during the tropical ocean global atmosphere (TOGA) coupled ocean atmospheric response experiment (COARE) Intensive Observation Period (IOP). Strong changes can be noted in the surface latent heat flux. All of the changes are of the same sign (i.e., none of the changes cancel if the net surface heat flux is being evaluated). Therefore, a 1 °C change (or error) in sea surface skin temperature would result in a change (or error) of 27 W m^{-2} in the net surface heat flux. In many instances, an error of this magnitude would be large enough to change even the sign of the net surface heat flux and could significantly modulate the atmospheric boundary layer and convective processes. Therefore, care is needed in using the correct skin SST to evaluate the surface fluxes and in determining the diurnal variation of SST in the TWP (Webster et al. 1996).

Table 3.3. Changes in surface heat flux components associated with a 1 °C change in SST for average conditions during the TOGA COARE IOP. (After Webster et al. 1996)

	Flux Change	
Component	W m^{-2}	Percentage
Upwelling long wave	6.3	1.3
Sensible heat	2.4	23.3
Latent heat	18.7	16.2

3.11
Marine Plankton and Clouds

There are some indications (see Ayers and Gras 1991; Charlson et al. 1987; Gras 1995) that the marine biosphere may influence, or possibly control, climate (Fig. 3.17), in view of the fact that dimethylsulfide (DMS) is a major source of cloud condensation nuclei (CCN) in remote marine regions (Table 3.4). The mechanism probably involves the following processes:

1. Some marine algae produce dimethylsulfoniopropionate (DMSP), the precursor of DMS. DMSP is released from the algal cells into the ocean where it breaks down to DMS which escapes to the atmosphere (Harvey et al. 1996).
2. In the atmosphere DMS molecules survive for less than a day. They react with hydroxyl radicals and become oxidized to sulfur dioxide, and finally non-sea salt sulfate. NSS-SO$_4^{-2}$ either condenses on existing particles, forms new ones (homogeneous nucleation) or is processed in cloud droplets (heterogeneous processes) to form particles in the evaporating parts of clouds.
3. Immediately below the cloud-base, NSS-SO$_4^{-2}$ particles are the most numerous components of the aerosol. Particles of sea-salt are present but are not so common.
4. The largest sizes of the aerosol act as the CCN from which cloud droplets grow in the updrafts of air near the cloud-base. These are believed to be dominated by NSS-SO$_4^{-2}$.
5. The amount of CCN influences the number and size of droplets that make up a cloud, which in turn affects cloud appearance and radiative properties. A cloud made of many small droplets looks whiter (i.e., has higher albedo) than a cloud made of fewer large droplets (Harvey et al. 1996).
6. CCN numbers are perhaps also related to the lifetime and amount of cloud. The more numerous and whiter the clouds, the more sunlight they reflect back to space and the more potential they have to cool the planet.

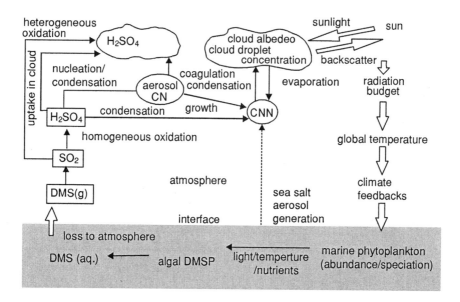

Fig. 3.17. The DMS-cloud-climate feedback cycle in the remote marine boundary layer. (After Harvey et al. 1996)

7. The influence of CCN on cloud albedo is known as the "indirect aerosol effect", as opposed to the "direct effect" of particles scattering and absorbing radiation.
8. There may exist a feedback loop whereby an increase in atmospheric DMS leads to more reflective clouds and a cooler planet. Lower temperatures will feed back to planktonic production of DMS in an unknown way, either increasing or reducing emissions.

It is estimated that, to counteract the warming effect due to doubling CO_2 in the atmosphere, an approximate doubling of CCN would be needed. Both natural sources (DMS emission) and anthropogenic sources (pollutant emissions) can influence the indirect effect of CCN.

Figure 3.18 shows that globally averaged changes in radiation ("radiative forcing") due to aerosol are comparable (and opposite in sign) to those due to additional greenhouse gases in the atmosphere today compared with preindustrial times. This forcing due to gas is termed the enhanced greenhouse effect and is fairly well quantified. In contrast, the aerosol effects, especially the indirect effect, are not well quantified, and make up a major source of uncertainty in attempts to model climate.

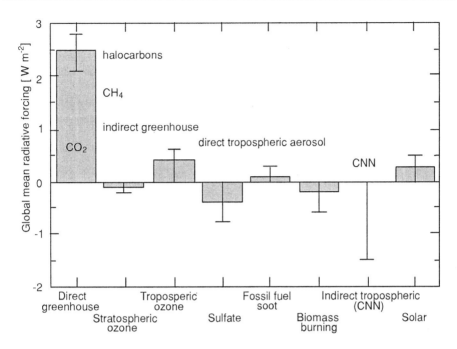

Fig. 3.18. Estimates of globally and annually averaged anthropogenic radiative forcing (in W m^{-2}) due to changes in concentrations of greenhouse gases and aerosols from preindustrial times to the present day (IPCC 1996)

3.12
Ocean and Climate

Oceans have assumed an ever greater significance in our understanding of climate. Ramanathan and Collins (1991) claim that there is a natural thermostat which prevents sea surface temperature (SST) from rising above 305 K. The mechanism which relies on extensive cirrus cloud cover over the warmest waters to reflect solar radiation has been challenged by Fu et al. (1992). According to Fu et al., changes in cirrus cover are related more to changes in atmospheric circulation than to SSTs, and changes in surface evaporation are more important than Ramanathan and Collins allow. This evaporation provides continuous air conditioning of the tropical SSTs, reducing the need for any thermostat (Stephens and Slingo 1992). A comparison of the global mean sea level with the global mean surface temperature is shown in Fig. 3.19.

Table 3.4. Linkages connecting DMS produced by phytoplankton to clouds. (After Harvey et al. 1996)

Process	Notes
DMS flux	Uncertainty (factor of 2) in areal distribution of DMS_{sea} and in the sea-air magnitude exchange coefficient; there are few measurements in some regions of the South Pacific.
DMS oxidation	There are no good models; environmental factors influence the aerosol (CN) efficiency of gas-to-particle conversion (OH, pre-existing aerosol).
GN to CCN	There are no good models; clouds have a role in CCN conversion production. The vertical profile of CCN is not well known.
CCN to cloud	Understanding of cloud microphysical processes and their droplet conversion modeling is further advanced than other areas.
Cloud micro-physics/cloud amount	More information is required on cloud microphysics versus cloud albedo in some regions; remote sensing and cloud albedo from satellite will improve and is a powerful tool for gathering statistics of cloud cover and albedo.

The topic of climate feedbacks is highly controversial. Although there is a tendency to use observations to study feedbacks, it is often impossible to separate cause and effect. With innovative use of observations obtained before and during the 1987 El Niño climate fluctuation in the Pacific, Ramanathan and Collins tried both to isolate and to quantify the cloud feedback. They proposed a simple relationship between changes in sea surface temperatures, cirrus clouds and solar radiation, and assume that other processes are of secondary importance. But we do not know whether cause and effect can be established so clearly in the equatorial Pacific, where the conventional theorem is that SSTs are maintained by many coupled air-sea interactions and ocean-atmosphere transport processes, operating on a variety of time and space scales (Stephens and Slingo 1992).

Some recent estimates of sea-level rise (EPA 1995; see Fig. 3.20) are lower than earlier estimates, primarily because of lower temperature projections that take into consideration the cooling caused by aerosols in the upper atmosphere. EPA (1995) made the following observations:

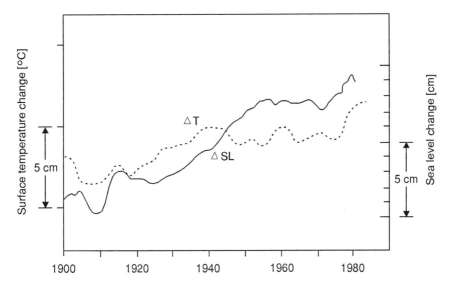

Fig. 3.19. Comparison of the global mean sea level change (Δ SL) with the global mean surface temperature change (Δ T)

1. Global warming is most likely to raise the sea level 15 cm by the year 2050 and 34 cm by the year 2100,
2. There is a 1% chance that global warming will raise the sea level by 1 m in the next 100 years and 4 m in the next 200 years,
3. By the year 2100, climate change can increase the sea-level rise by 4.2 mm/year,
4. Stabilizing emissions by the year 2025 could cut the rate of sea-level rise in half, and
5. Along most coasts, factors other than anthropogenic climate change will cause the sea to rise more than the rise resulting from climate change alone.

Global CO_2 surveys of the Indian, Pacific, and south Atlantic Oceans have been completed recently. About every 100 to 150 km along cruise tracks, a vertical profile of several water samples has been analyzed for total dissolved inorganic carbon (DIC) and at least one other carbonate-system parameter. The cruises have continuously measured near-surface pCO_2. The anthropogenic CO_2 content of seawater can, in principle, be calculated directly from DIC, total alkalinity and dissolved-oxygen measurements made in the interior of the ocean together with knowledge of the carbon:oxygen stoichiometry for the respiration of marine organic matter and the distribution of CO_2 at the sea surface (Wallace 1995).

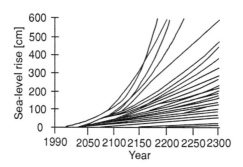

Fig. 3.20. Cumulative contribution of climate change to sea level as calculated from a wide variety of simulations of future climate. (From EPA 1995)

Attempts have been made to quantify the transport and storage of heat and freshwater within the world's oceans and to examine the large-scale, north-south transport of CO_2 by the ocean circulation. Such information, when combined with estimates of atmospheric CO_2 transport, is used to infer the large-scale distribution of sources and sinks of CO_2 (both natural and anthropogenic).

It is well known that the absorption by cloud particles (water droplets) increases with the increase of the fraction of diffuse radiation and the mean multiplicity of scattering, as well as with the decrease of the single scattering albedo. Only the first two of these factors depend on cloud type. For fixed pressure, temperature, and concentration, the absorption by atmospheric gases is determined by the value of photon mean free path in clouds. Radiation may escape through the sides of cumulus clouds, so that the mean multiplicity of scattering and the photon mean free path in cumulus are less than in stratus. When the sun is in zenith, the fraction of diffuse radiation is the same for both cloud types, thus the absorption in stratus exceeds the absorption in cumulus. As the optical thickness of atmospheric aerosol is much smaller than that of the clouds, any absorption difference between cumulus and stratus clouds primarily results from multiple scattering in clouds.

Several processes are known to play important roles in determining the SST distribution. Large-scale motion also has a significant role in governing the observed distributions of cloudiness and SST. The evolution of El Niño is thought to be influenced principally by shifts in the circulation patterns of both the atmosphere and oceans. This strong dynamical influence is perhaps the most fundamental difference between the theories of Ramanathan and Collins and Fu et al. who believe that large-scale dynamical processes spread local influences to other regions of the tropics and that analyses like that of Ramanathan and Collins need to consider this larger domain. According to them the correlation between SST and cloudiness breaks down on the largest scales, which they believe are the most important. Ramanathan and Collins argue that their thermostat operates locally, over the highest SSTs, with atmospheric motions invoked to export the long wave cloud forcing from the region. Another important process relevant to tropical SSTs is precipitation, which exceeds evaporation in the west Pacific and provides an influx of freshwater into the ocean. This reduces the salinity of the upper few meters of the mixed layer, creating a stable, lighter layer of water at the surface. This stable layer resists wind-driven mixing with the colder underlying water and is heated by the absorption of solar radiation. Such a warmed layer promotes further convection, more rainfall and further warming. Perhaps Ramanathan and Collin's thermostat operates only over such regions of significant freshwater input (Stephens and Slingo 1992).

We know very little about the processes that maintain the warmest SSTs on the planet. Compared with the tropical eastern Pacific, the western Pacific is a region where the winds are generally light, the evaporative heat flux is small, the structure of the ocean mixed layer is complex and the SST is highest (Stephens and Slingo 1992). So far, it has been possible to expose only the simplest aspects of these coupled processes and it appears that the long wave cloud forcing, the precipitation and evaporation as well as dynamical influences on cloud cover must also be assessed to determine whether Ramanathan and Collins's thermostat can function. It may be impossible to isolate a given feedback unambiguously from other coupled processes using available observations.

Figure 3.21 shows a tentative model of a part of the global cycle of carbon emphasizing interactions between the riverine and ocean systems. The sum of all inputs is made equal to the sum of all outputs in this model. Fluxes are in units of 10^{12} moles C yr^{-1}. For some flux estimates, the range is also stated. Note that organic carbon accumulation in sediments being less than 11% of total annual productivity, is difficult to estimate. Some of today's carbon fluxes may be greater than in olden times, prior to 1000 years ago. Very tentative calculations indicate that today's fluxes of organic carbon associated with river transport and accumulation in continental sediments may be double that of the ancient fluxes, and present-day organic carbon storage in marine sediments may be two to three times greater than in the past. Flux values given above are only tentative estimates. Total organic carbon production is taken as 9000×10^{12} moles C yr^{-1}; 60% of this being terrestrial and 40% marine. Total production is known to a precision of about ±15%.

Surface oceanic waters are typically saturated with methane whose origin has remained unclear mainly because methane producing organisms need an anaerobic environment whereas the surface ocean is well-oxygenated. Karl and Tilbrook (1994) have shown that methane produced by bacteria in the reducing environment provided by sinking particles of organic matter can result in the observed methane saturation.

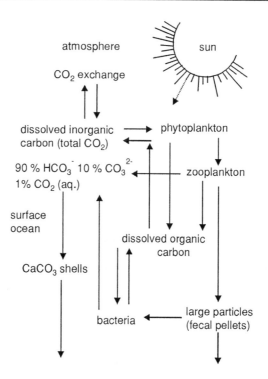

Fig. 3.21. A tentative model of a part of the present-day carbon cycle in oceanic systems

3.13
El Niño and Related Phenomena

The normal patterns of temperature and circulation in the Pacific involve warm waters and a deep thermocline in the west, cold water, shallow thermocline and upwelling in the east and strong trade winds maintaining high sea levels in the west. In vertical cross section, the Pacific may be likened to an asymmetrically heated basin with rising air and precipitation in the west and the reverse in the east. When this system is perturbed, anomalies in ocean forcing by the winds may initiate Kelvin and Rossby waves. The former, travelling eastward along the equatorial wave guide, deepen the thermocline in the eastern Pacific, interrupting normal upwelling and producing the anomalous warm sea-surface temperatures characteristic of El Niño; the latter may sometimes propagate to higher latitudes, with implications for temperate zone climate. There is a strong seasonality in the effects; it is necessary to understand the annual cycle in order to understand El Niño. Effects are found in precipitation, winds, river flow, lake levels, and indeed almost every climatic variable. Moreover, effects are seen not only in South America but around the world. However, there is some instability in many of the observed relationships. The impacts of El Niño on biological systems, particularly fisheries, have attracted some recent study. As a historical anomaly, El Niño was indeed originally described purely as a local biological anomaly. There is no strong evidence that El Niño alone has been responsible for major long-term changes in the highly productive Humboldt Current ecosystem in recent decades. Rather, a complex set of interactions among physical, biological and socio-economic factors may have been responsible. Between the 1960s and the 1970s, the fisheries saw a major collapse in anchovy landings, and a shift from an anchovies-dominated system to a mixed system of sardines and other pelagic fishes. El Niño has clearly demonstrable effects on the fisheries, e.g., a 50% decline in yield during moderate to strong events may have $ 300 M impact. The impact of El Niño seems to depend on the existing state of the ecosystem and may have highly significant second, third or higher order effects.

The extreme variability of tropical weather can be anticipated from the widely different terrain found between the tropics of Capricorn and Cancer. Despite such variability, however, there exist interconnections between different regions. For instance, the atmospheric circulation over India and Indonesia is closely related to the barometric pressure changes over the southern Pacific Ocean. This phenomenon is called the southern oscillation (SO). The SO is, in turn, related to El Niño and the sea temperatures off the western coast of northern South America. During otherwise normal periods, the water temperature at the surface of the western Pacific is warmer than toward Indonesia and South Asia. This difference comes from the upwelling of colder water from the depths off the coasts of Peru and Ecuador, the upwelling being driven by the easterly trade winds. El Niño tends to recur every 2–7 years, occurring when the SO is weak and the eastern Pacific waters are warm. El Niño usually brings drought to India and Indonesia but when at last it ceases, the coasts of Peru and Ecuador, the rice farmers of the south China sea and the bay of Bengal can look forward to good monsoon rains to fill their rice fields.

3.14
Marine Eutrophication

Eutrophication of coastal waters is a growing problem, caused by airborne pollutants and by over-supply of nutrients. Whereas phosphorus generally sets the limit to plant growth in lakes and streams, nitrogen usually does so in the sea. When nitrogen enters in a form available to plants, phytoplankton and algae start to grow and proliferate.

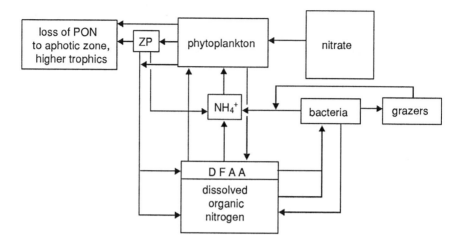

Fig. 3.22. A generalized nitrogen cycle in the euphotic zone of some coastal water columns. Arrows show direction of fluxes. The area of the enclosed boxes reflects approximate pool sizes. Open boxes denote chief sources and sinks of nitrogen, ZP Zooplankton, PON particulate organic nitrogen, DFAA dissolved free amino acids. (After Hollibaugh, 1980)

The water becomes less transparent, and because of the diminished light, large algal species such as bladder wrack can no longer survive at previous depths. In this way quite moderate additions of nitrogen can bring about extensive changes in the marine ecosystem.

Exceptionally favorable conditions can cause an explosive growth of phytoplankton. Oxygen deficiency can then result when dead organic matter from algal blooms sinks down to the sea bed and decomposes. Decomposition is a highly oxygen-consuming process. If the process continues, anaerobic bacteria will get to work, and hydrogen sulfide is formed. The input of nutrients is not beyond human control and eutrophication can be dealt with in large part through a drastic reduction of the emissions of nitrogen (Fig. 3.22).

There can be significant nutrient input via the atmosphere to the ocean. Nutrient-enriched rains markedly enhance marine bioproduction (Paerl 1985). Atmospheric input of nutrients to the ocean is one route where mankind's activities influence oceanic productivity, particularly in oligotrophic areas, and nitrogen limitation may be more prevalent than phosphorus limitation in euphotic layers of the ocean. Phosphorus limitation appears to be important in those coastal areas (e.g. estuaries) where direct waste drainage occurs, whereas nitrogen limitation tends to predominate in the open ocean, especially in oligotrophic areas. Zhang (1994) has given evidence of a correlation between harmful plankton blooms in the Yellow Sea (northwestern pacific) and episodic atmospheric depositions of nutrients in coastal oligotrophic zones. He noted that eutrophication may be induced by the atmospheric supply of nutrients and trace species. In the Yellow Sea, where the influence of waste runoff from land is very limited and/or completely absent, atmospheric deposition can sometimes become the major source of nutrient elements to the euphotic zone, particularly in regions where upward input (e.g. upwelling) is small. On average, episodic deposition of nutrient elements accounts for only a small (less than 10%) fraction of the concentrations in seawater (Owens et al. 1992). However, individual rain events sometimes result in temporal eutrophication of surface waters, which may cause harmful blooms to develop in the northwestern Pacific shelf regions.

Data from modern and ancient marine sediments suggest that burial of the limiting nutrient phosphorus is less efficient when bottom waters are low in oxygen. Mass balance calculations using a coupled model of the biogeochemical cycles of carbon, phosphorus, oxygen, and iron indicate that the redox dependence of phosphorus burial in the oceans provides a powerful forcing mechanism for balancing production and consumption of atmospheric oxygen over geologic time. The oxygen-phosphorus coupling further guards against runaway ocean anoxia (Van Capellen and Ingall 1996). According to Van Capellen and Ingall, phosphorus-mediated redox stabilization of the atmosphere and oceans may have been crucial to the radiation of higher life forms during the Phanerozoic (Fig. 3.23). The redox-sensitive recycling of nutrient phosphorus in the oceans regulates marine net primary production on time scales much longer than the oceanic residence time of reactive phosphorus (about 55 000 years).

Through its dependence on water column oxygenation, marine phosphorus cycling is also related to the oxygen content of the atmosphere. For this linkage, the couplings between phosphorus and all major redox elements, carbon, oxygen and iron need to be considered. Phosphorus-limited primary production and oceanic oxygenation depend on the global-scale vertical circulation of the oceans: nutrient phosphorus must be brought up to the photic zone and dissolved O_2 must be transported downward from the surface ocean. The distribution of iron between oxidized and reduced species and the burial efficiencies of organic phosphorus and Fe(III)-bound phosphorus depend on the oxygenation of marine bottom waters.

The inverse relation between oceanic water column oxygenation and surface primary production provides a negative feedback that helps stabilize atmospheric oxygen levels over geologic time. The feedback mechanism can also dampen the effects of decreased oxygen consumption by weathering. On geologic time scales, the fast responses of ocean oxygenation, water column recycling of phosphorus and net primary production cause near-instantaneous readjustments in the sedimentary burial of organic carbon plus iron sulfide to changes in the oxygen content of the atmosphere (Van Cappellen and Ingall 1996).

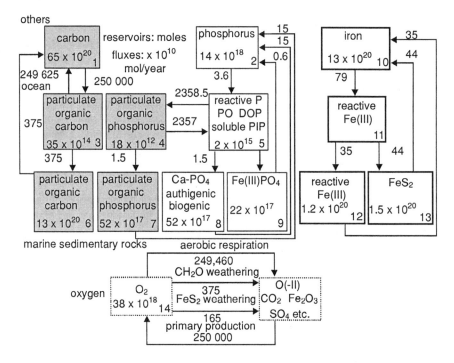

Fig. 3.23. Modern, prehuman steady-state biogeochemical cycles of carbon, phosphorus, iron, and oxygen. *Boxes* reservoirs, *arrows* material fluxes. Reservoir 1 contains all oxidized carbon in the exogenic cycle, as well as terrestrial organic carbon and marine dissolved organic carbon. Reservoir 2 includes phosphorus stored on land, plus oceanic unreactive particulate inorganic phosphorus. Terrestrial iron and unreactive marine iron are grouped in reservoir 10. Reservoirs are measured in moles, fluxes are measured x 10^{10} mol yr^{-1}. (After van Capellen and Ingall 1996)

Van Cappellen and Ingall concluded that oceanic phosphorus cycling plays a determining role in the long-term stabilization of redox conditions of the atmosphere-ocean system. The phosphorus-mediated coupling of ocean productivity and atmospheric oxygen provides effective safeguards against sudden depletion or buildup of atmospheric O_2 and against the irreversible eutrophication of the oceans.

3.15
Marine Litter

Litter is not only aesthetically unappealing, it has direct impacts on the marine ecosystem. It provides novel attachment sites for sessile organisms, kills or injures animals that become entangled in or eat litter, and can aid dispersal of terrestrial organisms. The amount of litter at sea is increasing despite control measures. Floating litter drifts throughout all the oceans after being dumped from ships or blown and washed from land. Plastic articles pose the greatest problem: once at sea, plastics are virtually immune to degradation and can drift for years, covering vast distances.

3.15.1
Coastal Dwellers

A doubling of atmospheric CO_2 could cause sea levels to rise by up to 1 m by 2100. Many of the world's coastal and delta regions are densely populated, and the flooding that would result from a sea-level rise would disrupt or destroy human settlements, agriculture and industry, and affect millions of people. Half the world's population lives in coastal regions, and the flood-prone delta regions of Egypt, India, Bangladesh and China are densely populated. A rise in sea level could cause the shore to move inland several kilometers in low-lying areas and displace millions of people.

In Bangladesh, for instance, about 10 million people live on coastal land less than 1 m above sea level, and a 1-m rise in sea level would be likely to flood 17% of the country. Storm flooding would also become more common and would particularly threaten areas in which hurricanes or severe storms already occur frequently, such as the Indian subcontinent, the western Pacific and the Caribbean islands. The loss of natural defenses, such as wetlands and mangroves, against flooding would increase the vulnerability of coastal regions (UNEP 1993). Many countries – often those that can least afford it, such as Bangladesh, China and Egypt – would also suffer from a loss of agricultural land if sea levels rise.

3.16
Disruption of Marine Ecosystems

Disruptive effects on marine food-producing ecosystems are caused by the use of water as a transport medium for organic waste, pathogenic microorganisms, and nutrients supporting growth of living organisms, emanating mainly from households in urban areas, and from agricultural activities.

The coastal zone is a region of intensive biogeochemical and physical activity comprising, particularly in the tropics, the most productive biosystems and the major fishing grounds that support two billion people with animal protein (Arrhenius 1992). Seagrass beds, coral reefs and mangroves exemplify such ecosystems. They are under increasing pressure from human activities and growing populations. About 60% of the world's population and most of the urban centers are situated in the coastal vicinity. By the year 2000, about 70% of the global population of over six billion are expected to live in these regions. Currently, the activities of growing populations are rapidly degrading the tropical coastal zones. About 70% of the mangrove forests in Southeast Asia, South Asia and the Caribbean have been destroyed in the last four decades and some 90% of the coral reefs in Southeast Asia have been severely affected by, inter alia, eutrophication and sedimentation. This reduces the productivity of the coastal areas, and most local fishing communities are now facing decreasing catches. Coastal marine pollution also reduces the biological diversity of coral reefs and other coastal marine habitats.

Several metals enter the sea from diverse sources. Figure 3.24 illustrates the fate of these metals. Many marine organisms absorb some of the metals discharged into the sea. Figure 3.25 illustrates the effect of marine environmental variables in such uptake. Thus, disruption of the coastal environment by pollution with untreated and improperly treated sewage, livestock waste and agricultural fertilizer leakage containing organic substances, nitrate and phosphate, pathogens and persistent toxic metals and organic compounds poses a growing threat to food security and the health of billions of people in the developing tropical countries (Arrhenius 1992). This problem is, in fact, a problem of rapid-scale growth of an inappropriately closed system. If properly closed, the system could change a grave pollutant into a valuable resource for the agricultural and aquacultural photosynthetic systems. Of the 9–12 gigatons (Gt) of carbon now fixed in agricultural systems each year, 3–4 Gt go into the human food system and a major part of this amount ends up as sewage and waste. With rapid population increase, this vast amount is sure to increase exponentially.

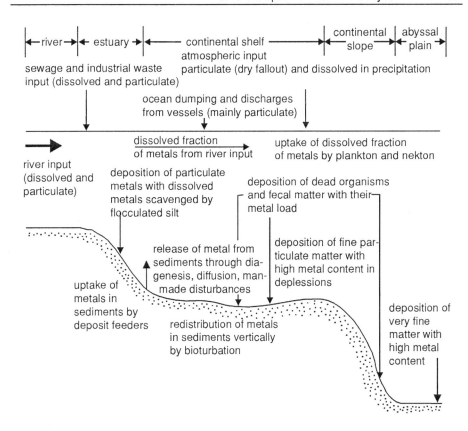

Fig. 3.24. The fate of metals entering the sea. (From Waldichuk 1986)

There is a need for integrated management of the environment in coastal zones, including land use in the river basins draining into coastal zones, and the airborne and waterborne discharges from households. However, the rapid disruption of tropical and subtropical coastal ecosystems is mainly due to improper control of material flow in a broad sense. The goal must also comprise an efficient closure of the nutrient cycles for carbon and nutrient minerals by shunting these flows to agricultural and aquacultural production systems, while avoiding contamination of these systems with pathogens and toxic substances. Pollution from humans and their livestock is an old environmental problem. There are many examples of the disastrous effects of the uncontrolled combined use of aquatic systems, as a recipient of biological waste and as a food source, for rural as well as urban areas (Arrhenius 1992). Such dangers are more serious in tropical coastal zones today than in the past.

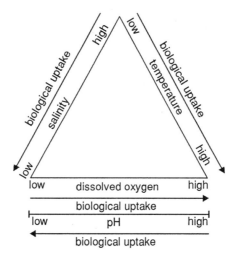

Fig. 3.25. The general effect of marine environmental variables on the uptake of metals by marine organisms. *Arrows* denote the direction of increasing uptake

Fortunately, not all of the present trends are negative. The intensive development of ponds used for aquaculture in some coastal regions holds back nutrients that would otherwise go to the sea. It also enhances production. This solution uses natural processes rather than high technology. Where socially acceptable, such intensification of coastal land use needs to be popularized.

The common technology currently used in many tropical areas is extensive investment in sewerage networks and large centralized municipal wastewater treatment plants: a technique transferred from the temperate regions. Properly used, it counteracts waterborne diseases. However, there are the following problems with these systems (see Arrhenius 1992):

1. Water of high quality, in many areas a scarce resource, is used as a transport medium for waste.
2. The fundamental material flow problem is not properly addressed.
3. Toxic substances, e.g., heavy metals and pathogenic microorganisms prohibit productive use of the nutrients in sludges.
4. Rapidly increasing investments and management costs greatly stress the weak economies of developing countries.
5. Intensive animal husbandry involves overproduction of manure. Current cultivation techniques lead to loss of nitrogen and phosphorus from this manure and from artificial fertilizers to surface waters and aquifers.

6. New agricultural technologies are urgently needed to stop current leakages, as the closing of the nutrient cycle in sewage systems should lead to additional application of large amounts of nutrients on agricultural land.

New approaches that are now available include: biogas and composting technology, natural or artificial soil filtration, multi-level biological systems based on natural or artificial wetlands, biological systems based on macro- and microphytes, bacteria and algae-based systems for selective uptake and concentration, ecological engineering, i.e. the use of self-designed or engineered ecosystems (Arrhenius 1992). Several nonconventional waste and wastewater treatment technologies are ready for a wider implementation. The following two strategies appear to be highly promising:

1. Closing the system near the waste and sewage source by water-saving techniques involving separation of different categories of the present sewage components. Such processes that allow productive use of nutrients do exist, and
2. Recovery by biomass production of large amounts of sewage nutrients which are currently released to tropical aquatic systems (Arrhenius 1992).

While constructing new systems, the action plans should focus on integrated material flow control and emphasize the use of organic matter and nutrients for production, rather than their disposal as wastes.

Top predators in the open-ocean (e.g. swordfish, tuna and shark) contain high levels of methylmercury in their muscles. The same is true for some fish-eating mammals like the pilot whale. In contrast, the muscles of seals contain predominantly inorganic mercury. It has hitherto been assumed that this mercury occurs naturally and that man is the victim rather than the cause of the high methylmercury levels found in the open oceans. Recent analysis of mercury in seawater has revealed rather low mercury levels in uncontaminated seawater in the range of 0.1–1 ng l^{-1} (Jernelov and Ramel 1994). Existing data on mercury levels in seawater and in the atmosphere suggest that open-ocean surface water has a residence time for mercury of only decades. This means that atmospheric fallout is important for mercury levels in water and biota. We also know that anthropogenic sources account for about 50% of atmospheric mercury; thus, there is a case for a reevaluation of human impacts as a cause of the high methylmercury concentrations in open-ocean top predators. Mercury may well turn out to be a transboundary air pollutant (Jernelov and Ramel 1994).

3.16.1
Hazards of Harmful Substances Carried by Ships

Attempts are underway to develop requirements for the maritime transport of individual hazardous substances and for operational discharges of their residues at sea, as well as considerations related to liability and compensation for hazardous and noxious substances in cases of accidental damage to property or the environment. The following issues are important in this context:

1. Copper-based antifouling paints are increasingly used to replace organotin-based paints and the hazards related to the release of copper and copper compounds from such paints into the marine environment need to be considered.
2. In the absence of data from tests to identify the potential of many chemicals to contaminate seafood, there is need for designing methods on how to evaluate thresholds of substances by using known or calculated sensory detection thresholds of chemicals in water and air, e.g. by estimating tainting properties of aliphatic ketones, which are in good agreement with values derived from tainting tests. The procedure also works well for straight chain alkanes.
3. There is no evidence of any damage to marine organisms caused by the bioaccumulation of fluoride compounds.
4. More attention needs to be given to substances of low acute toxicity, low volatility and high viscosity which, after release from chemical tankers, may float on the sea surface, affecting coastlines and damaging wildlife. They need to be suitably regulated. The hazards of these substances, including many of vegetable and animal origin, should also be evaluated.

3.16.2
Sea Surface Microlayer

The sea surface microlayer is a source of contaminant accumulation and can cause modification of biological processes and air/sea exchanges. There is great need for an in-depth review of physical, biological and chemical processes that occur at the sea-surface microlayer and their relevance to global change and effect on the marine environment and its living resources.

Natural surface-active substances are often enriched in the sea surface compared with subsurface water. Amino acids, proteins, fatty acids, lipids, phenols and many other organic compounds accumulate at the surface. The biota of the water column below is the source for most of the enrichment of natural (nonpollutant) chemicals. Plankton produces dissolved compounds as products of its metabolism. Air bubbles, rising through the water column, scavenge these organic materials and bring them to the surface. Also, as plankton die and disintegrate, some particles and many of the breakdown products (oils, fats, and proteins, etc.) float to the surface.

The accumulation of natural organic chemicals modifies the physical and optical properties of the sea surface. Thin organic films, invisible to the naked eye, are ubiquitous in aquatic systems. In areas where currents converage, thicker films accumulate.

Under light to moderate wind conditions, areas of accumulated film dampen small waves and become visible as "surface slicks". Strong surface tension forces exist, creating a boundary region where turbulent mixing is much reduced (IMO 1993).

Growing population and industrialization have resulted in increasing atmospheric transport of pollutant materials over the ocean. Atmospheric deposition of this material and of naturally occurring substances constitutes an important source of various chemicals to the sea surface microlayer. Many of these substances are surface active and contribute to increased concentrations in the surface microlayer and could result in increasing incidence of coherent films or slicks in both coastal and open ocean regions. High concentrations of toxic chemicals are also often found in the surface microlayer compared to the subsurface bulkwater.

Microbiological studies indicate that microlayers are generally greatly enriched in the abundance (density) of microorganisms compared with subsurface water. A few studies suggest high biochemical activity in surface films. However, the effects of this microbial activity on air-sea exchange rates of radiatively important gases or other materials are unknown. Phytoneuston (microalgae) occur in high densities compared with phytoplankton in most ocean areas examined. The rates of photosynthetic carbon fixation are often higher in microlayers compared with subsurface water. Blooms of *Trichodesmium* (possibly fixing atmospheric N_2) are common in the tropics. However, the overall regional or global importance of phytoneuston on CO_2 (or other gas) exchange from the atmosphere to the ocean is not known (IMO 1993).

As regards ichthyoneuston, many pelagic, commercially-important fish species have floating eggs and/or larval stages, that develop in contact with the microlayer. A contaminated microlayer may exert toxic effects on fish embryos and larvae. Also, neustonic fish embryos, collected in surface skimming nets from contaminated areas, have a higher incidence of chromosome abnormalities compared with those from less contaminated areas. However, the effects of microlayer contamination on fisheries recruitment at the population level are mostly unknown.

There is some evidence that the lower temperature of the sea-surface microlayer compared with the bulk water can lead to a significant increase in our estimates of the ability of the oceans to take up atmospheric carbon dioxide. Further, the current controversy over the magnitude of the transfer rate for gas exchange across the sea surface may be resolved by the existence of specific catalysts for carbon dioxide transfer occurring at enhanced concentrations in the microlayer. The following are some additional issues which need to be considered:

1. Temporal and spatial variability of chemical and biological enrichment in the microlayer,
2. The actual exposure hazard in situ of indigenous neuston species to microlayer contamination,

3. Quantitative estimates of how natural microlayers alter air-sea transfer compared with models which do not include a microlayer,
4. Biology and chemistry of surface layers in freshwater environments,
5. Exposure of aquatic surface layer communities to ultraviolet-B radiation and its implications for global change,
6. The horizontal transport of surface slicks and their deposition in coastal zones, and
7. Physico-chemical data on the behavior of micelles and data on the engineering floating processes (see IMO 1993).

3.17
Oil Spills

Oil spills at sea differ from each other in regard to their fate and behavior. Spill behavior determines the effects that the oil will have either on marine life or the shoreline. The precise fate of an oil spill depends on the type of oil spilled and the sea and weather conditions. Figure 3.26 shows the percentage of oil spills from various sources, and Fig. 3.27 illustrates the incidence of oil spills from tanker accidents in terms of number of spills and the total amount spilt. The behavior of oil spills is governed by the chemistry of oil, e.g., the contents of saturates, aromatics, asphaltenes and resins. The most important behavioral feature of oil spills is their evaporation. It is this aspect that is responsible for the largest changes in mass balance after a spill and affects other behavioral parameters. The second most important factor is emulsification, which depends on oil composition. Oils that emulsify undergo a rapid and pronounced increase in viscosity and stop spreading and evaporating. Resins and asphaltenes in the oil tend to stabilize water in the emulsion whereas aromatics dissolve the asphaltenes and thus delay emulsification. Some oils lose some aromatics by evaporation so that they emulsify. Natural dispersion is the third most important behavior of spilled oils. The saturate component of oils disperses in highly turbulent conditions. Asphaltenes, however, retard natural dispersion. Evaporation is probably the best understood and modeled behavior. Emulsification is also well understood while natural dispersion is poorly understood. Crude and refined oils are composed of a large number of different compounds, each with their own behavior and properties. Most crude oil consists of saturates or alkanes. Alkanes are the desirable end-product of crude oil because the C_8 to C_{18} fraction is directly used as gasoline and diesel fuel, the largest use for crude oil. Larger alkane compounds ($> C_{24}$) are known as waxes and can be treated by petroleum companies to convert them into smaller compounds. The chemistry of alkenes is similar to the alkanes, except that they contain one or more double bonds and are less desirable as a source for fuel. They can be treated in refineries and converted to alkanes.

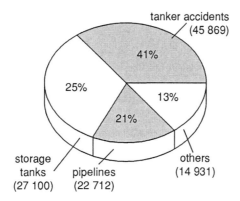

Fig. 3.26. Oil spills from accidents (tonnes, 1990)

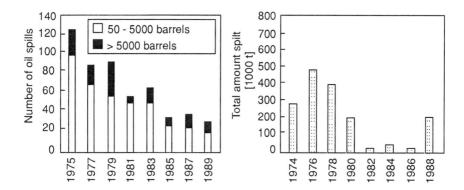

Fig. 3.27. Incidence of oil spills from tanker accidents

Aromatics are cyclic compounds containing double bonds that are conjugated or that are spaced two-apart and in rings. These properties confer special stability. Large aromatic compounds, known as polycyclic aromatic hydrocarbons (PAHs), occur in crude oils and cause frequent environmental concern resulting from combustion sources. Cycloalkanes or naphthenes are like aromatics except that they lack double bonds. The naphthenoaromatic compounds contain both aromatic and naphthenic rings. Resins or polars are compounds of the other classes having heteroatoms or molecule(s) such as sulfur, oxygen or nitrogen. These compounds have polarity and are partially soluble in water and oil. Several other classes of compounds are also present but do not influence the behavior of oil to a large extent. These are exemplified by mercaptans, metal-containing compounds, steroids and even biologically-derived compounds. Several properties are important to the understanding of oils and their behavior. The most important is viscosity, i.e. the shear resistance (force) divided by the shear rate (velocity). The oil viscosity determines its spreading rate. The density of oil determines how an oil will float on water. It is intimately related to viscosity.

Water content is an important parameter for oils that have emulsified or formed water-in-oil emulsions. This gives some indication of the expansion of spill volume resulting from water incorporation and also of the stability of the resulting emulsion. Water content is not important for those oils that have not emulsified. Surface and interfacial tension need to be considered. Interfacial tension is the energy difference between the oil layer and air or water, which is not very important because it cannot be used in isolation to predict any behavioral aspects of the oil and varies little between oils. The pour point is the temperature below which the oil ceases to pour as a fluid. The pour point is difficult to measure and also tends to vary. It is not a precise physical measure for a multi-component oil such as crude and many of the distilled products.

An oil can be divided into its chemical constituents using precipitation methods. The saturates or alkanes constitute the backbone of the commercial desire for oil. Saturates are largely dispersed chemically or naturally in water and this behavior correlates with saturate content. There is a negative correlation of natural dispersibility with increasing aromatic, asphaltenic and resinic content. Larger saturates or waxes (alkanes with chains longer than C_{24}) at concentrations greater than about 3% in oils, affect viscosity, adhesion and weathering. Resins and asphaltenes stabilize water-in-oil emulsions. The higher the resin and asphaltenic content, the more viscous the oil and the less it will spread. Saturate content is also related to biodegradability in the same way as natural dispersion.

To determine the long-term fate and behavior of oil, one useful parameter is the total amount of BTEX (Benzene, Toluene, Ethylbenzene and Xylenes). This fraction is largely evaporated soon after a spill, but it is also the fraction that is responsible for the short-term aquatic toxicity of the oil. A number of biomarkers in the oil such as hopanes are useful measures for studying biodegradation.

3.17.1
Behavior of Oil in the Environment

The behavior of oil in the environment mostly depends on its chemical composition. Gasoline and diesel fuel are strongly and rapidly evaporated. Both disperse naturally to a large extent in turbulent water. A light crude oil evaporates to about 50% in 2 days and disperses naturally in turbulent seas. A medium crude oil evaporates to about one third in 2 days. After evaporation, water-in-oil emulsions form, depending on contents of asphaltene, resin, and aromatics.

Evaporation is usually necessary before emulsification occurs and competes with natural dispersion. If emulsification occurs, the spill dynamics and behavior of the oil change greatly. Emulsification means the formation of water-in-oil emulsions ("chocolate mousse" or "mousse"). These emulsions greatly change the properties and characteristics of oil spills. Stable emulsions contain 50–80% water, thus increasing the volume of spilled material. The density of the resulting emulsion can be as high as 1.05 g ml^{-1} compared to an initial density of about 0.80 g ml^{-1}. Most significantly, the viscosity of the oil changes markedly. A liquid product changes to become a heavy, semi–solid material. Emulsification is the second most important behavioral characteristic of oil after evaporation. Emulsification slows down evaporation and the oil rides lower in the water column, showing different drag with respect to the wind. Sinking of oil can be divided into sedimentation of oil droplets and sinking en masse. The latter is further divided into over-washing by water and simple sinking. Biodegradation slowly changes the behavior and fate of oil spills only slightly.

Oil spills at sea always pose a potential risk of causing environmental damage, associated with drilling and production activities and transportation of oil by tankers or pipelines. The high level of public concerns followed by the Exxon Valdez, Mega Borg and Haven oil spill incidents have resulted in renewed international interest in research on spill response techniques.

When oil is spilled on sea, it is subject to several processes including spreading, drifting, evaporation dissolution, photolysis, biodegradation, and formation of both oil-in-water and water-in-oil emulsions. The weathering processes can lead to marked changes in the properties of the oil and also in oil behavior with time at sea. The main factors influencing the rate and extent of weathering are waves, wind, sunlight, temperature and salinity. Spreading, evaporation and formation of water in oil emulsions, can greatly increase oil viscosity and is critical for decision makers concerned with the use of the different spill response techniques during an oil spill combat operation; these parameters should therefore be assessed in relation to the time after an oil spill. Figure 3.28 illustrates the various aspects of oil spill processes. Figure 3.29 illustrates the process of photolysis of the oil film.

3.17.2
Spill Response Techniques

The choice of spill response techniques depends on weather conditions, changes in the properties of the oil and on any delayed response time following an oil spill. Major spill response techniques during an oil spill combat operation include use of mechanical recovery equipment, in situ burning and chemicals or "soaps" as dispersants and demulsifiers. Figure 3.30 compares the relative efficiency of various response techniques. Ignition and combustion of an oil layer having a thickness less than approximately 2 mm is extremely difficult. The spreading of oil from a source on the open seas will, within a short period of time, cause a reduction of the thickness to less than 1 mm making ignition and combustion impossible.

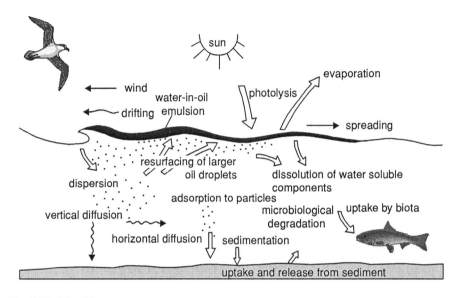

Fig. 3.28. Oil spill processes

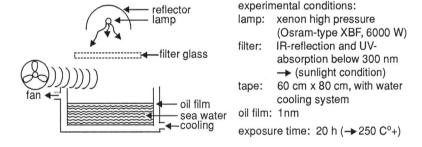

experimental conditions:
lamp: xenon high pressure
         (Osram-type XBF, 6000 W)
filter: IR-reflection and UV-
         absorption below 300 nm
         → (sunlight condition)
tape: 60 cm x 80 cm, with water
         cooling system
oil film: 1 nm
exposure time: 20 h (→ 250 C°+)

Fig. 3.29. Photolysis of the oil film

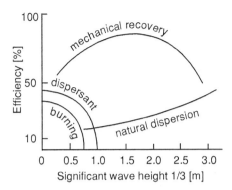

Fig. 3.30. Effectiveness of response techniques

To allow ignition and in situ burning at sea, fire resistant or fireproof booms are needed to create a sufficient oil thickness. Due to the heavy weight of fireproof booms, however, they have poor wave following properties and are ineffective in average waves higher than 3 m.

In wind velocities above approximately 2 m s^{-1} or 4 knots, emulsification occurs within minutes and may increase the water content up to about 50–60% within a few hours. The efficiency of in situ burning can vary from 50%–80% depending on oil type and weathering properties. For emulsions, the efficiency is lower (10–30%).

Simply adding inorganic nutrients (e.g. nitrate, phosphate) can stimulate the natural rate of oil biodegradation on polluted shorelines. Hydrocarbon-degrading microorganisms are ubiquitous in most ecosystems, and over 30 different genera of oil-degrading bacteria and fungi have been identified. For many years, scientists have hoped that enhanced microbial decomposition of oil by converting hydrocarbons largely to recyclable products might decrease the long-term effect of a spill on shoreline biota. However, although bioremediation did increase oil biodegradation rates on contaminated beaches it certainly did not turn out to be an instant magic bullet. Some positive reports of bioremediation have been balanced by a few negative reports (failures). But attitudes changed as a result of the clean-up operations following the spill of crude oil from the Exxon Valdez in Prince William Sound, Alaska, in 1989. Part of the response strategy for this oil spill was the use of an oleophilic fertilizer (Inipol EAP22) to stimulate biodegradation of the crude oil. Biodegradation is effective only on beaches where high levels of pore water nitrogen are maintained.

3.17.3
In Situ Burning of Oil Spills

In situ burning of spilled oil has certain advantages ever other countermeasures. It has the potential to rapidly convert large amounts of oil into carbon dioxide and water, with a small percentage of smoke particulates and other unburned and residue byproducts. Burning of spilled oil from the water surface reduces the chances of shoreline contamination and damage to biota by removing the oil from the water surface before it spreads and moves. In situ burning requires minimal equipment and is less laborious than other techniques. Because the oil is mainly converted to airborne products of combustion by burning, the need for physical collection, storage, and transport of recovered fluids is reduced to the few percent of the original spill volume that remains as residue after burning (Walton et al. 1993).

Burning of oil spills forms a smoke plume containing smoke particulates and other products of combustion which often persist over several kilometers downwind from the burn. This affects public health. Air quality is also affected by evaporation of large oil spills that are not burned. Volatile organic compounds (VOC) e.g. benzene, toluene, and xylene and polycyclic aromatic hydrocarbons (PAH) are found in the air downwind of an evaporating crude oil spill.

In situ burning of oil is a most important technique for rapid removal of large quantities of oil at sea, particularly in ice infested waters. Much work has been done on burning of fresh or unemulsified oil but some studies have dealt with in situ burning of emulsions. Oil at sea tends to emulsify and after one day the water content can be as high as 70%. If in situ burning is to be a practical and efficient measure against spilled oil at sea, more attention has to be given to the burning of emulsions. If emulsions cannot be burned in situ, burning will be limited by a response time of only a few hours.

In situ burning of emulsions differs from burning of unemulsified oil. The controlling process seems to be the removal of the water in the emulsions. Two different measures may be utilized for this, viz., boiling the water out or breaking the emulsion thermally or chemically.

In order to ignite an emulsion, the igniter must have the following properties: (1) it should burn long enough and with sufficient heat so as to form an oily layer on top of the emulsion – a layer thick enough to burn and then ignite the layer, and (2) it should ignite oil in an area that is big enough to ensure a self sustained burn. Emulsions can be burned; however, for oils evaporated more than 18%, and with a water content of more than 20%, the gelled gasoline is insufficient as an igniter. Gelled or liquid crude oil often proves to be a better igniter for such emulsions.

Particulate size not only has important health implications but also impacts the dynamics of smoke settling. Particulates having an aerodynamically effective diameter of less than 10 μm are respirable and may be drawn into the lungs with normal breathing.

In general, small particle sizes have the greatest resistance to settling and can be expected to be carried much further from the burn site than larger particles. It is not possible to directly translate the observed irregular shape of smoke particles into aerodynamically effective diameters. The aerodynamically effective diameter of a particle is defined as the diameter of a smooth spherical particle with a unit density of 1 g cm^{-3} that has the same settling velocity in air. Hence, the aerodynamically effective diameter of a particle depends on the size, shape and density of the particle. Cascade impactors measure particle size distribution by the amount of particulate deposited on a series of plates (Walton et al. 1993). The particulate laden air is drawn through the cascade impactor which consists of a series of stages each having a nozzle and plate. Aerodynamic forces determine the size ranges deposited on the plate in each stage and the sizes that pass through to other stages downstream. The fraction of the total deposition collected by each stage of the device determines the distribution of the aerodynamically effective diameter of the particles.

The visual appearance of the smoke plume from the burning is a factor that influences the acceptance of in situ burning as an oil spill response method. A highly visible smoke plume elicits a stronger public concern than a barely visible smoke plume, regardless of the chemical composition.

3.17.4
Persistence of Oil in Subtidal Sediments

Oil may reach the sea bottom in diverse ways. It may sink by itself, adsorb to sediment particles and sink, or be ingested and then evacuated by organisms as a fecal pellet and sink. Shore-stranded oil may weather and be washed into the subtidal zone, sometimes associated with sediment. Once it has become part of the benthos, oil can weather by loss of components to the water column, may be ingested by organisms or may be degraded by microbes. It is often buried by subsequent sedimentation, bioturbation or reworking of the sediment by waves, currents or ice. After burial, it can continue to weather or may be resuspended into the water column on sediment particles by waves, currents or ice. The fate and persistence of oil in the subtidal zone depends upon the combination of all these processes. The effects upon biota and its recovery depend in turn on the fate and persistence of the oil (Humphrey 1993).

Oil may be removed from subtidal sediments through several processes. The fate of oil in a subtidal environment is determined by the amount of oil reaching the sea floor and on the physico-chemical conditions there. Fine sediments retain oil longer than coarse sediments. Microbial degradation and loss to interstitial water constitute important routes for oil out of the sediment. Both are increased with increased flushing of coarser sediments. Microbial degradation requires nutrients and oxygen and is increased with increased flushing (Humphrey 1993). In anoxic sediments, which are usually fine grained, oil can sometimes persist for many years. Lighter hydrocarbons usually disappear very rapidly, but when conditions are anoxic, they tend to persist.

Oil reaches the sea floor by sedimentation. While on the sea floor, subtidal sediment may be covered by further sedimentation and may be reworked by bioturbation and physical processes. The rates of these processes depend on sediment grain size distribution, bottom currents, depth and the composition of the sea water above the sediment. Large pads of oil become heavier than the supporting water by incorporating mineral particles or flotsam, and then sink. Often, oil reaches the sea floor by sedimenting with other materials. Small oil particles incorporate mineral material and can also be ingested by animals and be evacuated in fecal pellets. Once on the sea floor, the oil particles will be buried by continued sedimentation and by reworking of the sediments by bioturbation or physical processes. Once buried, the oil tends to come into an anoxic zone, where degradation is very slow. The same processes which can bury oil particles sometimes also bring the particles back to the surface. If reworking of the bottom sediment is an active process, it is likely that the oxygen concentration in the sediment is also high. Reworking can be caused by bioturbation or by such physical processes as bottom current scouring and suspension, and wave and ice action. As for biodegradation, the oil will be degraded if any low molecular weight components are present. The rate of biodegradation decreases as the oil ages and fewer light ends remain. Biodegradation in oxic zones is much faster than in anoxic zones. It has been observed that low molecular weight polar components of oil and the products of biodegradation are often relatively soluble, and dissolve in interstitial water, being carried to the sediment surface with flushing (Humphrey 1993).

Many of the processes and effects are related; for instance, fine grain sediments occur in low current regimes and have shallow oxic depths; oil may not penetrate this type of sediment, but if it does, it degrades only slowly. Instead of specific process algorithms, a table of half-lives can be used (see Table 3.5).

3.17.5
Containment of Oil in Flowing Water

Booms containing oil generally fail if towed at speeds greater than about 0.25 m s^{-1} or if deployed on rivers moving faster than this speed. The problem of containing oil on moving water has been recognized for many years (see Brown et al. 1993). Few, if any containment studies have addressed the problem of trapping viscous, neutrally buoyant oils. Brown et al. (1993) have undertaken a laboratory study of the hydrodynamic properties of variously shaped objects and also a meso-scale flume study of several containment concepts to determine if these can be used to contain oil on fast flowing water. The laboratory study showed that stable vortices are difficult to generate and that spilled oil is not easily trapped by them. Several filter materials were tested in an outdoor flowing channel with both floating and neutrally buoyant oil. Although some of these materials trapped and held heavy oil, they were not a significant improvement over nylon fishing nets which had been tested previously. A hydrofoil device showed some promise at trapping floating oil.

Table 3.5. Environmental factors influencing the half-life of oil sediment. (Humphrey 1993)

Process or effect	Short half-life	Long half-life
Ice cover	Mostly ice free	Mostly ice covered
Sediment depth	Shallow	Deep
Wave energy	High	Low
Bottom currents	High	Low
Grain size	Coarse	Fine
Nutrients	Unlimited	Limited
Oxygen	High	Low
Sedimentation	Low	High
Sediment reworking	Active	Rare

3.17.6
Oil Sinking

Spilt oil normally floats, but occasionally it sinks below the sea surface, reaching the sea bed or remaining at some intermediate level. Floating oil can be seen, but once it is out of sight it also goes out of mind. Hence it is possible that the frequency of oil sinking and the amount of material involved may have been underestimated in past oil spills. Just occasionally, circumstances allow a closer examination of the fate of oil leaving the sea surface.

The density of oil in relation to that of the surrounding sea water is clearly the major factor involved in oil sinking. Because of the turbulent nature of the sea and the influence of waves and currents, slightly buoyant oil can become submerged. Also, larger patches of oil that become swamped in breaking seas can be carried below the surface by the orbital motion of waves and often take a considerable time to resurface. Some oils, e.g. weathered residual fuel oils, can prove extremely difficult to track either visually or by remote sensing.

3.17.7
Bioremediation of Oil Spills

Bioremediation is the exploitation of the natural ability of living microbes to reduce, remove or transform (biodegrade) contaminating organic chemicals in the environment. This requires the establishment of conditions in the contaminated environment suitable for the growth of pollutant-degrading micro-organisms. Some pollutants amenable to bioremediation include crude oil, oily sludges and refined petroleum products, such as jet fuel, diesel, gasoline, phenolic wastes and creosotes. Bioremediation is not normally used to replace traditional physical cleanup of crude oil spills, but rather as an adjunct procedure (Foght and Westlake 1992). Some advantages include:

1. Low environmental impact in environmentally sensitive or inaccessible areas such as aquifers,
2. Relatively low cost compared to excavation and safe landfilling or incineration of contaminated soil,
3. Detoxification resulting (ideally) in transformation of the pollutant into carbon dioxide, water and cell mass, and
4. Cost-effective treatment at oil concentrations too low for efficient physical cleanup techniques.

The disadvantages include:

1. Longer time frame for treatment compared to physical cleanup,
2. Site-specific limitations that require adaptation of the methodology to meet local conditions (e.g. temperature), and
3. Unidentifiable residues tend to persist even after the most efficient bioremediation treatment, leading to uncertainties regarding detoxification (Foght and Westlake 1992).

Bioenhancement, biostimulation and bioaugmentation are the main classes of bioremediation technologies. Bioenhancement relies on indigenous microbial flora present at the polluted site. The innate biodegradative abilities are then accelerated by adding chemicals such as nutrients (fertilizer) or surface active agents (dispersants). In bioaugmentation, in contrast, pollutant-degrading deficiencies of the natural flora are overcome by introducing some microbes into the environment, often accompanied by growth nutrients. The added microbes may simply be strains isolated from the polluted site, grown in the laboratory on the pollutant and then reintroduced after multiplication, or they may be strains specifically adapted, selected, or genetically engineered in a laboratory. To be effective, strains used for bioaugmentation treatment must be non-pathogenic and must be able to compete with the natural flora.

No single method exists for bioremediation: different techniques are used for different environments. Simple composting, landfarming, and the use of windrows have been successful in cases of soil contamination. Such sites represent essentially contained systems in which either bioaugmentation or bioenhancement are theoretically possible. Aquifer decontamination, a semicontained system, can be achieved with bioenhancement techniques using injection wells. The use of chemical fertilizers to stimulate natural flora on oil-polluted marine beaches has received much attention lately, but neither bioenhancement nor bioaugmentation appear to be feasible in the open sea because of dilution and migration effects. Bioremediation has also been successfully attempted in completely closed systems, such as oil tanker bilges and holds; however, this does not result from oil spilled in the environment.

Bacteria, fungi and yeasts can biodegrade almost all naturally occurring organic chemicals, which they use as growth substrates. An individual strain may have a relatively limited substrate range and alone may be incapable of completely degrading such a complex substrate mixture as crude oil, but a community of degradative strains (consortium), each possessing its own specific degradative ability can effectively biodegrade oil. In the 1980s, small scale trials had suggested that nitrogenous fertilizers might stimulate microbial growth and oil degradation following an oil spill. However, when the Exxon Valdez incident polluted 2000 km of the Alaskan coast in 1989, the technique of bioremediation was viewed rather skeptically. A critical analysis of the aftermath of Exxon Valdez has since shown that bioremediation can indeed be quite effective and holds good potential for tackling similar incidents in the future (Bragg et al. 1994).

In an unpolluted environment there exists an equilibrium among the diverse microbes present. Microbial growth in most natural environments is sometimes limited by the availability of biodegradable carbon sources. Oil, containing biodegradable hydrocarbons, creates a flux in that some organisms will grow, most will be unaffected and others will be killed. Microbes having the ability to degrade the oil will briefly flourish, by outcompeting others. Very often, bacteria are most important in bioremediation.

In bioaugmentation, the approach is to lessen the adaptation period necessary for enrichment of the indigenous species by introducing pre-adapted microbes. The normal degradative abilities of the natural population may be enhanced by introducing strains with specific degradative capabilities. Any introduced microbes, however, must be able to survive and compete with the resident microbial community which is already adapted to the local conditions. This constraint means that bioaugmentation is a relatively untested technology.

Carbon, nitrogen and phosphorus are necessary for microbial growth to occur. Their ideal proportion is 100:10:1. N and P are usually growth-limiting, and this is particularly true when biodegradable carbon such as spilled crude oil is introduced. When oil-polluted soil is incubated under laboratory conditions and nitrogen and phosphorus are added, the soil hydrocarbons disappear. Addition of nitrogen and phosphorus helps improve conditions for the growth of hydrocarbon-degrading micro-organisms.

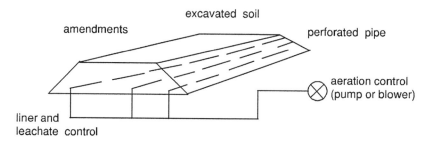

Fig. 3.31. Simplified bioremediation of excavated contaminated soil using a windrow. (After Foght and Westlake 1992)

Many biostimulation agents are indeed fertilizer formulations. Oleophilic fertilizers associate with the spilled oil and slowly release nutrients in marine environments. Some microbes (aerobes) require oxygen for growth, while others require the exclusion of oxygen for growth to occur. In practical terms, only the aerobes participate significantly in short-term bioremediation of spilled oil. The initial attack by hydrocarbon-degrading micro-organisms is an oxidative attack on the oxygen-deficient oil hydrocarbons and hence air is required for efficient biodegradation (Fig. 3.31).

The anaerobic degradation of hydrocarbons can occur via a slow anaerobic growth on several n-alkanes. Oxygen can be limiting and has to be supplied by aeration. This can be accomplished in soil by either positive or negative pressure using a windrow and perforated pipe. Nutrients and moisture may be added at the surface, which is then covered. Perforated pipe internal to the windrow allows for aeration. A liner may be used for control and optional recovery of leachate. In aquifers, oxygen is supplied using downhole air pumps or by adding hydrogen peroxide. In coastal marine areas, tidal action aerates the shoreline and water column. It is very difficult to aerate anything below the surface of anaerobic sediments.

3.17.7.1
The Crude Oil Component

Crude oils are complex, heterogeneous substrates often containing numerous different chemical species present in different proportions (Fig. 3.32). Oils may be fractionated into four classes of chemicals (Fig. 3.33): aliphatics (saturated hydrocarbons, including straight, branched and cyclic alkanes), aromatics (unsaturated hydrocarbons, some with alkyl side chains), resins (polar nitrogen-, oxygen-, and sulfur-containing heterocyclic aromatics) and asphaltenes (a solubility class comprised of large sheets of condensed aromatic hydrocarbons).

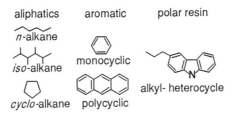

Fig. 3.32. Examples of compounds in three classes of petroleum chemicals

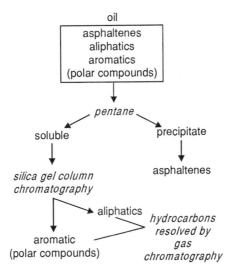

Fig. 3.33. Fractionation of oil into asphaltenes, aliphatics, and aromatic hydrocarbons (analyzed using a different detector)

The latter two classes are relatively recalcitrant to biodegradation because their hetero-geneity and large size offer too complex a variety of chemical bonds for microbes to attack effectively. Generally, increasing size and complexity decreases degradability, and some compounds are biodegraded very slowly, if at all. Most aliphatics and low molecular weight aromatics are readily degraded by a variety of microbes. The lighter (that is, low molecular weight) fractions generally contain the most biologically toxic compounds in oil (for example, benzenes), but fortunately these are also the first to volatilize after a spill, so usually their presence is transient.

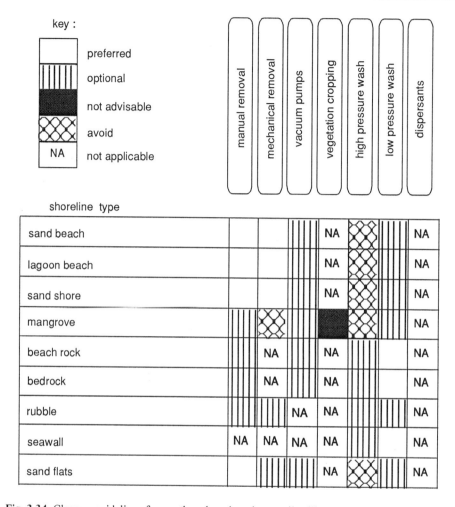

Fig. 3.34. Cleanup guidelines for weathered crude or heavy oil spills

In a natural environment, incomplete microbial attack yields oxidized products that are more water-soluble than the hydrocarbons and these may serve as substrates for other microbes that are not primary hydrocarbon-degraders. With most oils, complete biodegradation is not feasible and residues will result from bioremediation processes. Such residues should be less toxic than the initial material, although the polar products of incomplete oxidation may be more mobile and therefore difficult to manage. Different types of oil are differentially susceptible to biodegradation and require different time frames for significant biodegradation. Microbial activity may have played a role in the conversion of conventional crude oils to heavy oils and bitumens.

The concentration of oil in the environment influences its biodegradation. Minute amounts of oil may be insufficient to induce the necessary degradative enzymes, whereas excessive oil concentrations may overwhelm the microbial flora with toxic compounds or physically prevent adequate aeration. The physical state of the oil also has an effect on its biodegradability. Oil is inherently water-insoluble which means that degradative microbes must function at oil-water or oil-surface interfaces. Therefore, on water, the thickness of the oil slick determines the surface area available for microbial attack. The formation of emulsions, whether by wave action or resulting from the application of bioremediation dispersants, tends to ameliorate biodegradation through increased contact between microbes and oil droplets or increased partitioning of toxic components into the water column. Interestingly, many microbes naturally produce surface active agents (biosurfactants) in response to crude oil, resulting in dispersion of oil droplets in water. This phenomenon, however, is not always linked with degradative ability. At the other extreme, the formation of stable water-in-oil or oil-in-water emulsions ("mousses") retards biodegradation because of the low surface area presented (Foght and Westlake 1992). At low temperatures, oil becomes more viscous and spreads less, reducing the available surface area. Its volatilization decreases. In contrast, "tar balls" composed predominantly of high molecular weight materials are formed from oils that have been excessively "weathered" (i.e., the light ends have been volatilized); these tar balls are highly resistant to microbial action.

Various permutations of microbial community, environmental conditions (such as temperature, nutrient availability and oxygenation), and oil type, provide a diversity of possible combinations for biodegradation, and each site requires its own evaluation for engineering the most efficient bioremediation plan. Figure 3.34 illustrates some cleanup guidelines for weathered crude or heavy product spills.

Table 3.6. Nomenclature of dispersants

Common name	Dispersant characteristics	
	Solvent	Use
Industrial cleaner	Light aromatic hydrocarbon	Neat from vessel
Conventional dispersant	Hydrocarbon base, low aromatics contents	Neat from vessel
Concentrate dispersant	Oxygenates (e.g. glycol ethers) and hydrocarbon base	Dilute from vessel neat from vessel or aircraft

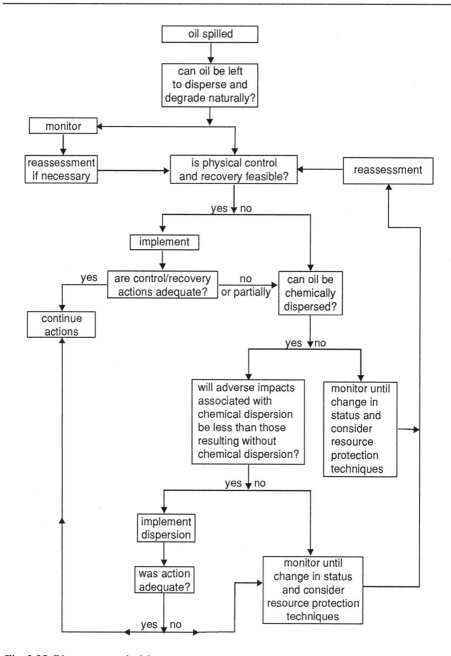

Fig. 3.35. Dispersant use decision tree

Note the following points:

1. The "industrial cleaners" are no longer used as oil spill dispersants
2. Concentrates are sub-divided into two groups, those capable of water dilution and those meant to be applied without dilution. In practice, most concentrates can be placed in both groups but the more recent "high efficiency" products have been developed specifically with undiluted application in mind.

The effectiveness values of dispersants used in the field are in the range of 20 to 70%, but most values are below 35%. The effectiveness of dispersants depends on such factors as oil properties, application method, droplet size, oil thickness, dosage rate or dispersant-to-oil ratio, wind and wave conditions, salinity, temperature, viscosity and emulsification (see Daling and Brandvik 1990).

Mechanical equipment based clean-up is the commonest response method for oil spills. Mechanical recovery equipment includes a large number of booms and skimming devices. The equipment tends to be designed for operation in calm, protected or open ocean areas, and skimmers are often designed to operate within certain viscosity ranges. Their performance capabilities and effectiveness depend on the area of operation and the weathering properties of the oil, in particular the sea state, oil type, oil thickness and degree of emulsification (see Etkin 1990).

Energy has often been stated as the reason for the varying results of dispersant effectiveness tests. Difficulties have been faced in varying and measuring energy. There are differences of opinion as to what is meant by energy in the context of oil dispersion. The differences in energy levels and the way these have been applied to the oil/water mixture, result in effectiveness values that are unique. Fingas et al. (1993) have studied the relationship of dispersant effectiveness with the factors of dispersant amount and mixing energy. They varied energy by changing the rotational speed of a specially-designed apparatus. They found that effectiveness goes up linearly with energy, expressed as flask rotational speed. Natural dispersion was also measured and showed similar behavior to chemical dispersion except that the thresholds occur at a higher energy, and effectiveness rises more slowly with increasing energy. The effect of dispersant amount at higher energies was the same as was found in previous studies for lower energies. The effectiveness increased exponentially with increasing dispersant amount. Although there exists a trade-off between dispersant amount and energy to achieve high effectiveness values, energy is the more important factor (Fingas et al. 1993).

Release of oil into coastal waters often results in stranding of oil along the shoreline. Chemical cleaning agents are one option that can be used to mitigate detrimental effects of stranded oil on natural shorelines under appropriate circumstances. Such agents are used because of the biological sensitivity of indigenous fauna and flora to stranded oil, amenity considerations of a shoreline, or concern about refloating of oil and subsequent stranding on adjacent shorelines. However, prior to the use of cleaning agents at a spill site, information regarding the performance of available cleaning agents must be obtained (e.g., the relative performance of agents for removing stranded oil from surfaces). Through standardized laboratory testing procedures, it is possible to evaluate the relative performance of cleaning agents.

3.17.7.3
Emulsification

Emulsification is the process of the formation of water-in-oil emulsions often called "chocolate mousse" or "mousse". These emulsions greatly change the properties and characteristics of oil spills. Stable emulsions have between 50 and 80% water, increasing the volume of spilled material from two to five times the original volume. The density of the resulting emulsion can be as great as 1.03 g ml^{-1} compared to a starting density as low as 0.80 g ml^{-1}. Most significantly, the viscosity of the oil can often increase 1000-fold. This changes a liquid product to a heavy, semi-solid material. Emulsification is the second most important behavioral characteristic after evaporation. Emulsification has a strong effect on the behavior of oil spills at sea. As a result of emulsification, evaporation slows significantly; spreading slows by similar rates, and the oil rides lower in the water column, showing different drag with respect to the wind. Emulsification also has significant effects on other spill aspects; spill countermeasures are quite different for emulsions. Emulsions are hard to recover mechanically, treat or burn (Fingas et al. 1993).

The Canadian environmental agency, Environment Canada, conducted a series of studies on the physics of emulsion formation (Bobra 1992). This study has provided experimental results to show that emulsion formation is a result of surfactant-like behavior of the polar and asphaltene compounds. The latter are similar compounds and both behave like surfactants when they are not in solution. The aromatic components of oil solubilize asphaltene and polar compounds. When there are insufficient amounts of these components to solubilize the asphaltenes and polars, these precipitate and are available to stabilize water droplets in the oil mass (Fingas et al. 1993).

3.17.8
Remote Sensing

Remote sensing is an important part of oil spill combat strategy. The public expects that the government or the spiller know the location and extent of the contamination. Remote sensing can be used to increase spill cleanup efficiency. Furthermore, recent advances in electronics technology have made the instrumentation much cheaper and more capable. The definition of remote sensing implies that a sensor is used to detect the target of interest from a distance. The most common form of remote sensing as applied to oil spills is aerial remote sensing – that is using aircraft or satellites as platforms. Visual observation, irrespective of the platform used, is by definition not remote sensing. Optical techniques constitute the commonest means of remote sensing. Cameras are particularly common because of their low price. Aerial mapping is very common and many companies now use aircraft equipped with video cameras.

Oil has a higher surface reflectance than water in the visible range. Heavy oil appears brown, showing up in the 600 to 700 nm region. Mousse shows up in the red-brown or closer to 700 nm. Sheen appears silvery and reflects light over a wide spectral region up to the blue. There is no strong information in the visible region from 500 to 600 nm, so often this region is filtered out, to give better contrast. Experimenters have found that one technique for giving high contrast to visible imagery is to set the camera at the Brewster angle (53 degrees from vertical) and use a horizontally aligned polarizing filter which passes only that light which is reflected from the water surface. It is this component that contains the information on surface oil.

Oil on shorelines is difficult to identify because weeds can have similar appearance, and oil on darker shorelines cannot be detected. Use of the visible spectrum for oil detection appears to be limited; it does, however, offer economical means of documenting spills and means of providing baseline data on shorelines or relative positions.

Oil which is "optically thick" absorbs solar radiation and re-emits the radiation as thermal energy largely in the 8 to 14 μm (8000 to 14000 nm) region. This enables oil detection by infrared sensing. This oil appears to be cool in the infrared (compared with the surrounding water). Thin oil or sheens are not detected by infrared. Thick oil appears hot or white in infrared data, intermediate thicknesses appear cool and black, and thin oil is not detectable. The thicknesses at which these transitions occur are not known, but the transition between the heated and cooled layer lies between 50 and 150 μm and the minimum detectable layer between 10 and 70 μm.

Infrared cameras are now commonly used; earlier, scanners with infrared detectors were largely used. Infrared detectors suffer from the disadvantage that they require cooling to avoid thermal noise, which would destroy any useful signal. The traditional cooling agent has been liquid nitrogen, which gives about 4 h of service. New, smaller sensors can use electric thermal coolers or Joule–Thompson coolers. Most infrared sensing takes place at what is known as the thermal infrared at the wavelengths of 8 to 14 μm.

Oil slicks show high reflectivity of ultraviolet radiation even at thin layers (0.01 μm). This may be used to map even thin sheens of oil. Overlapped ultraviolet and infrared images are often used to provide a relative thickness map of spills.

Fluorosensors rely on some compounds in the oil absorbing ultraviolet radiation and re-emitting part of this energy in the visible. Since very few other compounds show this tendency, fluorescence is a strong indication of oil presence. Natural fluorescing substances such as chlorophyll, fluoresce at sufficiently different wavelengths to avoid confusion. There is, however, a strong natural fluorescent material known as "gelbstoff" which consists of a broad spectrum of fluorescing materials. Some laser fluorosensors operating in the ultraviolet region are also being used now.

Radars optimized for oil spills can also be usefully applied to remote sensing, in particular for large area searches and for night-time or bad weather work.

Oil on the ocean is a stronger emitter of microwave radiation than water, and thus appears as a bright object on a darker sea. This differential emission can help detect oil.

Another instrument that holds potential for detecting neutrally-buoyant oil is the lidar bathymeter. This uses a laser in the green region to map bottom contours. Presence of massive amounts of oil floating beneath the surface could be detected. Oil can be visibly detected on ice. It appears black or brown against the white background of the ice. Sediment has the same appearance and hence interferes.

3.18
Use of Indicators to Judge the Condition of Marine Ecosystems

A phased approach to the use of biological indicators in the measurement of the condition of the marine environment is desirable. The initial phase involves detection of a problem. The second phase is the assessment, definition and characterization of the problem, and the final phase is managerial activity to solve the problem. In each phase, the aims and uses of bioindicators may vary.

There is a need for use and application of various indicators of the condition of marine ecosystems. Although much recent progress has been achieved in the assessment of chemical impacts on ecosystems, these should not be considered in isolation. Biological and physical impacts such as the introduction of exotic species, overexploitation, sedimentation and dredging also have profound impacts. These may interact with effects of contaminants in marine ecosystems. In addition, globally, the level of concern about one type of issue in comparison to another may vary substantially from country to country and from region to region.

The approach to environmental assessment has two core components. The first is the recognition that the study has different phases which lead to determination of the goals, needs and uses of indicators of the condition of an ecosystem. The second component is that there are tiers of hierarchical sets of indicators that might be examined, depending on the defined goals of the study.

A tiered structure may include a range of biological organizations from the molecular level, extending to the cellular and organismal levels, populations and assemblages. Investigations at one level of organization may be appropriate for one type of problem, whereas another type of problem may require investigations at all levels.

A suitable mixture of indicators is needed for sampling to detect changes in components and functions of ecosystems over time. This is because the range and scale of potential stresses acting at any time in a given area is quite large. The assessment of the well-being of a system needs to include indicators of chemically-induced effects (e.g. biomarkers) in the biota. Also, several measures should be used to detect responses to disturbances such as organic or nutrient enrichment, changes in sedimentation, erosion or hydrography, exploitation and other processes that influence whole organisms, populations, assemblages and the processes and interactions operating at these levels of biological organization. Finally, there must be indicators of the risk of, or the actual, change of habitat due to such processes as dredging, clearing, landfill and development of the shoreline that will lead to alterations or losses of populations or sub-systems in an area (see GESAMP 1995).

In field studies, indicators should be in a tiered approach, so as to meet the needs of the three phases of an investigation. The composition of each phase and the number of tiers appropriate for a given situation would depend on the specific objectives of the research and the characteristics of the field site. The first phase is to identify a response. This may happen because of some observed biological effect (a fishery fails or a series of algal blooms occur). The second phase will be to characterize and understand the cause.

A single conceptual model elucidating potential linkages among phases of investigation and tiers of study would amount to an oversimplification of the diversity of threats that currently challenge marine habitats. Some knowledge of the variability in the natural system is an essential prerequisite for assessment of anthropogenic effects. A good example is the occurrence of "unusual" plankton blooms caused by variations in hydrographic conditions, which differ from the "normal" blooms and yet are not due to anthropogenic influences.

It has been previously believed that clear cause and effect relationships should be developed among the tiers of investigation for a technique to be considered valuable for broad application. Evidence of impacts on cells, for example, must be clearly related to the consequence for the functioning of populations of individuals. It has now been realized that this is an unachievable as well as unnecessary objective. Rather, measurements of effects on cells, tissues or individuals all contain different information, expressing different aspects of response to a stress and all are valid and important in the evaluation of environmental impact. Molecular biomarkers of exposure help to quantify the links between additional/increased chemical exposure and the first stages of biological response. These measurements may not give information on the performance of the individual organism, or a change in the reproductive potential within the target population. But in some applications it is important to consider whether a particular effect is more or less useful in establishing connections among tiers of investigation. As different indicators may respond to different stresses, it is wise to employ a suite of indicators to assess the condition of different components of the ecosystem.

3.19
Scales of Effects

Chronic effects of contaminants, eutrophication, or changes in sedimentation require measurement of the levels of variables after the onset of the stress, or the steady rate of change of the variable as the effects of the stress accumulate. Acute scenarios are more likely to be measurable in the context of "pulse" responses, hence attention is to be focused on reliable measures of rates of change in the various indices. Most stresses act together simultaneously. Many stresses are diffuse and of uncertain timing and frequency. All these are difficult to measure accurately.

Weak, diffuse impacts do not affect populations or assemblages quickly. Suborganismal measures are the only possibilities of detecting them. For these chronic problems, precise estimation of the magnitude of the chosen indicator(s) relative to undisturbed background or standard measures is needed. In contrast, many acute events are a result of accidents (e.g. oil-spills) for which it is difficult to design an advance program of monitoring and detection. This type of environmental disturbance requires monitoring of very general indicators so that a range of disturbances is detected. It also requires widespread and frequent sampling. Unpredictable episodic events pose special tricky problems for environmental assessment. The most practicable way to determine what happens as a result of episodic disturbances, such as oil-spills, destruction of an area of mangrove forest, accidental discharges of chemicals and so forth, may be to use "event-driven" sampling.

Spatial scales of environmental disturbances are no simpler (Underwood 1993). The crucial point is to have good understanding of the disturbance and the processes that disseminate or constrain its influence. Barring the most widespread disturbances that affect the whole world, or whole ocean basins, all scales of disturbance are amenable to investigation by suitable measurements. Large-scale stresses cause some stress responses at any hierarchically chosen smaller scale. But at a small scale the detection of large patterns is sometimes difficult because of the multitude of small-scale events that can mask the large-scale signal. Detecting large-scale stresses requires examination of processes at a sufficient scale to filter out local stress responses. But the problem here is that very large-scale disturbances cannot be detected by any comparisons of disturbed and control or reference locations. The problem of scale and scaling will, therefore, have to be addressed in each case, using applicable scaling laws (Levin 1993).

There are only two possible methods for detecting deterioration of ecosystems due to very large-scale disturbances. One is the comparison of temporal trends, and the knowledge of the existence of a particular disturbance. Major forcing functions for marine ecosystems are physical structures and processes related to the hydrology and hydrodynamics of the system. Any major changes in these are likely to bring about changes in the ecosystem (GESAMP 1995). It is also often advisable to measure the conditions prevailing before, or in the absence of, an impact. As virtually all systems influenced by human activities experience local changes, the crucial question is when these changes could be regarded as symptoms of a deterioration of the system. Merely assessing the intensity of the change is not sufficient. The second method for studying ecosystem-wide changes is to compare measures of stress against some absolute standard.

3.20
Scenarios

Some adverse effects/disturbances in marine ecosystems include: (1) "point source" chemical contamination, (2) loss of habitat on local and regional scales, (3) eutrophication caused by excessive nutrients or organic material (see Fig. 3.36), and (4) the effects of UV-B radiation as a consequence of depletion of stratospheric ozone. Different bioindicators are needed for studying the above four types of cases.

3.20.1
Human Impacts

Mankind influences marine habitats by biological, chemical and physical interactions over different temporal and spatial scales. The spatial distribution of an impact may be local, regional or global. The spatial scale of the impact may not necessarily be related to the physical scale of the stress because a range of mechanisms such as hydrodynamic and atmospheric transport processes or the migratory patterns of organisms may extend local effects regionally or globally. Multiple local effects can become regional when they collectively cover a large area. Some other relevant aspects of human impacts are their magnitude, frequency and reversibility. Human activities interact with the characteristics of habitats, making habitats differentially vulnerable to human stresses.

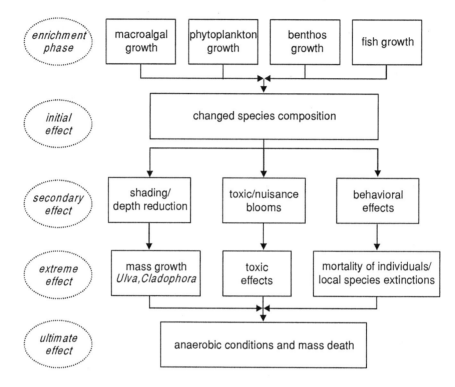

Fig. 3.36. General model of effects of eutrophication on marine systems (From Gray 1992)

3.21
Shallow Water Habitats

The marine environment can be classified into shallow and deep-water habitats, which generally represent near shore and offshore environments, respectively. Near shore habitats can be further distinguished as being estuarine (i.e. subject to freshwater influence from the land) or non-estuarine. Within these broad groupings, habitats may comprise either hard or soft substrata or biologically structured habitats such as seagrass beds or mangrove forests. This habitat classification does not define physical boundaries. The continuity imposed by water means that different marine habitats are interconnected and can rarely be treated in isolation. This physical continuity also explains why stresses can cause impacts on different spatial scales.

3.21.1
Estuarine

Estuarine habitats are heavily influenced by human activity. There is much concern about loss of habitat due to diverse activities such as filling of wetlands, silviculture, the diversion of water, the introduction of exotic species; aquaculture and other stresses which develop in the catchment of rivers. Near shore environments under the influence of freshwater from land are often under additional stress caused by contaminants in river runoff. The contaminants may be organic, such as untreated domestic sewage or chemical (e.g. industrial wastes or pesticides washed out from agricultural areas). Sedimentation associated with land-based activities such as agriculture, silviculture, mining and estuarine and marine dredging can also degrade near shore habitats.

Important ecosystems situated in estuarine areas include wetlands, (e.g. tropical mangrove forests), seagrass beds, and seaweed assemblages. Besides the above impacts on specific ecosystems, soft-bottom habitats, including seagrass, mangrove and algal assemblages and the expanses of sediment on the seafloor which lack conspicuous epibenthic organisms, are vulnerable to forces affecting sediment movement and alteration of sediment characteristics. Processes of erosion, deposition and alteration of grain size harm organisms in these soft sediments. Hard bottom assemblages in estuaries can likewise be affected by diverse impacts associated with runoff from land and such factors as excessive exploitation.

3.21.2
Non-Estuarine

Coral reefs are among the most important non-estuarine ecosystems in shallow water and are hard bottom structures. They line the coastlines of parts of the tropical and subtropical belt where environmental conditions generally allow sufficient penetration of light for photosynthesis by their autotrophic symbionts, and fluctuations in temperature and salinity do not exceed the physiological limits of the reef inhabiting organisms. These ecosystems are important sources of food and other products for nearby human populations. The rapid increase in human population and the heavy dependence on reefs (especially in developing countries) has led to overexploitation of resources. Certain species of finfish, mollusks, crustaceans and echinoderms are harvested on a selective basis. This can change the structure and species composition of communities. Coral reefs are also being destroyed physically in some parts of the world as a result of destructive fishing, sedimentation from land runoff, dredging, mining and filling (as in land reclamation). Globally, some impacts that occur in near shore marine environments include organic enrichment of kelp forests, disposal of dredge spoil and discharge of thermal effluents into rocky embayments.

3.22
Recycling in the Deep Ocean

Lots of sediments are washed down into oceans every day. Sediments from terrestrial sources are transferred via submarine channels to the deep ocean, and abyssal currents can transport the material thousands of kilometers around the submarine margin, then return it to land via the "subducting" or crust-consuming trenches (see Carter 1995).

Rising mountain ranges, because of their inherent instability, the erodible nature of their rocks and climatic factors, shed huge volumes of detritus first into the rivers and then onto the continental shelf. The volumes delivered offshore account for about 3% of the world's sediment input to the oceans (Carter 1995).

3.23
Deep Water Habitats

The open ocean constitutes the largest part of the marine environment. Because of its longer distance from human settlement it is less subjected to anthropogenic impact than coastal areas. The open ocean is divided into the benthic realm and the overlying water column (pelagic realm). The water column encompasses thousands of cubic kilometers and is also structured by physical processes. It can therefore be subdivided into a number of different regions which include the ocean surface microlayer, the euphotic zone, intermediate waters and abyssal depths and from oceanic gyres to frontal systems. Offshore pelagic assemblages are subject to chemical contamination which may lead to effects on biological systems. The primary concern has, however, been with overexploitation of their living resources. Bottom habitats, both hard and soft, are subject to continuing deposition and movement. Deep water assemblages are subject to long-distance influences including low levels of contaminants, especially those associated with transported sediments. Additional threats are posed by deep-sea dumping and the prospects of mineral exploitation. Some deep water fish stocks are being heavily exploited without adequate knowledge of the stocks or their productivity.

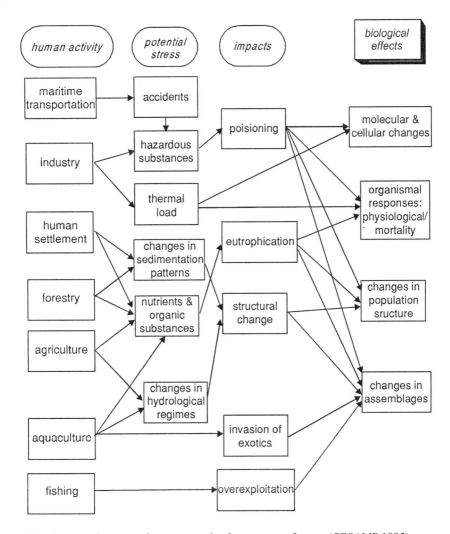

Fig. 3.37. Some major, general stress scenarios for mangrove forests. (GESAMP 1995)

Long-distance transport of chemical contaminants and nutrients can influence open ocean productivity. Effects of global climate change on ocean circulation and weather patterns, coupled with other stresses, may also cause significant changes. Figure 3.37 is an example of an impact scenario for mangrove habitats but which could as well be applicable to other habitats.

3.24
Indicators of Exposure and Effect

Estimating the extent to which contamination may adversely affect the biota within marine ecosystems is an important goal of a program that monitors environmental quality. Bioassays (i.e., toxicity tests) are experiments in which organisms are exposed to different substances or combinations of substances to determine concentrations that adversely affect them (cf. Chap. 1). The period of exposure may be short (acute tests) or long (chronic tests). Measured effects include mortality and sublethal biological responses such as impaired growth, irregular development, abnormal behavior, motility or reduced reproductive output. Different bioassays may be used for specific purposes, e.g.:

1. Comparison of the toxicity of effluents from industry and treatment plants,
2. Comparison of the toxicity of chemicals or chemical mixtures,
3. Comparison of field samples from different locations,
4. Comparison of the sensitivities of different species to the same substances, or
5. Identifying new problems.

The short-term nature of most bioassays, which only consider single species under controlled conditions, rather than many species under natural conditions, limits their predictive value as indicators of complex chemical and biological interactions (GESAMP 1995). Precise exposure condition is a very important variable that affects organism-toxicant contact and partitioning within the organism (McCarty and Mackay 1993). Subsequent chemical uptake, organ-system chemical transfer, toxic response and metabolism all depend on exposure, concentration and composition. Hence, bioassays on fresh field-collected samples, or conducted in situ are better indicators of field exposures and help identify available toxicants (GESAMP 1995).

Numerous single-species marine bioassays are available as standard methods or protocols (e.g. ASTM 1993). The standardized bioassays cover the span from acute sublethal and lethal assays to chronic sublethal assays, measuring the bioaccumulation and effects of contaminants. Commonly used tests measure mortality in marine fish, bivalve mollusks, copepods, microalgae, rotifers, amphipods, echinoid gametes and crustacean larvae. End points include mortality, abnormal behavior, slowed or delayed growth, inhibited photoluminescence, inhibition of fertilization, abnormal or delayed embryonic development, decreased molting success, etc.

Recent applications of toxicity testing have included assessments of the effects of contaminants in the sea surface microlayer (Karbe 1992). The microlayer was sampled and then thoroughly mixed with clean water and toxicity tested on this water. Although such tests have indicated significant toxicity, there is some uncertainty as to the extent organisms in nature are actually exposed to the surface microlayer.

3.25
Biomarkers

The term "biomarker" refers to a biological response which can be specified in terms of a molecular or cellular event, measured with precision, and which gives information on either the degree of exposure to a chemical and/or its effect upon the organism or both. Biomarkers are influenced by the bioavailability of contaminants, the routes of exposure and the level and time of exposure. The first response of organisms takes place at molecular and cellular levels of target organs and tissues before effects become visible at higher (individual, population or assemblage) level.

Several biomarkers at the tissue, cellular and molecular levels are available to evaluate the status of an organism and to detect exposure to contaminants (Huggett et al. 1992). As molecular biomarkers are tightly linked to the insult, they tend to be sensitive indicators of specific types of exposure (GESAMP 1995).

Biomarkers may fall into several categories (see Table 3.7) depending upon their diagnostic value. These categories are: (1) biomarkers of exposure only, (2) biomarkers of exposure with uncertain eventual consequence and (3) biomarkers of known deleterious consequence based on mechanistic understanding.

Table 3.7. Categorization of selected biomarkers (relating to DNA integrity) based on their diagnostic value

Biomarker	Category (see text)	Remark
Strand breakage	1	Potential genotoxic insult
Adducts	2	Exposure to specific class of chemicals (mutagens, carcinogens)
Photoproducts	3	Exposure to UV-B radiation
Chromosomal aberrations	2	Detected by flow cytometry and cytogenetic analysis

Biomarkers of exposure represent either general or specific responses. General ones include those responses that are not compound- or class-specific but indicate merely that exposure to some exogenous agent might have occurred (e.g., DNA strand breaks or chemically-induced oxidative stress). Specific biomarkers are used to indicate or confirm the class of chemicals or toxic agent involved. Examples include organophosphate/carbamate inhibition of serum cholinesterase activity, polyaromatic hydrocarbon (PAH) adducts with DNA and protein, lead inhibition of aminolevulinic acid dehydratase (ALAD), and DNA photoproducts caused by ultraviolet radiation (see Huggett et al. 1992).

Biomarkers of effect detect a wide variety of responses to pollutants. Some response indicators measure the effects of contaminants on the whole organism and include physiological, behavioral and ecological variables.

3.26
Physiology

Physiological responses of marine organisms to pollutant exposure depend on the biological availability, uptake and distribution of the chemicals within the body. Physiological responses integrate subcellular and cellular processes and can be chosen to be representative of the fitness of the whole organism. The most important physiological changes associated with contaminant exposure tend to harm the organism's growth, reproduction and survival. Physiological indices linked to the survival and growth or reproductive potential are therefore potentially highly effective in assessing the effects of contamination gradients.

3.27
Ecology

Several indices of population structure or species assemblages measure ecological responses. Broadly, some of these are intrinsically univariate measures – i.e. a single, measurable, variable response to an environmental stress. Others are contrived univariate measures of the collective properties of more than one variable. Four types of univariate indices are commonly used to assess changes in populations and assemblages:

1. Loss of species over time,
2. Diversity indices tend to simplify the expression of biological complexity,

3. Changes in the type or abundance of dominant species. Under severe stress, dominance by a few species increases; these are small-sized, rapidly growing, so-called opportunist species (e.g. Olsgard and Hasle 1993). On coral reefs, macroalgal dominance increases in response to nutrients, and

4. Reductions in the size of harvested species are indicators of population change due to overharvesting or overexploitation, e.g. of fisheries.

Such ecological characteristics of populations (as fecundity, age- or size-specific patterns of mortality, rates of birth or recruitment) and measures of activity or behavior are useful for determining whether populations are stressed. But much better knowledge of how different types of stress are likely to affect characteristics of populations is urgently needed.

Multivariate statistical analyses are valuable techniques for assessing change due to stress in complex assemblages of many species. Analyses based on matrices of species abundance in several sites have been shown to detect subtle effects of stress on natural systems (see Warwick 1988; Warwick and Clarke 1991).

3.28
Marine Pollution in India

India has a coastline of about 7500 km, and about 25% of its population lives in coastal areas. Many major cities including Calcutta, Madras and Bombay are located along the coast. There are 11 major, 16 medium and 78 smaller ports in India. One of the wettest countries in the world, India has an annual average rainfall of about 1000 km^3. The country has 14 major, 44 medium and 162 small rivers with a mean annual runoff of 1645 km^3, not all these rivers discharge into the sea. About 500 million tonnes of sediment annually discharge into the seas off India. In the Arabian Sea, the southwest monsoon results in intense upwelling along the west coast of India, accounting for the high productivity and the high fisheries potential in this area. Both the Arabian Sea and the bay of Bengal see large semi-diurnal tides of amplitudes of 1–8 m and are also affected by the biannual reversal of the monsoon winds. These two factors produce strong flushing of the coastal areas, effectively dispersing pollutants. The open oceans around India have so far remained fairly pollution free but coastal pollution has been increasing.

The major pollutant in coastal areas is domestic sewage. Over 4 km^3 of such wastes are discharged into the seas off India annually (Glasby and Roonwal 1995). Bombay discharges 365 million tonnes of sewage effluent to the sea annually, Calcutta about 400 million tonnes. Only a small proportion of this sewage is treated before discharge.

Upgrading sewage treatment facilities might lead to substantial reduction of coastal pollution.

Mahim bay in Bombay covers 64 km^2 and once used to be a healthy ecosystem with fisheries, oyster beds, mangroves and migrating birds, but is now one of the most polluted areas in the country. The dissolved oxygen levels in the bay water often drop to 1 mg l^{-1}. The water is rich in nutrients (eutrophic), has high biological oxygen demand and high coliform counts. The near shore waters show severe organic pollution during ebb tide. Mahim bay is now virtually an open sewer. The benthic fauna of Bombay is badly depleted, affecting fishing. Fisheries are retreating from the shore, and fishermen must go over 10 km offshore to get a worthwhile catch. The main health hazards for humans are gastrointestinal diseases resulting from the consumption of contaminated sea food.

About 0.5 km^3 of industrial waste is annually discharged into the seas around India. Industries such as paper, textile, chemical, pharmaceutical, plastic, food, leather, jute, pesticide and oil contribute to this waste. Particular consideration should be given to metals such as lead, zinc, cadmium, mercury and chromium. So far the concentrations of these elements in seawater, sediments and biota in offshore areas have by and large not posed any serious problem. Discharges of these elements into the rivers have also not caused serious problems to the marine environment. Over 50% of the metals discharged into the Ganges settle out in the estuarine and river mouth regions and only 15% reach the bay of Bengal. Nonetheless, there are problems in coastal areas near big cities. In Thane creek, the concentrations of heavy metals in marine organisms are so high that some marine species have become unfit for human consumption. The Ulhas river, which drains into Thane creek in Bombay, is quite rich in heavy metals, as revealed by an estimate based on monitoring of 18 of the 48 major industries. About 11 tonnes of Cu, 400 tonnes of Zn, 7 tonnes of Hg and 0.5 tonnes of Cr are discharged annually into the Ulhas river (see Glasby and Roonwal 1995). Fly ash from power stations is an important source of metal pollution in India. Over 100 million tonnes of fly ash are produced each year, much of it being transported in the atmosphere before deposition.

An estimated 380 000 tonnes of pesticides and other halogenated hydrocarbons are used in India each year, of which 55 000 tonnes are used in agriculture. The total annual consumption of DDT and its isomers is 107 000 tonnes per year (see Glasby and Roonwal 1995). These chemicals are used as pesticides, herbicides and fungicides in agriculture, and for controlling such vector-borne diseases as malaria. The seas are the ultimate repository of these highly persistent chemicals. Pesticide concentrations are much higher on the east coast than on the west coast because they are transported to the east coast by the large rivers. These compounds result in reproductive failures in marine birds and fishes and inhibit photosynthetic activity of algae.

The Arabian sea is the main tanker route for the transport of oil from the Gulf to the Far East. Tanker accidents, oil well blowouts, ballast discharge and bilge washings are the main sources of oil pollution in the marine environment. Oil slicks and tar balls are common along this tanker route. The tar particles have a residence time of 30–45 days before they start sinking, and tar-like residues are washed up on the west coast beaches of India, particularly during the southwest monsoon.

3.29
The Indian Ocean Law and the Environment

About one third of the world's population live in the Indian Ocean region and depend heavily on the resources of the sea to meet their needs. But, from eastern Africa to western Australia, from the Antarctic to the Indian subcontinent, the waters of the Indian Ocean already show the adverse impact of human pressures. The oceans are under stress from growing coastal populations, rapid industrialization and overuse of resources. Many people in the coastal zone depend on its waters for their livelihood. Paradoxically, coastal waters suffer most from the effects of pollution (Wickremeratne 1991).

The increasing marine degradation from human activities has contributed to the rapid development of international ocean law over the past two decades, and the law of the sea was agreed upon in 1982. It is a global instrument dealing with all ocean space and with all the many ocean uses – shipping and navigation, mineral development, fisheries, scientific research, environmental protection as well as with settlement of disputes. This convention has drawn up new boundaries of state sovereignty and jurisdiction. It gives landlocked and geographically disadvantaged states new rights. New universal standards have been set. More than 35 Indian Ocean states have so far signed the convention.

The law of the sea convention encourages states to cooperate, directly or through relevant international organizations, in the protection and preservation of the marine environment. The regional seas program plays an increasingly important role in implementing the Convention. It involves five regions of the Indian Ocean: Eastern Africa, the Red Sea and Gulf of Aden, the Kuwait action plan region, the South Asian Seas and the East Asian Seas. The programs for these regions are planned by the countries concerned, while keeping certain basic concepts in view such as helping the countries to assess marine pollution, promoting the control of marine pollution, coordinating efforts in management, and supporting education and training.

Under the convention, a state can claim a 12-mile territorial sea. Most states in the Indian Ocean region have implemented these provisions. States have sovereignty over these waters, but the exercise of the sovereignty is subject to observing certain commitments. All ships and aircraft are given right of "innocent passage" through or over territorial waters without hindrance from the coastal state so long as they do not violate its sovereignty.

States have the right and duty to oversee that all ships passing through their territorial waters observe the generally accepted international rules with respect to safety at sea and the control of pollution from ships. Where ships come into port, a country can exercise even greater controls as a port state. Some states lie adjacent to important navigational routes, which are considered international straits. The convention provides for the right of transit passage through such straits. The Indian Ocean has six important international straits: the Bab el Mandeb in the Red Sea, Hormuz in the Gulf, Lombuk, Macassar, Malacca and Singapore in the East Asian Seas.

3.29.1
Exclusive Rights

The new concept of the exclusive economic zone (EEZ) has been devised to enable coastal states to claim exclusive rights to the living and non-living resources across a belt of 200 nautical miles from the shore. The seas beyond national jurisdiction are known as the high seas. In this area all states enjoy freedom of the seas.

Most Indian Ocean states have very narrow continental shelves – the submerged extension of land into the ocean. However, under the convention, coastal states are entitled to claim a shelf of up to 200 nautical miles even though the extension may not in fact be that wide. A nation has sovereign rights over the wealth of its continental shelf, even where this extends beyond 200 miles. The convention does not permit a state to claim rights beyond 350 miles.

Several Indian Ocean States have benefited from this provision, especially those in the Southern part of the bay of Bengal which have gained sovereignty over a large area of extended seabed (Wickremeratne 1991). The international seabed area refers to the ocean floor and subsoil beyond the limits of national jurisdiction. The Area is mostly in waters 3 to 5 km deep and extends over largely uncharted ocean floor. The area and its resources (such as the valuable "manganese" nodules; these nodules contain, in addition to Mn, some Co, Ni, and Cu) have been declared the common heritage of mankind. India was the world's first country to be granted rights to seabed mining in the Indian Ocean. At present it is the only country of the region expected to be commercially involved in mining the seabed for manganese nodules. Exploitation of the nodules can potentially be a boon to Indian Ocean economies. Under the terms of the convention, even nations which are not involved in mining will be entitled to share the benefits of this "common heritage". The regional seas programme has helped Indian

Ocean states to develop certain common environmental standards with regard to hazardous effluents.

3.30
Ship Pollution

A fairly common problem is ship-generated marine pollution. The Indian Ocean is the busiest tanker traffic route in the world. Tankers carrying almost 500 million metric tons of crude oil pass through this ocean region annually to meet the fuel needs of the industrialized countries. Discharges from vessels are the source of approximately 40–50% of all the petroleum introduced into the ocean. Oil slicks caused by the flushing of tanks and dumping of oily ballast are frequent occurrences along the tanker routes.

Several accidental oil spills have also occurred in the region, among them the sinking of the Ennerdale off the Seychelles, the grounding of the Tayub on the reefs of Mauritius, the grounding of the tanker Transhuron in the Laccadives, the grounding of the Showa Maru at Buffolo rock in Indonesia, and the sinking of a number of vessels in the Iran-Iraq War. These have spilled large amounts of oil into the Indian Ocean. Most is washed onto the coasts by the ocean currents. Consequently, some beaches have become heavily contaminated with tars (the solid residue of crude oil). Coral reefs and mangrove forests have been affected by oil because of the nearby tanker routes.

The danger of accidental spills as a result of collision at sea is growing with the ever increasing size of the supertankers carrying oil across the Indian Ocean. The rise in offshore oil production in the Indian Ocean has also increased the potential for major oil spills from blow-outs or leakage. The law obliges states to develop contingency measures to deal with pollution incidents in the marine environment. Until recently Indian Ocean states were quite unable to effectively deal with any major oil spill, but today many of the states have been able to develop fairly good capabilities to respond to such emergencies. The convention puts a responsibility on all states to prevent, reduce and control ocean dumping of wastes, even though dumping is not the most severe nor the most extensive of the ways in which the ocean is polluted (Wickremeratne 1991).

3.31
Waste from Land

A substantial portion of all ocean pollution originates on land. The inability to cope with increasing amounts of municipal wastes has magnified the problem of coastal pollution. The majority of states in the Indian Ocean region lack basic sewage treatment facilities and discharge raw sewage directly into the ocean. Industrialization has resulted in highly toxic effluents being discharged directly into bays and lagoons, later reaching the ocean. Fertilizers and pesticides also end up in streams that flow into the ocean. Algal blooms are increasing in frequency and scale, causing great concern especially in the Kuwait region. These blooms use up most of the oxygen in the water when the organisms decay, killing fish and other animals. Sedimentation, industrial effluent, and domestic garbage are the most serious sources of pollution from land.

3.31.1
Wanton Destruction

A very serious problem is being created by the systematic destruction of the "critical marine habitats" which are areas that serve as feeding, breeding, resting and nursery grounds for marine organisms. They are major sources of nutrients, are particularly rich in species, and are therefore highly productive. Such habitats are essential for the productivity and survival of commercial marine species and help to stabilize the coast against erosion by waves and storms (Wickremeratne 1991). Over 50% of the mangrove forests within the Indian Ocean region have been destroyed in the last two decades. Mangroves are being felled for firewood, charcoal and building materials and cleared for shrimp farms, housing and commercial development. They are also being affected by land-fill activities, freshwater flooding and pollution.

Coral reefs have much ecological significance and economic value, but are being degraded and even totally destroyed throughout the region by sedimentation from land-fill and coastal construction work, dredging and upstream erosion. Forest clearing, dam building and construction of harbors, sea walls and breakwaters contribute to soil erosion and siltation. Coral reefs are also being destroyed by siltation. Marine parks and reserves have been created and their coverage is likely to be extended.

3.32
Animals in Peril

Whales, dugongs, turtles and some other marine animals are threatened by direct exploitation, pollution and habitat destruction. The international whaling commission (IWC) has already declared the Indian Ocean a sanctuary. Marine mammal conservation is attracting attention in the Indian Ocean region. Some protection has been gained for the tiny Indus river dolphin in Pakistan and other cetaceans in the Indian Ocean.

Five species of marine turtles are found in the Indian Ocean, four of which are endangered, being threatened by poaching, encroachment of tourism into their habitat and accidental entanglement in fishing nets. Breeding of seabirds is threatened by human activities, and the numbers of pelicaniforms (boobies and frigate birds) are declining. In view of this, some Indian Ocean states have established reserves for seabird colonies. Major seabird nesting islands in Indonesia, Malaysia, the Philippines, Tanzania and Saudi Arabia are protected.

At least half of the total animal protein comes from the ocean to feed the growing population. Traditionally, fisheries in the Indian Ocean have been small-scale and concentrated in coastal areas. India, Pakistan, and Thailand have recently developed large industrial fisheries but even today, small fisheries account for more than 90% of the total fish catch of about 4 million metric tons.

Foreign vessels are legally allowed into the exclusive economic zone of many Indian Ocean states to fish tuna under bilateral agreements or through joint venture projects, but with due thought for conservation. Distant water vessels operating in the Indian Ocean now have to take on board observers from the coastal states, and the vessels can catch only a certain amount of fish.

3.33
Desalination of Seawater

In late 1993, the total supply of freshwater from desalination plants from seawater and brackish water reached 18 700 000 m^3 per day worldwide (Goto 1994). This is equivalent to the daily water demands of 62 300 000 people, assuming that 300 l are used per person per day. (Actual consumption varies from region to region depending on living standards, habits and customs.) Many of these desalination plants are located in the oil-rich countries of the Middle East: 27% in Saudi Arabia, 11% in the United Arab Emirates (UAE), 8% in Kuwait. Advanced countries with water shortage problems, e.g. the USA, operate only about 15% of these plants. Three main methods used for desalting of seawater are distillation, reverse osmosis and electrodialysis in combination with reverse osmosis. This last method is applicable where brackish water runs underground. In reverse osmosis, a semipermeable membrane is used to produce freshwater out of seawater. Under 50 to 60 kg cm^{-2} of pressure, freshwater permeates this membrane while salt does not. In electrodialysis, freshwater is made by applying an electric current to seawater in which two special membranes, cation (positive ion) exchange membranes and anion (negative ion) exchange membranes, are lined up alternately.

It usually costs 1–2 US$ to produce 1 m^3 of freshwater in a desalination plant with a capacity of 4000 m^3 per day. In the Middle East countries, where oil is inexpensive and is used to generate electricity rather cheaply, the distillation process is most frequently used since it can utilize low pressure steam from steam-power stations. The reverse osmosis process is more popular in the rest of the world. The desalination of brackish water is, on the whole, less expensive than that of seawater (Goto 1994).

References

Arrhenius E (1992) Protecting tropical and subtropical coastal waters: a resource for future generations. Ambio 21:488–90

ASTM (1993) American society for testing and materials. ASTM standards on aquatic toxicology and hazard evaluation. ASTM, Philadelphia

Ayers GP, Gras JI (1991) Seasonal relationship between cloud condensation nuclei and aerosol methanesulphonate in marine air. Nature 353:834–835

Barbault R (1995) Biodiversity: stakes and opportunities. Nature and Resources 31:3

Berthelot B, Deschamps P-Y (1994) Evaluation of bio-optical algorithms to remotely sense marine primary production from space. J Geophys Res 99:7979–7989

Bobra MA (1992) A study of water-in-oil emulsification. Environment Canada Manuscript Report EE-132

Bragg JR, Prince RC, Harner EJ, Atlas RM (1994) Effectiveness of bioremediation for the Exxon Valdez oil spill. Nature 368:413–418

Brown HM, Nicholson P, Goodman RH, Berry BA, Hughes BR (1993) Novel concepts for the containment of oil in flowing water. Proc. 16th Arctic and Marine Oil Spill Program. Tech Seminar, Calgary (Canada), 7–9 June, 1993, Vol I. Environment Canada, Ottawa, pp 485–496

Bruland KW, Donant JR, Hutchins DS (1991) Interactive influences of bioactive trace metals on biological production in oceanic waters. Limnol Oceanogr 36:1742–1755

Cappellen PV, Ingall ED (1996) Redox stabilization of the atmosphere and ocean by phosphorus-limited marine productivity. Science 271:493–496

Carter L (1995) What goes down must come up: recycling in the deep ocean. Water and Atmosphere (NIWA) 3:23

Charlson RJ, Lovelock JE, Andreae MO, Warren SG (1987) Oceanic phytoplankton, atmospheric sulfur, cloud albedo and climate. Nature 326:655–661

Daling P, Brandvik PF (1990) Characterization of crude oils for environmental purposes. Oil and Chemical Pollution 7:199–224

Davies-Colley R, Kirk J (1995) Ocean productivity and water color. Water and Atmosphere (NIWA) 3:20–22

EPA (1995) The probability of sea level rise (EPA 230-R-95-008), U.S. Environmental Protection Agency, Washington, DC

Etkin DS (1990) Cold water oil spills. Cutter Information Corp, Arlington, Mass

Fingas M, Fieldhouse B, Bier I, Conrad D, Tennyson E (June 7–9, 1993) Development of a test for water-in-oil emulsion breakers. Proceedings of the 16th Arctic and Marine Oil Spill Program (AMOP) Technical Seminar, Calgary, Alberta, Canada, Vol. 2, pp 909–924

Foght JM, Westlake DWS (1992) Bioremediation of oil spills. Spill Technol. Newsletter 17 (3):1–10

Fu R, DelGenio AD, Rossow WB, Liu WT (1992) Cirrus-cloud thermostat for tropical sea surface tested using satellite data. Nature 358:394–397

GESAMP (IMO-FAO/UNESCO/WMO/WHO/IAEA/UN/UNEP) (1995) Joint group of experts on the scientific aspects of marine environmental protection. 1995: biological indicators and their use in the measurement of the condition of the marine environment, Report No. 55

Glasby GP, Roonwal GS (1995) Marine pollution in India: an emerging problem. Current Science 68:495–497

Goto T (1994) Freshwater from the sea. UNEP-IETC Newsletter (Fall, 1994), p 4

Gould SJ, Eldrcdgc N (1993) Punctuated equilibrium comes of age. Nature 366:223–227

Graneli E, Harldsson C (1993) Can increased leaching of trace metals from acidified areas influence phytoplankton growth in coastal waters? Ambio 22:308–311

Gras JL (1995) CN, CCN and particle size in southern ocean air at Cape Grim. Atmospheric Research 35:233–251

Gray JS (1992) Eutrophication in the sea. In: Columbo G (ed) Marine eutrophication and population dynamics. Proc 25th European Mar Biol Symp Olsen and Olsen, Denmark, pp 3–15

Harvey M, Sturrock G, Swan H (1996) Can marine plankton affect clouds? NIWA Water and Atmosphere (New Zealand) 4:9–12

Hollibaugh JT (1980) Amino acid fluxes in marine plankton communities contained in CEPEX bags. In: Freeland HJ, Farmer DM, Levings CD (eds) Fjord oceanography. Plenum Press, New York, pp 439–445

Hong X, Raman S (1996) Coupled interactions of organized deep convection over the tropical western Pacific. In: Proc 5th Atmospheric Radiation Measurement Science Team Meeting, March 19–23, 1995. San Diego, California, Published by US Dept of Energy, Washington, DC, pp 139–141

Huggett RJ, Kimerle RA, Mehrle PM, Bergman HL (1992) Biomarkers: biochemical, physiological and histological markers of anthropogenic stress. Lewis Publishers, Boca Raton, Florida

Humphrey B (1993) Persistence of oil in subtidal sediments. Proc 16th Arctic and Marine Oil Spill Program Tech-Seminar, Calgary, Canada

IMO (1993) Report No 51 of the 23rd Session, London, 19–23 April, 1993. International Maritime Organization, London

IPCC (1996) Climate Change 1995: The science of climate change. Contribution of WG1 to the Second Assessment Report of the Intergovernmental Panel on Climate Change. Houghton JT et al (eds) Cambridge Univ Press, Cambridge

Jackson JBC (1994) Constancy and change of life in the sea. Phil Trans R Soc Lond 344 B:55–60

Jernelov A, Ramel C (1994) Mercury in the environment. Ambio 23:166

Karbe L (1992) Toxicity of surface microlayer, subsurface water and sediment elutriates from the German bight: summary and conclusions. Mar Ecol Progr Ser 91:197–201

Karl DM, Tilbrook BD (1994) Production and transport of methane in oceanic particulate organic matter. Nature 368:732–734

Kirk JTO (1994) Light and photosynthesis in aquatic ecosystems. 2nd edition. Cambridge Univ Press, Cambridge

Kolber ZS, Barber RT, Coale KH, Fitzwater SE, Greene RM Johnson KS, Lindley S, Falkowski PG (1994) Iron limitation of phytoplankton photosynthesis in the equatorial Pacific Ocean. Nature 371:145–148

Koski RA (1995) The making of metal deposits. Nature 377:679–680

Levin SA (1993) Concepts of scale at the local level. In: Ehleringer JR, Field CB (eds) Scaling physiological processes: leaf to globe. Academic Press, New York, pp 7–19

Levinton J (1988) Genetics, paleontology and macroevolution. Cambridge Univ. Press, Cambridge

MacDonald IR (1998) Natural oil spills. Sci Am 279, Nov. 30-35

Martin JH, Coale KH, Johnson KS, Fitzwater SE, Gorton RM, Tanner SJ, Hunter CN, Elrod VA, Nowicki JL, Coley TL, Barber RT, Lindley S, Watson AJ, Scoy KV, Law CS, Liddicoat MI. Ling R, Stanton T, Stockel J, Collins C, Anderson A, Bidigare R, Ondrusek M, Latasa M, Millero FJ, Lee K, Yao W, Zhang JZ, Frienderich G, Sakamoto C, Chavez F, Buck K, Kolber Z, Greene R, Ralkowski P, Chisholm SW, Hoge F, Swift R, Yungel J, Turner S, Nightingale P, Hatton A, Liss P, Tindale NW (1994) Testing the iron hypothesis in ecosystems of the equatorial Pacific Ocean. Nature :123–129

Martin JH, Fitzwater SE (1988) Iron deficiency limits phytoplankton growth in the northeast Pacific. Ambio 17:341–43

Martin JH, Gordon RM, Fitzwater SE (1990) Iron in Antarctic waters. Science 345:56–158

McCarty LS, Mackay D (1993) Enhancing ecotoxicological modelling and assessment: body residues and modes of toxic action. Environ Sci Technol 27:1719–1728

Mullin MM (1993) Biological oceanography. Tree 9:452

Olsgard F, Hasle JR (1993) Impact of waste from titanium mining on benthic fauna. J Exp Mar Biol Ecol 172:185–213

Overpeck JT, Webb RS, Webb T (1992) Mapping eastern North American vegetation change of the past. Geology 20:1071–74

Owens NJP, Galloway JN, Duce RA (1992) Episodic atmospheric nitrogen deposition to oligotrophic oceans. Nature 357:397–399

Paerl HW (1985) Enhancement of marine primary production by nitrogen-enriched acid rain. Nature 315:747–749

Piazena H, Häder D-P (1995) Vertical distribution of phytoplankton in coastal waters and its detection by backscattering measurements. Photochem Photobiol 62:1027–1034

Piazena H, Häder D-P (1997) Penetration of solar UV and PAR into different waters of the Baltic Sea and remote sensing of phytoplankton. In: Häder D-P (ed) The effects of ozone depletion on aquatic ecosystems. Ch. 5. Environmental Intelligence Unit, Academic Press and RG Landes Comp., Austin, pp 45–96

Ramanathan V, Collins W (1991) Thermodynamic regulation of ocean warming by cirrus clouds deduced from observations of the 1987 El Niño. Nature 351:27–32

Sheldon PR (1993) Making sense of microevolutionary patterns. In: Evolutionary patterns and processes. Less DR, Edwards D (eds) (Linn Soc Symp Vol 14), Academic Press, London pp 19–31

Stanley SM, Yang X (1987) Approximate evolutionary stasis for bivalve morphology over millions of years: a multivariate, multilineage study. Paleobiology 13:113–139

Stephens G, Slingo T (1992) An air-conditioned greenhouse. Nature 358:369–370

Sullivan CW, Arrigo KR, McClain CR, Comiso JC, Firestone J (1993) Distributions of phytoplankton blooms in the southern ocean. Science 262:1832–1837

Uitto JI (29 Oct 1993) Management of marine areas for sustainable development: agenda 21 and the role of the UN university. No 4. EMECS Newsletter, Hyogo, Japan,

Underwood AJ (1993) The mechanics of spatially replicated sampling programmes to detect environmental impacts in a variable world. Aust J Ecol 18:99–116

UNEP (1993) The impact of climate change. UNEP/GEMS Environment Library No 10. Nairobi

Vincent A, Clarke A (1995) Diversity in the marine environment. TREE 10:55–56

Waldichuk M (1986) Biological availability of metals as a factor in monitoring marine sediments for pollution. pp. 30–61. In: Yu A, Israel AV, Tsyban AS, Sarkisyan SP, Basinova OE, Geroimenkova, Mamayev VO (eds) Integrated global ocean monitoring. Proc I Internat Sympos Vol 2, Leningrad Gidrometeoro Rizdat, USSR, p 442

Waldichuk M (1990) The contamination of seas. Proc. 4th Conf. on Toxic Substances. Montreal April 90. Environment Canada, pp 1–33

Walton WD, Evans DD, McGrattan KB, Baum HR, Twilley WH, Madrzykowski D, Putorti AD, Rehm RG, Koseki H, Tennyson EJ (1993) In situ burning of oil spills: mesoscale experiments and analysis. In: Proc 16th Arctic and Marine Oil Spill Program. Tech Seminar, Calgary (Canada), 7–9 June, 1993, Vol II Environment Canada, Ottawa, pp 679–734

Wallace DWR (1995) Monitoring global ocean carbon inventories. Ocean Observing System Development Panel, Texas A & M Univ, College Station, Texas

Warwick RM (1988) Analysis of community attributes of the macrobenthos of Frierfjord/Langesundfjord at taxonomic levels higher than species. Mar Ecol Progr Ser 46:167–170

Warwick RM, Clarke KR (1991) A comparison of some methods of analysing changes in benthic community structure. J Mar Biol Ass UK 71:225–244

Watson AJ, Law CS, Scoy KV, Millero FJ, Yao W, Friederich GE, Liddicoat MI, Wanninkhof RH, Barber RT, Coale KH (1994) Minimal effect of iron fertilization on sea-surface carbon dioxide concentrations. Nature 371:143–145

Webster PJ, Clayson N, Curry JA (1996) Clouds, radiation, and the diurnal cycle of sea surface temperature in the tropical western Pacific. In: Proc 5th Atmospheric Radiation measurement Science Team Meeting, March 19–23, 1995, San Diego, California, Published by US Dept of Energy, Washington, DC, pp 349–353

Wickremeratne S (1991) The law of the sea and the Indian Ocean. UNEP, Nairobi

Wilkinson M (1995) The intertidal zone. Water and atmosphere. NIWA (NewZealand) 3:17–19

Zhang J (1994) Atmospheric wet deposition of nutrient elements: correlation with harmful biological blooms in northwest Pacific coastal zones. Ambio 23:464–468

4 Ozone Changes

4.1
The Ozone Shield

Between about 15 and 50 km above our heads lies a unique, lifesaving layer of poison. If this layer had not developed, Earth would have been a small, unremarkable sphere, tucked away in a far corner of one among billions of galaxies – peopled by no more, at best, than the most primitive underwater life forms. If it were to disappear, the sun's ultraviolet radiation would sterilize the surface of the globe, destroying all terrestrial and much aquatic life. This layer is made of ozone (Fig. 4.1) which is a poison; any animal that inhaled more than a trace of it would die. Near the Earth's surface, ozone is an increasingly troublesome pollutant, a constituent of photochemical smog and of the cocktail of pollutants popularly known as acid rain (UNEP 1989). But safely up in the stratosphere (15 to 50 km above the Earth's surface) ozone is as important to life as oxygen itself. If concentrated and compressed under atmospheric pressure at the bottom of the atmosphere, the whole ozone layer would be only 3–5 mm thick. This thin layer of ozone in the stratosphere is at its thickest between about 20–40 km up.

Normally, the overall concentration of stratospheric ozone remains fairly constant. High above the stratosphere, the density of gases is so low that oxygen atoms rarely come across other molecules to collide with and consequently not much ozone forms there. Below the ozone layer, too little short wavelength solar radiation penetrates to allow any significant amounts of ozone to form. Thus, most of the ozone is found in a stratospheric layer at altitudes from about 12 to 45 km (see Fig. 4.2).

Near the ground in the troposphere (0–15 km), ozone formed through several chemical reactions, involving hydrocarbons and nitrogen oxide emissions from motor vehicles and industries, constitutes a potent greenhouse gas. It also has adverse impacts on human health at high concentrations. Thus, ozone plays two remarkably different roles in global environmental change, viz., (1) in the stratosphere as a shield against UV-B radiation and (2) closer to the ground as a greenhouse gas and a health hazard.

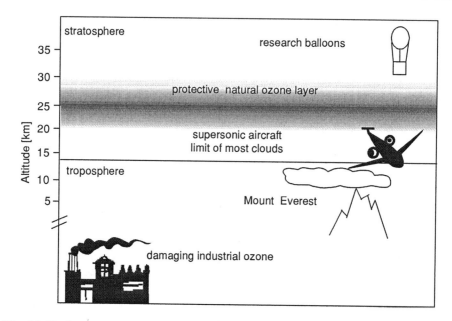

Fig. 4.1. Vertical distribution of O_3 in the atmosphere

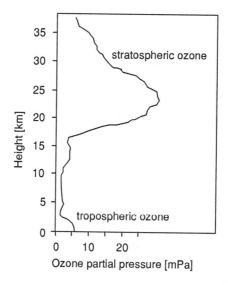

Fig. 4.2. Ozone levels at different altitudes in the troposphere and the stratosphere

Ozone absorbs short wavelength ultraviolet solar radiation; it protects plants, animals, humans and other life forms from much of this harmful radiation. Stratospheric ozone totally prevents the shorter and lethal wavelengths of UV radiation (UV-C, less than 280 nm) from reaching the Earth's surface. It also filters out up to about 90% of the mid wavelengths (280–315 nm, the so-called UV-B) which are less damaging than the UV-C. The ozone layer neither affects the transmission of the longer wavelengths of UV (UV-A, 315–400 nm) through the atmosphere nor the visible range (400–700 nm).

4.2
Ozone Formation

The first photochemical theory of stratospheric ozone formation, involving four oxygen-only reactions, was proposed by Chapman (1930a,b). He proposed initiation by the photolysis of O_2 in the region of the Schumann–Runge bands at $\lambda < 200$ nm. The resulting oxygen atoms form ozone by termolecular combination with molecular oxygen. In turn, ozone may be removed by photolysis in the Hartley continuum between 200 and 350 nm or by reaction with oxygen atoms. The basic Chapman mechanism is as follows:

$$O_2 + h\nu \rightarrow O + O, \tag{4.1}$$

$$O + O_2 + M \rightarrow O_3 + M, \tag{4.2}$$

$$O_3 + h\nu \rightarrow O + O_2, \tag{4.3}$$

$$O + O_3 \rightarrow 2\ O_2. \tag{4.4}$$

By considering the intensity and spectral distribution profile of sunlight, the concentrations of O_2 and N_2 (M being an arbitrary third body), the absorption cross sections and photolysis quantum yields of O_2 and O_3, and the rate coefficients of termolecular ozone formation and bimolecular oxygen atom attack on O_3, one may indeed predict the altitude and concentration of the ozone layer, but such calculations lead only to semi-quantitative agreement with the measured ozone profiles (Atkinson et al. 1989).

Rate constant measurements in the 1950s showed that reaction (4.4) was in fact much slower than was believed at the time of Chapman's theory, which resulted in the ozone concentrations calculated from this reaction sequence being higher than that observed in the stratosphere. This led to the postulation by Hunt (1966) that cyclic reactions of HO_x radicals (H, HO, HO_2), originally expounded by Bates and Nicolet

(1950), were responsible for some of the "missing" ozone removal arising from the slow rate of reaction (4.4). For example,

$$H + O_3 \rightarrow HO + O_2, \qquad (4.5)$$

$$HO + O \rightarrow H + O_2, \qquad (4.6)$$

Net: $O_3 + O + 2 O_2$.

The next developments in the photochemical theory of ozone took place in the early 1970s when Crutzen (1970) proposed the catalytic cycle involving NO_x and Johnston (1971) suggested that nitrogen oxides from the exhaust of supersonic aircraft flying in the stratosphere deplete the ozone layer.

$$NO + O_3 \rightarrow NO_2 + \qquad (4.7)$$

$$NO_2 + O \rightarrow NO + O_2, \qquad (4.8)$$

Net: $O_3 + O = 2 O_2$.

4.3
Miracle Substances

For half a century the chlorofluorocarbons (CFCs) that are doing most to damage the ozone layer were accepted as extremely useful substances, uniquely useful to both industry and consumers and harmless to human beings and the environment alike. CFCs are totally inert, fully stable and unreactive. They are neither flammable nor poisonous. They can be produced cheaply and stored very easily. They do not damage the land, sea or air and they neither react with other substances in the biosphere nor dissolve to pollute the rain. All these properties have contributed more and more to their immense popularity and utility. Invented more or less by accident in 1928, they were first developed as the working fluid for refrigerators. Since 1950, they have been used as propellants in aerosol cans. The computer revolution proved their usefulness as solvents, because they clean delicate circuitry without damaging its plastic mountings, and the fast food revolution enlisted them to blow up the foam for polystyrene cups and hamburger cartons. About 30% of world CFC production is used in refrigerators, freezers and air conditioners, about 25% in spray cans (this percentage has decreased substantially since), another 25% in blowing foams for various uses from buildings and cars to fast food containers, and the remaining 20% for cleaning and other purposes (UNEP 1989, 1994).

Their very stability, so useful on Earth, enables them to attack the ozone layer. Unchanged, they slowly drift up to the stratosphere, where intense UV-C radiation (wavelengths shorter than 280 nm, chiefly about 240–280 nm) splits their chemical bonds, releasing chlorine which strips an atom from the ozone molecule, turning it into ordinary oxygen. The chlorine acts as a catalyst and hence can go on repeating the process. In this way, every CFC molecule destroys thousands of molecules of ozone.

Some halons, related substances primarily used as fire extinguishers, are even more damaging, destroying ozone up to ten times as effectively as the most destructive CFCs. Concentrations of halons – though very small – are doubling in the atmosphere about every 5 years. The most damaging CFCs are also increasing rapidly; concentrations of CFC-11 and CFC-12 (the most widespread CFCs) double every 17 years and concentrations of CFC-113 every 6 years (UNEP 1989). Most of the these dangerous chemicals are quite long-lived. CFC-11 lasts for an average of 75 years in the atmosphere, CFC 12 for some 110 years, CFC-113 for 90 years and Halon-1301 for an average 110 years. This gives them enough time to drift up into the stratosphere and stay there to deplete ozone. Some CFCs have shorter lifetimes, and cause much less damage. Carbon tetrachloride, also used in fire-fighting, is slightly more destructive than the most damaging CFCs. Methyl chloroform, largely used for cleaning metal, is less virulent, but poses a threat because its use is rising. Other substances are less significant. Nitrous oxides, released by nitrogenous fertilizers and by burning coal and oil, are also long-lived and can destroy ozone, but only a tiny proportion reaches the stratosphere. Supersonic aircraft and space shuttles release nitrogen oxides and chlorine, respectively, into the stratosphere; but there are so few flights that these have little impact. Figures 4.3 and 4.4 illustrate the 1986 global consumption of chlorofluorocarbons and halons. Since 1989 there has been a downward trend in CFC production (see Fig. 4.5) largely because of efforts to limit emissions of fully halogenated CFCs.

Fig. 4.3. 1986 global consumption of CFCs, by use and by region

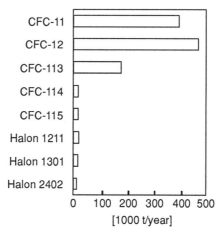

Fig. 4.4. Estimated world consumption of main CFCs and halons

Table 4.1 shows the extent to which the CFC global market in 1991 has changed in comparison to the 1986 market. Chlorofluorocarbons (CFCs) and other man-made chemicals have already created a hole in the ozone layer over Antarctica. They have also been clearly implicated in limited ozone losses over the Arctic. And ozone has been disappearing over mid to high latitudes for more than a decade. However, there may still be worse to come. The possibility of appearance of an ozone hole over the Arctic exists and it could slide south to affect densely populated areas of Europe. Because ozone concentrations are normally relatively high over the Arctic, even extensive ozone destruction is unlikely to create a "hole" there.

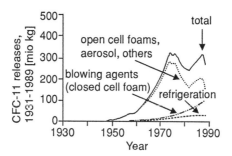

Fig. 4.5. Trends in the releases of CFC-11 since 1930 (Boden et al. 1992)

Ozone losses are also expected in other parts of the world to be more rapid than had been predicted. It appears that the 9% decline in ozone at mid-latitudes in the summer can cause about a 12% increase in UV-B. These forecasts have been made for the upper atmosphere (see Kerr 1992).

4.4
Catalytic Ozone Depletion

A major surprise came in 1985 when large springtime depletions of ozone over Antarctica - the "Ozone Hole" – were reported. Intensive studies in the field and in the laboratory revealed the occurrence of novel chemical processes in the lower stratosphere. These included heterogeneous reactions occurring on the surface of polar stratospheric clouds which form by condensation of nitric acid and water. These reactions lead to conversion of most of the chlorine to catalytically active forms. Certain man-made chemicals, such as chlorofluorocarbons (CFC), destroy the ozone in the stratosphere. During the last decade there has been about an 8% loss in the ozone layer in winter in mid and high northern and southern latitudes, and up to about a 60% loss in Antarctica during the southern spring. (There has apparently been no or little loss of ozone in tropical or equatorial areas.) Every spring, up to 95% of stratospheric ozone over Antarctica is destroyed at a height of about 12–25 km above the Earth's surface – creating the so-called "ozone hole". Throughout the entire stratosphere (up to 50 km) about one-half of the ozone layer is being destroyed each spring. Man-made chlorine and bromine compounds appear to be responsible for these ozone losses. Under certain conditions in the Antarctic, one molecule of catalytic chlorine monoxide can destroy thousands of ozone molecules.

Although it was suspected as early as the 1960s that human activities were damaging the ozone layer, the issue was not taken seriously until Molina and Rowland (1974) hypothesized that chlorine atoms, produced by photochemical degradation of chlorofluorocarbons (CFCs) in the upper stratosphere, may lead to large depletions of ozone via the catalytic cycle:

$$Cl + O_3 \rightarrow ClO + \qquad\qquad\qquad (4.9)$$

$$ClO + O \rightarrow Cl + \qquad\qquad\qquad (4.10)$$

Net: $O_3 + O = 2\,O_2$.

Table 4.1. The CFC global 1991 market compared with the 1986 market

Use	Percent decline
Propellants	58
Cleaning agents	41
Blowing agents	37
Polyurethane	33
Phenolic	65
Extruded polystyrene	
Sheets	90
Boards	32
Polyolefin	31
Refrigerants	7
Total market	40

Chlorofluorocarbons were rapidly accumulating in the atmosphere and because of their long lifetimes, any effects of these molecules would be impossible to reverse on effective time scales. This development has focussed on chlorine as the most serious threat to the ozone layer, and has resulted in much effort to replace the long-lived CFCs by molecules with similar properties, but which are more readily removed from the atmosphere, for example by reaction with HO radicals in the troposphere (Atkinson et al. 1989).

The stratosphere's haze of tiny particles sets the stage for ozone destruction by catalyzing chemical reactions that, directly and indirectly, help the conversion of the chlorine in CFCs into its ozone-destroying form, chlorine monoxide. Until now, only the stratospheric ice particles that form in polar regions had been definitively implicated in ozone destruction. But recent results show that particles elsewhere are fostering similar ozone-destroying chemistry worldwide. They include the natural aerosols of sulfuric acid in mid-latitudes and perhaps ice particles in the tropics also. The finding that particles are engaged in ozone destruction outside the polar regions has been known for some time to atmospheric chemists. But the efficiency with which chlorine monoxide is now forming in mid-latitudes is indeed alarming. With cold winter temperature accelerating the chemistry, summertime levels of 0.025 parts per billion (ppb) had quadrupled to 0.100 ppb by December 1991.

One cause for the jump in chlorine monoxide is obvious. Nitrogen oxide protects ozone by tying up chlorine in a harmless form. A reaction catalyzed by sulfuric acid aerosols immobilizes the nitrogen oxide, blocking its protective role. Certain aerosols can lock up large amounts of the protective compound. The aerosols are also weakening another crucial component of the ozone defenses, a reaction between methane and chlorine that forms hydrogen chloride. That reaction locks up chlorine that might otherwise form destructive chlorine monoxide. In 1991, less hydrogen chloride could be detected than aerosol-free air should produce. Then there is strong and surprising ozone-depleting chemistry in the tropics. Flying south out of Bangor, Maine, an ER-2 flight encountered totally unexpected thin sheets of air particularly rich in chlorine monoxide as far south as the latitude of Cuba. The sheets might conceivably have spun off the chlorine monoxide rich vortex of swirling winds over the Arctic. But it appears far more likely that the sheets originated in the tropics. Indeed, the most dramatic example was found near the Caribbean, not near the Arctic vortex (Kerr 1992). Thus it appears that, like the polar regions, the tropics may be a center for intensified production of chlorine monoxide. The crucial factor, as at the poles, might be the presence of ice crystals. That may not be so far-fetched; when we consider that because of the cooling effect of the exceptionally strong updrafts in the tropics, the lower stratosphere there is the second coldest region on Earth (the coldest being the stratosphere over the poles). The detection of chlorine monoxide-rich gases near the tropics may well turn out to be the tip of an iceberg. It could be an early warning of more troubles to come. All that is needed now for some serious ozone destruction at northern latitudes is more sunlight, which turns on the catalytic destruction of ozone by chlorine monoxide.

Most of the chemicals released into the atmosphere become converted to other forms or are completely removed from the atmosphere within a few years because these molecules react with the major oxidants in the atmosphere or are photolyzed at wavelengths longer than 190 nm. In contrast, some chemicals of very low reactivities do not react with the oxidants in the stratosphere and the troposphere. If an industrially produced chemical has an atmospheric lifetime of a few thousand years or more, it accumulates in the atmosphere for very long periods. Best examples are fully fluorinated organic compounds that are chemically very inert and unusually long-lived atmospheric constituents. Even when some of the fluorine is replaced by chlorine, the stability remains quite high.

The atmospheric lifetime of each of these species, a measure of its accumulation tendency, is of particular importance because it is used in the calculation of the global warming potential (GWP), which depends on the infrared absorption spectrum and the lifetime of atmospheric gases. The GWP of CF_4, for example, is as high as that for chlorofluorocarbon-11 (CFC-11 or $CFCl_3$; Ravishankara et al. 1993).

The simplest molecule in this class, CF_4, is also a by-product of aluminum production. CF_4, C_2F_6, and SF_6 have been found to have atmospheric lifetimes of hundreds of years. All these molecules are solely anthropogenic; CF_4 is nearly inert in the stratosphere and troposphere and has a lifetime of over 10 000 years.

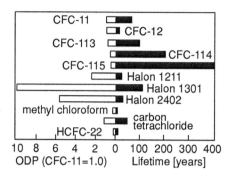

Fig. 4.6. Ozone depletion potentials (ODP) and lifetimes of halocarbons, assuming the ODP of CFC-11 = 1.0

The atmospheric lifetime of the fluorinated gases CF_4, C_2F_6, $c\text{-}C_4F_8$, $(CF_3)_2$, $c\text{-}C_4F_6$, C_5F_{12}, C_6F_{14}, C_2F_5Cl, $C_2F_4Cl_2$, CF_3Cl, and SF_6 are of serious concern because of the effects that these long-lived compounds acting as greenhouse gases can have on global climate. Figure 4.6 compares the lifetime in years and the ozone depleting potential of some halocarbons.

There has also been some concern about the growing use of halons (bromochlorofluorocarbons) and the high catalytic efficiency for ozone destruction by BrO_x. There have been several studies of the reaction between BrO and ClO, which assumes added importance because of its role in polar ozone depletion. The kinetics and temperature dependence of the three product channels of this reaction are now well determined:

$$BrO + ClO \rightarrow Br + ClOO, \tag{4.11}$$

$$BrO + ClO \rightarrow Br + OclO, \tag{4.12}$$

$$BrO + ClO \rightarrow BrCl + O_2. \tag{4.13}$$

It has now been established that the partitioning of chlorine in the polar springtime stratosphere is highly perturbed and that this is due to heterogeneous reactions such as the reaction given below. These reactions occur on the surface of polar stratospheric clouds resulting from the condensation of water and nitric acid at the low prevailing temperatures (<190 K). These processes also lead to considerably reduced concentrations of NO_x in the polar winter stratosphere.

$$ClONO_2 + HCl \rightarrow Cl_2 + HNO_3. \tag{4.14}$$

The subsequent photochemistry in the polar spring gives rise to ozone depletion by a completely different set of reactions involving the ClO dimer, Cl_2O_2 (Molina and Molina 1987):

$$Cl_2 + hv \rightarrow 2 \; Cl, \tag{4.15}$$

followed by:

$$Cl + O_3 \rightarrow ClO + O_2, \tag{4.16}$$

$$ClO + ClO + M \rightarrow Cl_2O_2 + \tag{4.17}$$

$$Cl_2O_2 + hv \rightarrow Cl + ClOO, \tag{4.18}$$

$$\underline{ClOO + M \rightarrow Cl + O_2 + M,} \tag{4.19}$$

Net: $2 \; O_3 = 3 \; O_2$,

4.5
The Family of Organic Reactions

Tropospheric photochemistry has developed rapidly since the realization that relatively large steady state concentrations of HO and HO_2 radicals are present in the sunlit troposphere. Because of the high reactivity of HO radicals, this provides an important sink for many trace gases. Further developments in tropospheric chemistry have been in the area of photochemical oxidant ("smog") formation, in formulating the mechanisms for hydrocarbon oxidation in the presence of NO_x, leading to ozone production in polluted air. The column density of O_3 through the troposphere is a small fraction (~10%) of that in the stratosphere, with the presence of O_3 in the troposphere being attributed to a combination of downward transport from the stratosphere and in situ photochemical production being balanced by deposition and destruction at the Earth's surface (Logan 1985).

However, despite its small contribution to the total atmospheric column density of O_3, tropospheric O_3 plays a crucial role in the chemistry of the atmosphere through the production of the $HO^\bullet$ radical.

The formation of $HO^\bullet$ radicals in the troposphere leads to the degradation of organic chemicals introduced into the troposphere, thus limiting their transport to the stratosphere (where they photolyze and/or react with $HO^\bullet$ radicals and ^1O atoms), and acts essentially as a low temperature combustion process. This tropospheric removal of organic compounds by the $HO^\bullet$ radical (or the lack of this process) is central to the issue of the chlorofluorocarbons, the depletion of stratospheric ozone and the current search for alternative hydrofluorocarbons and/or hydrochlorofluorocarbons which might react sufficiently rapidly with the HO$^\bullet$ radical so as to be largely degraded in the troposphere (Atkinson 1989). Figure 4.7 illustrates the roles of the oxidants, O$_3$, OH$^\bullet$ and H$_2$O$_2$ in atmospheric photochemical reactions.

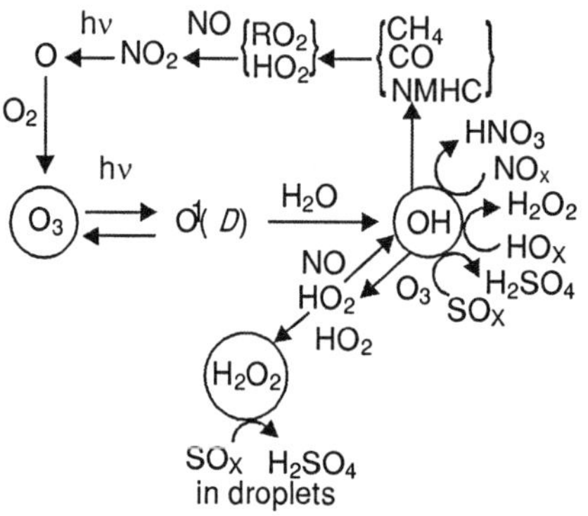

Fig. 4.7. Roles of oxidants in atmospheric photochemical reactions. (After Thompson 1992)

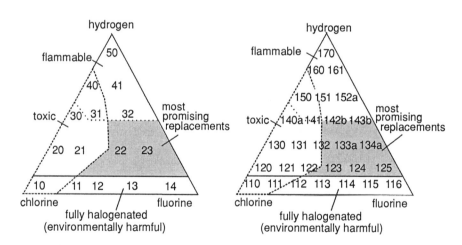

Fig. 4.8. Relation between chemical composition and properties of CFCs, HCFCs and HFCs. (UNEP 1992)

The relationships between chemical composition and properties of CFCs, HCFCs and HFCs are illustrated in Fig. 4.8. Hydrogen increases flammability. An absence of hydrogen increases atmospheric lifetime and ozone depleting potential (ODP). Chlorine increases toxicity. HCFC-22, HCFC-134a, and other substances in the center-right of the triangles are the most suitable substitutes for CFC refrigerants.

4.5.1
Organochlorine Compounds in Polar Regions

Organochlorine (OC) chemicals, including chlorinated pesticides and polychlorobiphenyls, are found at appreciable concentrations in the polar regions, presumably as a result of long-range atmospheric transport. Concentration data in Arctic and Antarctic air, snow, atmospheric deposition, fish and seals, measured by various investigators, have been compiled and interpreted to determine latitudinal and temporal trends. The often surprisingly high concentrations have been explained partly by the temperature dependent partitioning of these low volatility compounds. A process of global fractionation may be occurring in which organic compounds become latitudinally fractionated, "condensing" at different ambient temperatures depending on their volatility. Wania and Mackay (1993) suggested that compounds with vapor pressures in a certain low range can preferentially accumulate in polar regions and possibly exert adverse and toxic effects on the indigenous population and on the Arctic ecosystem. There is a need to control or even ban certain chemicals which have a tendency to fractionate into the polar ecosystems (Wania and Mackay 1993).

It appears that the concentrations of the organochlorine compounds in air throughout the Arctic are quite homogeneous. Absolute concentrations and relative abundance of the different compounds are fairly similar at most sites. The hexachlorocyclohexanes are generally the dominant OC in Arctic air. Hexachlorobenzene is also present and shows little spatial variability.

For the Norwegian Arctic, the main source of OCs appears to be Eurasia. Polluted air masses also contribute most of the pollutant burden to the North American Arctic by crossing the polar ice cap. The predominance of Eurasian sources and the pronounced seasonal variation may be explained by distinctive meteorological patterns which favor the transport of polluted air from Eastern Europe northward during winter. Attempts to pinpoint distinctive sources and pathways for the OCs have been more successful than similar attempts for aerosols and sulfate. Atmospheric residence times of aerosols are generally short compared with hemispheric mixing times; thus, aerosols tend to be heterogeneously distributed in the atmosphere. Dispersion is mainly controlled by the location of the source areas and the general circulation pattern. Although OCs occur in air associated with particles, they are also found in gaseous form and are consequently subject to quite different deposition processes. Precipitation scavenging is not as effective with vapors as it is with particles.

According to Wania and Mackay (1993), the physical-chemical properties of these chemicals, and certain factors characterizing cold environments significantly contribute to the long-term spatial distribution patterns of OC chemicals. The volatility of a compound and the ambient temperature strongly influence the distribution patterns. Vapor pressure and solubility in water play a strong role in influencing or even controlling the environmental fate of chemicals. All are temperature-sensitive. With a fall in temperature, there is a tendency for chemicals present in the atmosphere to "condense" onto soil, water, aerosols, snow and ice. OCs absorbed to aerosol particles are deposited more rapidly than those in the vapor phase due to dry and wet particle deposition.

Other factors also influence OC concentrations in polar ecosystems: reaction rates are slower at lower temperatures. Photolysis is reduced by the low sun angle. Productivity may be lower in aquatic systems resulting in reduced organic carbon deposition and hence reduced scavenging of adsorbed chemicals to sediments (Wania and Mackay 1993). In total, OC behavior in polar ecosystems may be quite different from the relatively well characterized behavior in temperate ecosystems.

The more volatile a compound, the more easily and farther it tends to travel to remote polar regions. A nonvolatile, immobile substance such as benzo-a-pyrene remains in the region of its emission because of rapid deposition. With increasing volatility the range of transport increases and the compound condenses and accumulates at higher latitudes. Highly volatile compounds (e.g. the CFCs) probably do not face sufficiently low environmental temperatures to condense appreciably (Fig. 4.9; Wania and Mackay 1993). Spatially, this phenomenon should cause a shift in the composition of the OCs towards the more volatile compounds in progressively colder regions. Likewise, while the deposition and concentration of the least volatile chlorinated organics should decrease with latitude, the more volatile ones may be expected to have a more uniform or even reversed deposition profile, i.e. higher levels in colder regions.

Differences in volatility can explain part of the difference in abundance of the OCs in various environmental compartments. The dominant compounds in Arctic air, HCHs and HCBs, are also the most volatile, while the less volatile PCBs and DDTs are more important in the biotic samples. Other factors such as degradability, bioaccumulative potential and uptake efficiency also explain the dominance of the latter compounds in biotic samples (Wania and Mackay 1993).

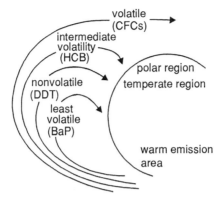

Fig. 4.9. Latitudinal fractionation of organic compounds according to their volatility as they condense at different ambient temperatures. (After Wania and Mackay 1993)

4.6
Methyl Bromide

Fire alters habitats, mobilizes nutrients and drives evolution. It is used in agriculture to clear fields of vegetation for planting or pastures and to burn post-harvest residues. Biomass burning is a form of nonindustrial pollution produced mostly in developing countries. One notable impact of biomass burning is that large amounts of methyl bromide (CH_3Br) are emitted.

Gases released by biomass burning change atmospheric chemistry. Nitrogen oxides (NO and NO_2) and carbon monoxide (CO) photochemically produce ozone in the troposphere downwind from the burning – just as in polluted urban air. Biomass burning also produces CH_3Cl, most of which is destroyed photochemically in the troposphere but some survives the upward journey to the stratosphere where the ozone-destroying Cl atoms are released.

In recent years there has been increasing pressure to declare methyl bromide a controlled ozone-depleting substance. Despite some uncertainties it appears that methyl bromide does deplete ozone. Most of the methyl bromide emitted as a result of human activities comes from fumigation of soils. Some comes from quarantine and commodity fumigation and structural fumigation. In 1990, over 65 000 tonnes of methyl bromide were produced commercially – about 40% in North America, about 30% in Europe, and 20% in Asia (UNEP 1992).

Bromine is, per atom, far more efficient than chlorine in destroying stratospheric ozone, and methyl bromide is the single largest source of stratospheric bromine. The two main previously known sources of this compound are emissions from the ocean and from the use as a soil fumigant in agriculture. Methyl bromide is also emitted in the smoke from various fuels (Mano and Andreae 1994). It is also present in smoke plumes from wildfires in savannas, chaparral, and boreal forest. Global emissions of methyl bromide from biomass burning are estimated to be in the range of 10 to 50 Gg (1 Gg = 10^9 g) per year, which is comparable to the amount produced by ocean emission and pesticide use and represents a major contribution (about 30%) to the stratospheric bromine budget (Mano and Andreae 1994).

Although stratospheric ozone losses observed so far are mostly attributable to Cl from CFC gases, in the Antarctic spring, up to 25% may be attributable to Br from synthetic halon compounds (and CH_3Br, even though stratospheric inorganic Br amounts are less than 1% of inorganic chlorine concentrations, Cicerone 1994). Free Br atoms catalyze stratospheric ozone destruction mainly through the reaction sequence

$$Br + O_3 \rightarrow BrO + O_2, \tag{4.20}$$

$$Cl + O_3 \rightarrow ClO + O_2, \tag{4.21}$$

$$\underline{BrO + ClO + light \rightarrow Br + Cl + O_2,} \tag{4.22}$$

Net: $2\ O_3 + light \rightarrow 3\ O_2$.

The Br atoms reach the stratosphere while bound in relatively stable organic compounds such as Halon-1301 (CF_3Br), Halon-1211 (CF_2BrCl) and CH_3Br. Atmospheric lifetimes of the halons are 65 and 20 years, respectively, so that almost 100% of the halons released reach the stratosphere intact. The 1990 worldwide release rates of halons imply a Br flux into the stratosphere of 7.2 x 10^6 kg yr^{-1}. Methyl bromide is highly reactive and its atmospheric lifetime is about 2 years. However, only a small fraction of CH_3Br emitted at the surface reaches the stratosphere, and there are natural sources as well as anthropogenic ones (Cicerone 1994). There is some probability that the oceans may serve as a buffer in the global CH_3Br system. Any decrease in anthropogenic emissions may be compensated by a flux from the ocean to the atmosphere.

Between 1984 and 1990 the annual use of the compound rose from 42 000 to 66 000 tonnes. The greatest area of use is soil disinfection (80%); agricultural products and materials in store demand 15% and building 5%. The use of the compound (in thousand tonnes, approximately), is distributed over the world approximately as follows: North America 28%, Europe 19%, Asia 15%, Africa 2%, South America 1.5% and Australia 1%.

Since methyl bromide depletes the ozone layer, it is a controlled substance under the Montreal protocol. It is being used as a fumigant for soils, horticultural fumigation, pest control, beekeeping, quarantine, and for treatment of products prior to export. Under the Protocol, quarantine and preshipment use is not restricted but a permit is needed to import methyl bromide. The first option (quarantine) puts responsibility for permits in the hands of the importers/wholesalers, who then distribute the gas to final users. The second option places the permits in the hands of the final users, who would then arrange for importers to supply the gas on their behalf. There is a need to reduce the consumption of methyl bromide substantially in the near future.

In December 1993, the countries of the European Union decided that their use of methyl bromide in 1995 would have to be reduced to 1991 levels, and that it must lie 25% lower than this in 1998. Denmark has decided to discontinue its use entirely by 1998. The USA expects to cease using methyl bromide by 2001. Its use as a soil disinfectant has been banned in the Netherlands since 1992.

Not all the methyl bromide which is used ends up in the atmosphere. In soil disinfection, for instance, it is partially decomposed in the ground, the bromide remaining in the soil. The emission factor is the ratio between the amount of methyl bromide which is emitted to the atmosphere and the amount that is used (van Haasteren 1994). The percentage can vary between 20 and 80% when the compound is used as a disinfectant in greenhouses; when used in the open field, the emission lies between 60 and 80%. The temperature of the soil is particularly important in the decomposition of methyl bromide. If methyl bromide is used for the disinfection of buildings or ships, the emission is 100%. When stored materials and other products are disinfected, the emission will be 80 to 90%. Methyl bromide also comes from certain oceanic processes.

The combustion of biomass also gives rise to the compound. No accurate estimates of these emissions are available. The share of anthropogenically produced methyl bromide, used as a fumigant, has been estimated at 25%. Model calculations reveal that the anthropogenic emissions of methyl bromide have probably been responsible for one twentieth to one tenth of the total ozone loss (UNEP 1992).

In principle, some replacements are already available for several applications of methyl bromide. For quarantine treatment and the disinfection of some stored materials there is, however, either no alternative or only a very limited selection available. There is need to intensify the search for suitable alternatives to methyl bromide (van Haasteren 1994).

Near total depletion of the ozone in surface air is often observed in the Arctic spring, coincident with high atmospheric concentrations of inorganic bromide. Barrie et al. (1988) suggested that the ozone depletion was due to a catalytic cycle involving the radicals Br and BrO; however, these species are rapidly converted to the nonradical species HBr, $HOBr$ and $BrNO_3$, quenching ozone loss. McConnell et al. (1992) proposed that cycling of inorganic bromine between aerosols and the gas phase could maintain sufficiently high levels of Br and BrO to destroy ozone, but they did not specify a mechanism for aerosol-phase production of active bromide species. Other authors have proposed such a mechanism, based on known aqueous-phase chemistry, which rapidly converts HBr, HOBr and $BrNO_3$ back to Br and BrO radicals. This mechanism should be particularly efficient in the presence of the high concentrations of sulfuric acid aerosols observed during ozone depletion events.

The main sinks of methyl bromide are believed to be atmospheric oxidation by hydroxyl radicals and photolysis and uptake by the oceans. CH_3Br also seems to be consumed in soils. Shorter et al. (1995) have shown that upon contact with soils, CH_3Br is rapidly and irreversibly removed to below the levels found in the global atmosphere. The uptake process is bacterially mediated. The global annual soil sink is estimated to be 42 ± 32 Gg; coupled with other removal mechanisms, this suggests an atmospheric lifetime for CH_3Br of less than a year and also a smaller depletion potential than believed up to now (Shorter et al. 1995).

Methyl bromide is the most abundant gaseous bromine species in the atmosphere, with an average atmospheric mixing ratio in the Northern Hemisphere of 10–15 parts per trillion by volume (pptv), with somewhat lower values in the Southern Hemisphere (UNEP 1992). Its important atmospheric sources include: biological production and emission from marine environments, biomass burning, exhaust from cars burning leaded petrol and atmospheric emission of synthetic CH_3Br following its use as a soil, commodity or structural fumigant (UNEP 1992). Soil fumigation accounts for over 80% of synthetic CH_3Br use, and certain soil pests cannot be controlled without using methyl bromide. The principal loss process for atmospheric CH_3Br is its reaction with photochemically produced hydroxyl radicals. The chemical destruction rate of CH_3Br in salt water is well characterized, supporting an ocean uptake lifetime ranging between 3.7 and 2.7 years, with the lower estimates arising from the most recent evaluation. On the land surface, methylation of organic matter is its major chemical sink. CH_3Br is destroyed by reaction with sulfide and chloride anions in anaerobic salt marsh soil. But these chemical sinks seem to saturate at the high concentrations used in soil fumigation, so much of the applied CH_3Br escapes to the atmosphere. Aerobic methanotrophic bacteria probably consume CH_3Br (see Shorter et al. 1995). The regulation of methyl bromide used widely as a fumigant for soil, grain stores and buildings has been a very controversial issue. Annually some 66 000 tonnes of the chemical are released into the atmosphere, but there are scientific uncertainties about the atmospheric chemistry of methyl bromide. By the year 2000, man-made methyl bromide releases could be responsible for about 15% of predicted ozone depletion if production continued unrestricted.

Methyl bromide is a short-lived gas, with an atmospheric lifetime of about 2 years and high ozonedepletion potential. The high potential arises because bromine is about 40 times more efficient in destroying ozone as chlorine on a per molecule basis. There is a strong need to cut production and consumption of methyl bromide by 25% by the turn of the century. Production for quarantine fumigation may be exempted.

There exist several alternatives to the use of methyl bromide. For soil fumigation, there are organic growing practices, soil solarization and integrated pest management strategies. For building fumigation, alternative methods include heat, cold, borates and electricity. Identifying substitutes for commodity fumigation is rather difficult because many countries require methyl bromide fumigation as a part of their quarantine regulations, but there are alternatives as well. They include heat and cold processes and bioremediation. In the Netherlands, methyl bromide has been phased out already and replaced with steam as a fumigant.

Carbon dioxide may possibly be a substitute for methyl bromide for grain fumigation. In Kenya, cotton bags containing maize infected with the adult corn weevil were placed in silos. After the experimental silo was fumigated, the bags were analyzed for live and dead weevils, and the grain was incubated to see if new weevils hatched; they did so only in the control silo (not fumigated with CO_2). The fumigation also killed most of the insects. In contrast to methyl bromide, CO_2 may need to be applied only once, using the same ducting system that is used for methyl bromide. Providing the silos are properly sealed and a CO_2 rich atmosphere is maintained, the insects will not return. Also, if the CO_2 used comes from natural existing sources, it will not lead to the greenhouse effect (Banks 1995).

4.7
The Polar Stratosphere

Recent findings show that chemical perturbations, established in the Arctic stratosphere at the beginning of the winter, are responsible for ozone removal as sunlight returns at the end of winter. These measurements also indicate that stratospheric chemistry at mid-latitudes may be somewhat different from our previous understanding (Rodriguez 1993). Ozone levels are maintained by a delicate balance between production, transport and removal (Fig. 4.10). A decrease in ozone is observed at mid-high latitudes if ozone transport from the tropics is reduced or chemical ozone removal is increased. For example, ozone transport to the polar regions could be curtailed by vortices in atmospheric circulation which appear during the respective winter and spring.

Aircraft forays have established that, contrary to expectations, chemical removal of ozone by chlorine-catalyzed cycles is enhanced in the polar region during winter and spring. The main culprits for this aberrant behavior are seemingly innocuous solid nitric acid and ice clouds that are formed in the polar stratosphere during winter in the extremely cold temperatures. These polar stratospheric clouds (PSCs) have been known to exist for over a century but their role in stratospheric chemistry has only been known recently.

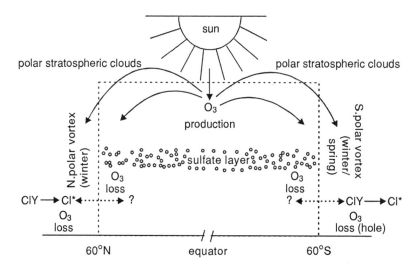

Fig. 4.10. Probing stratospheric ozone. Ozone is produced by sunlight in the tropics and transported to mid to high latitudes, where chemical removal occurs. Removal processes are catalyzed by different chemicals; chlorine has been of concern because of its anthropogenic origin. Heterogeneous reactions on the global sulfate layer or on the larger polar stratospheric clouds, can facilitate conversion of total inorganic chlorine to active forms (Cl), which destroy ozone. This process also needs sunlight and is particularly efficient within the Antarctic polar vortex. (After Rodriguez 1993)

So-called heterogeneous chemical reactions take place on the surface of these particles, strongly shifting the balance between inactive and active forms of chlorine and increasing the efficiency of chlorine in catalytic ozone removal (Rodriguez 1993). Processing of air by a PSC should leave a telltale footprint: enhanced concentrations of chlorine monoxide (ClO), which should persist in Arctic conditions for at least 1 to 2 weeks, even if the culprit PSC has since evaporated. Meteorological analysis can reveal whether the air was cold enough for the likelihood of PSCs forming sometime in the previous 10 days. Recent results have suggested a high correlation between measured concentrations of ClO and conditions for PSC formation, as determined from the temperature history. The seasonal behavior of ClO is also consistent with the meteorological analysis. Enhanced levels of ClO first appear in December, when the stratosphere gets cold enough for PSCs to form. The highest levels of ClO occur in January when the PSC coverage is sufficient for most of the vortex air to have been processed. The level of ClO starts declining in February when warming of the polar vortex leads to the disappearance of PSCs. This timing explains why no Arctic ozone hole has been seen: warming of the Arctic vortex occurs around February, before the arrival of sufficient sunlight for catalytic ozone removal to occur. In contrast, temperatures in Antarctica remain low until at least early to mid-October, which is equivalent to April in the Northern Hemisphere.

Footprints of heterogeneous processing by PSCs are also found in other species. In situ measurements of hydrochloric acid have already been made. This very stable and inactive form of chlorine had been calculated to be the major component of inorganic chlorine in the stratosphere. Reactions of HCl with $ClNO_3$ on PSC-like laboratory surfaces, however, have been shown to rapidly convert these species to active chlorine. There is now strong indication that heterogeneous reactions can occur outside the polar vortices. A ubiquitous layer of supercooled sulfuric acid aerosols, partly of volcanic origin, provides the necessary surfaces for heterogeneous reactions to occur.

The eruption of Mount Pinatubo in June 1991 substantially increased the global aerosol loading. The analysis of Wilson et al. (1993) indicates a footprint of heterogeneous reactions on ClO; however, these enhancements in ClO are not attributed to PSC appearance, but simply to higher concentrations of sulfate aerosols resulting from the spreading volcanic plume.

At sufficiently low temperatures, heterogeneous reactions on sulfate aerosolsmay be as efficient as those on PSCs. Toon et al. (1993) recorded measurements of HCl and $ClNO_3$ integrated column abundances. Reductions in both species were observed, despite the fact that meteorological analysis did not indicate formation of PSCs.

Browell et al. (1993), Salawitch et al. (1993) and Proffitt et al. (1993) have addressed the question of net ozone loss during Arctic winter. These workers used different sets of observations and methodologies to derive ozone losses but got similar results; chemical processes removed about 15 to 25% of the ozone at altitudes near 18 km during January and February, the months with the strongest perturbation in the chlorine chemistry.

Our understanding of polar ozone loss is based mainly on the heterogeneous conversion on particles of polar stratospheric clouds (PSCs) and of inorganic chlorine from its less reactive components hydrochloric acid (HCl) and chlorine nitrate ($ClNO_2$) to photochemically labile Cl_2 and HOCl. Conversion is thought to occur on surfaces of both type I nitric acid trihydrate (NAT) and type II (water ice) PSC particles that form in the lower stratosphere at temperatures at or below about 196 K and 188 K, respectively, when sufficient amounts of HNO_3 and H_2O condense on sulfate aerosol particles. The main reactions are

type I, II PSC

$$HCl + ClONO_2 \rightarrow Cl_2 + HNO_3,$$

(4.23)

type II PSC

$$ClONO_2 + H_2O \rightarrow HOCl + HNO_3, \tag{4.24}$$

type I, II PSC

$$HOCl + HCl \rightarrow Cl_2 + H_2O. \tag{4.25}$$

Simultaneous in situ measurements of hydrochloric acid (HCl) and chlorine monoxide (ClO) made by Webster et al. (1993) in the Arctic winter vortex have shown large HCl losses, of up to 1 part per billion by volume (ppbv), which were correlated with high ClO levels of up to 1.4 ppbv. Air parcel trajectory analysis showed that this conversion of inorganic chlorine occurred at air temperatures of less than 196 + 4 K. High ClO was always accompanied by loss of HCl mixing ratios equal to 1/2 (ClO + 2 Cl_2O_2). These data indicate that the heterogeneous reaction (4.23) on particles of polar stratospheric clouds establishes the chlorine partitioning, which, contrary to earlier notions, begins with an excess of $ClONO_2$, not HCl (Webster et al. 1993).

Toohey et al. (1993) reported in situ measurements of chlorine monoxide (ClO) at mid and high northern latitudes for the period October 1991 to February 1992. As early as mid-December and throughout the winter, significant enhancements of this ozone-destroying radical were noted within the polar vortex shortly after temperatures dropped below 195 K. Decreases in ClO observed in February were consistent with the rapid formation of chlorine nitrate ($ClONO_2$) by recombination of ClO with nitrogen dioxide (NO_2) released photochemically from nitric acid. Outside the vortex, ClO abundances were higher than in previous years as a result of NO_x suppression by heterogeneous reactions on sulfate aerosols enhanced by the eruption of Mount Pinatubo (Toohey et al. 1993).

Chemical reactions occurring on particles cause O_3 loss in the global stratosphere. At most latitudes, sulfuric acid aerosols drive heterogeneous chemistry. At high latitudes, PSCs dominate heterogeneous chemistry because of their large surface area. Sulfuric acid aerosols may also be important in the polar regions because chlorine-containing gases react on them at temperatures higher than those at which PSCs form. But it is very difficult to assess the role of sulfuric acid aerosolsin polar chemistry.

Toon et al. (1993) reported that on 19 January 1992, heterogeneous loss of HNO_3, $ClNO_3$, and HCl was observed in part of the Mount Pinatubo volcanic cloud that had cooled as a result of forced ascent. Portions of the volcanic cloud froze near 191 K. The reaction probability of $ClNO_3$ and the solubility of HNO_3 were close to laboratory measurements on liquid sulfuric acid. The magnitude of the observed loss of HCl suggested that it underwent a heterogeneous reaction. Such reactions could lead to substantial loss of HCl on background sulfuric acid particles and so be important for polar ozone loss (Toon et al. 1993).

Highly resolved aerosol size distributions measured from high-altitude aircraft have been used to describe the effect of the 1991 eruption of Mount Pinatubo on the stratospheric aerosol (Wilson et al. 1993). In some air masses, aerosol mass mixing ratios increased by factors exceeding 30 or more. Increases in aerosol surface area concentration were accompanied by increases in chlorine monoxide at mid-latitudes when confounding factors were controlled. This observation suggests that reactions occurring on the aerosol can increase the fraction of stratospheric chlorine that occurs in ozone-destroying forms (Wilson et al. 1993).

4.8
Polar Clouds and Sulfate Aerosols

Although heterogeneous chlorine activation reactions on polar stratospheric clouds (PSCs) play a key role in polar ozone depletion, the chemicalcomposition and the formation mechanism of the polar clouds is not settled. Koop and Carslaw (1996) suggested how one kind of PSC (type I) forms over and over again throughout the winter. Earlier, nitric acid trihydrate (NAT) was believed to be the main ingredient of type I PSCs. It was supposed that the global sulfate aerosols froze and formed sulfuric acid tetrahydrate (SAT) under the cold polar temperatures and that crystalline SAT then furnished a nucleating surface for NAT. Laboratory studies showed that neither of these two steps occurs readily. Besides, some type I PSCs are not composed of NAT. The current picture of PSC formation is that low-temperature liquid sulfuric acid aerosols absorb significant quantities of HNO_3, forming ternary solutions of H_2SO_4-HNO_3-H_2O. As temperatures fall further, the ternary solutions absorb more HNO_3 and H_2O, forming PSC particles that are essentially binary solutions of HNO_3-H_2O. These liquid PSCs seem to activate chlorine more efficiently than frozen NAT particles.

Although liquid sulfate aerosols do explain the presence of liquid PSCs, crystalline PSCs also exist at temperatures above the ice equilibrium temperature (ice frost point). Presumably, evaporation of crystalline PSCs (either ice or NAT) leaves a frozen sulfate core, probably in the form of SAT (Fig. 4.11). In the atmosphere, SAT remains frozen up to a temperature of 210 to 215 K but this high temperature probably never occurs during the polar winter; hence SAT remains frozen for the entire season. But SAT is not a suitable nucleating surface for PSCs. Thus, until recently, it was inferred that after crystalline PSCs formed once and left behind a frozen core, additional PSC growth would be hindered. PSCs could only form after either the temperature warmed to cause SAT melting or cooled to 189 K, so that ice could condense. Neither of these possibilities materializes so frequently as to explain the observations of PSCs throughout the winter (Tolbert 1996). SAT can melt upon moderate cooling in the atmosphere to form a supercooled ternary solution of particles of H_2SO_4-HNO_3-H_2O (Fig. 4.11), and the presence of HNO_3 in the atmosphere may regenerate liquid sulfate aerosols, allowing the cycle of PSC formation to restart. Adsorption of HNO_3 onto SAT makes the crystal melt at temperatures well below 210 to 215 K measured for SAT in the absence of HNO_3. This melting temperature depends only on the pressures of atmospheric water and nitric acid and occurs at a temperature several degrees above that at which NAT grows on SAT in laboratory experiments (Tolbert 1996). Thus, if SAT does really form in the atmosphere, one may predict that it will melt back to a ternary solution before NAT can directly condense, providing a clue to the mystery of how PSCs continue to form late in the polar winters.

PSCs are thought to consist of various nitric acid hydrates, sulfuric acid aerosols and water ice particles. These clouds tend to form at temperatures below 195 K. Water ice particles form at temperatures near 188 K and are large enough to fall out of the stratosphere on short time scales. The settling of these ice particles reduces both water and nitrogen concentrations in the stratosphere (dehydration and denitrification, respectively), further perturbing the stratospheric photochemical balance. Newman et al. (1993) have characterized the thermal hemisphere winter of 1991–1992 (AASE II) and contrasted this with the corresponding data for the winter of 1988–1989 (AASE I). Temperature analyses showed that nitric acid trihydrates (NAT temperatures below 195 K) should have formed over small regions in early December. Perturbed chemistry was found to be associated with these cold temperatures.The temperatures in the polar vortex warmed beyond NAT temperatures by late January (earlier than normal).

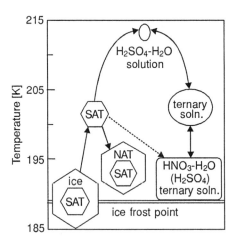

Fig. 4.11. Conversion of cloud particles from crystalline to liquid through the melting of SAT upon cooling (*dashed line*), which then permits PCSs to form again. (After Tolbert 1996)

In situ measurements of chlorine monoxide, bromine monoxide and ozone made by Salawitch et al. (1993) have been extrapolated globally, with the use of meteorological tracers, to infer the loss rates for ozone in the Arctic lower stratosphere during the AASE II in the winter of 1991–1992. The analysis indicated removal of 15 to 20% of ambient ozone because of elevated concentrations of chlorine monoxide and bromine monoxide. Observations during AASE II defined rates of removal of chlorine monoxide attributable to reaction with nitrogen dioxide (produced by photolysis of nitric acid) and to production of hydrochloric acid. Ozone loss ceased in March as concentrations of chlorine monoxide declined. Ozone losses could approach 50% if regeneration of nitrogen dioxide were inhibited by irreversible removal of nitrogen oxides (denitrification), as presently observed in the Antarctic or without denitrification if inorganic chlorine concentrations were to double (Salawitch et al. 1993).

4.9
Ozone in the Lower Stratosphere

The last decade has seen serious international efforts to better understand the physical and chemical processes which control the ozone level in the stratosphere. Most of the ozone depletion is observed in the lower stratosphere, and it is there that an acceleration of the ozone decline has been reported for the 1990 through 1995 period. The previous total ozone trend estimates were based on the Nimbus-7 TOMS (Total Ozone Mapping Spectrometer) data record. The ozone record in the 1992–1993 post-Pinatubo eruption has shown an anomalous loss of an additional 1–2% of total column ozone.

A very significant new development with the latest data sets is the small, barely significant trend in total column ozone in the tropics. This small tropical negative trend has been observed in both the satellite and ground-based total column ozone data.

The October monthly mean total ozone measured at the Antarctic Dobson stations continues to decrease. Ozone sondes launched from the South Pole in 1993 measured profiles with less than 100 DU (Dobson unit, 1 DU = 10^{-3} bar cm^{-1}), with essentially no measurable ozone between 14 and 19 km. The annual springtime ozone reductions at Halley Bay are now continuing into the late summer months. Since 1990, the lowest four January and the three lowest February monthly mean total ozone values have been recorded.

Although ozone destruction in the lower stratosphere is primarily due to the HO_x and ClO_x catalytic cycles, reactive nitrogen chemistry has an important role. It strongly affects these other cycles. With the NO_x species we have the most complete and longest-term set of measurements, the basic chemistry is well understood, and the heterogeneous chemistry is quite simple.

There is a continuous H_2O cross-tropopause transport up to 30 km, which maintains the tropopause values over a year or longer. This transport is relatively unaffected by horizontal mixing from subtropical regions. The tropical H_2O ascent data may account for the elevated hygropause. Another puzzling aspect is an observed hemispheric H_2O asymmetry thought partly to be a result of Antarctic dehydration.

Simultaneous determination of several radicals involved in the rate-determining steps has enabled a direct comparison of the relative importance of the various ozone loss cycles for the first time. Catalytic destruction by HO_x in the lower mid-latitudes accounts for over half of the total photochemical loss rate. At high latitudes, halogen catalytic cycles become quite important. NO_x tends to play a less direct role in ozone loss but is important in modulating HO_2 and ClO. The HO_x cycle is the largest contributor to odd oxygen loss in the lower stratosphere, but a major uncertainty in the HO_x cycle is that the rate coefficient for the reaction of HO_2 with ozone, together with the mechanism of this reaction, are poorly defined at the temperatures of the lower stratosphere.

NO_x is a kind of glue that holds the cycles together. Of all the reactions of NO_x catalytic ozone destruction cycles, the $O + NO_2 \rightarrow NO + O_2$ reaction is the most poorly understood. Fluorine and $C_xF_yO_z$ radicals are not such efficient ozone destroyers. One of the chief questions about ClO_x and BrO_x is the partitioning between active and unreactive forms. Some radical-radical reactions that can lead to reservoir species are quite hard to study.

Bromine is probably the most effective ozone destroyer on a per atom basis of the three most abundant halogen species, viz., fluorine, chlorine and bromine. However, there is much uncertainty as to the extent to which bromine radicals are converted to the reservoir HBr via reactions such as $BrO + HO$. The major products of the $HO_2 + BrO$ reaction are now known to be HOBr plus O_2, since the yield of HBr from this reaction has been shown to be very low at the stratospheric temperatures. Some ozone depletion probably occurs by coupling between IO_x, ClO_x and BrO_x cycles. The coupling of this iodine with ClO_x and BrO_x could possibly have some role in sudden episodic ozone removals in the Antarctic springtime troposphere.

4.10
CFC Substitutes

Recent debates have centered around some of the substances that are being proposed and used as alternatives for CFCs. The most frequently identified are some hydrochlorofluorocarbons (HCFCs) which can replace CFCs in many applications. However, because HCFCs also contain some O_3-destroying chlorine, HCFCs cannot be the final solution to the CFC problem. It is only for the time being that they remain good alternatives to CFCs. For use in refrigerators a mixture of propane and butane (two ozone-neutral substances) is being tested as alternative to CFC/HCFCs in Germany. Some researchers, however, feel that notwithstanding the example of the refrigerator, the case for ozone-neutral alternatives is not strong because hydrofluorocarbons (HFCs, chlorine-free) are strong greenhouse gases, and the more traditional chemicals such as ammonia, butane, and pentane can be toxic and flammable.

There is growing potential for CFC substitutes in the changeover from CFCs to environmentally friendly refrigerants. Hydrocarbon technology in particular has been readily accepted in projects and initiatives in some developing countries. In countries such as India and Pakistan, where summer temperatures often go up to 45 °C or higher, the main technical problem with refrigerators is that the compressors do not work properly, due to poor lubrication or fluctuations in the power supply voltage.

Hydrocarbon technology provides a cheap, efficient and safe substitute for CFCs. In some cases, cooking gas (LPG) is being tried as a substitute for CFC-12. LPG is much cheaper than CFC-12. The larger gas producers are planning to switch over from CFC-12 to HFC-134a which, however, is far more expensive. HFC-134a also has a high global warming potential. Although almost all manufacturers in Europe have switched to hydrocarbon technology, there is still at least partial resistance in many countries to accept this technology for refrigerator production. The USA and Japan have not accepted the use of hydrocarbons as coolant gases for safety reasons. Thus, hydrocarbons may be an option for Europe and the developing countries, but not for the USA and Japan.

Some ozone-friendly alternatives to CFCs as refrigerants are the following: butane and propane (highly inflammable, less efficient than CFCs, greater contribution to the greenhouse effect); ammonia (highly corrosive, danger to human health if it escapes, less efficient than CFCs, greater contribution to the greenhouse effect), HCFCs (efficient refrigerant, no toxic effects, destroys ozone but to a lesser extent than CFCs).

Small entrepreneurs in developing countries, who make a living by retrofitting refrigerators in the informal sector, are very much in favor of hydrocarbon technology (Dijkstra 1994). For local industries in these countries also, conversion to hydrocarbons is beneficial as they do not have to import the additional expensive lubricants for HCFCs and HFCs. It appears that the success of the Montreal protocol in the domestic refrigerator sector depends on hydrocarbon technology. HFC-134a is not an option and it should not be promoted in developing countries. In Europe, major sectors of industry have already abandoned it (Dijkstra 1994).

In the domestic refrigerator industry sector, an alternative to the ozone-depleting CFCs already exists. A cheap "eco-fridge" technology has been developed in Germany. As a high-volume utility product, the eco-fridge helps create environmental awareness among consumers. Plans are on their way to switch production to hydrocarbon technology at one or two production plants in China and India, with the following objectives:

1. To enable refrigerator manufacturers in developing countries to switch to the new technology
2. To set up an advisory service for companies in developing countries, and
3. To establish a German consortium so as to offer CFC substitution to refrigerator manufacturers and their suppliers.

In China, a reputed refrigerator manufacturer is considering converting one production line to work with hydrocarbon technology, in cooperation with a leading German maker. The German partner will plan, engineer, and convert the production lines and also train the Chinese technicians. Promotion of hydrocarbon technology in repair businesses also forms a part of the project. Hydrocarbon technology know-how will be passed on to a number of businesses, and technicians will be trained in retrofitting of the refrigerant liquids. In India, the Swiss Development Corporation (SDC) has been helping refrigerator manufacturers and craft businesses to find substitutes for CFCs since 1991. The aim of the project is to reduce CFC emissions and help improve the technological capabilities of the partner countries. Economic and ecological considerations need to be given equal emphasis.

Awareness about the adverse impact of chlorofluorocarbons on stratospheric ozone (WMO 1989; 1992) has prompted attempts to replace CFCs with environmentally acceptable alternatives such as hydrochlorofluorocarbons (HCFCs). Examples include HFC-134a (CF_3CFH_2), a replacement for CFC-12 (CF_2Cl_2) in domestic refrigeration and automobile air conditioning units, HCFC-22 (CHF_2Cl), a replacement for CFC-12 in industrial refrigeration units and HCFC-141b ($CFCl_2CH_3$), a replacement for CFC-11 in foam-blowing applications (Wallington et al. 1994). HFCs and HCFCs are volatile and insoluble in water. In the atmosphere they become oxidized into a variety of degradation products. While HFCs are ozone friendly, HCFCs have only a small ozone depletion potential. The global warming potentials of HFCs and HCFCs are also less than those of the CFCs. At the concentrations expected from the atmospheric degradation of HFCs and HCFCs, none of the oxidation products would be toxic. Since, unlike CFCs, HFCs and HCFCs contain C-H bonds, they are susceptible to attack by $OH^•$ radicals in the troposphere (Wallington et al. 1994). Reaction rates with $OH^•$ radicals determine the atmospheric lifetimes (2–40 years) of all HFCs and HCFCs which are much shorter than those of CFCs. The last stage in the removal of any species from the atmosphere is its heterogeneous deposition on the Earth's surface.

Hydrofluorocarbons (HFCs) or hydrochlorofluorocarbons (HCFCs) are attractive alternatives (substitutes) for CFCs because they have relatively short atmospheric residence times. HFCs and HCFCs are attacked by tropospheric hydroxyl radicals, leading to the formation of trifluoroacetate (TFA). Most of the atmospheric TFA is deposited at the Earth's surface, where it is resistant to bacterial attack. Therefore, use of HCFCs and HFCs may lead to accumulation of TFA in soils, where it could prove toxic or inhibitory to plants and soil microbial communities (Visscher et al. 1994). We know little about the toxicity of TFA but monofluoroacetate, which occurs at low levels in some plants and which is susceptible to slow attack by aerobic soil microbes, is highly toxic. Visscher et al. reported that TFA can be rapidly degraded microbially under anoxic and oxic conditions, implying that significant microbial sinks exist in nature for the elimination of TFA from the environment. They also showed that oxic degradation of TFA leads to the formation of fluoroform, a potential ozone-depleting compound with a much longer atmospheric lifetime than the parent compounds.

There exist certain destruction technologies that may be used to eliminate existing stocks of CFCs and other ozone-depleting substances, as well as creating a process which can approve destruction technologies in the future. These technologies have been endorsed by the Technology and Economic Assessment Panel of the UNEP. Incineration is one such technology. A few other prospective technologies may include more environmentally benign processes, such as electrochemical, physiochemical and dehalogenation. Recycling has been given a boost, as recycled ozone-depleting chemicals have been exempted from production or consumption categories.

4.11
Causes of Ozone Depletion

The stratosphere region of the atmosphere is characterized by a more or less continuous increase of temperature with height, caused mainly by absorption of ultraviolet radiation by a layer of ozone molecules with a peak abundance near 20 km. Ozone is a minor atmospheric constituent and its "mixing ratio" (the ratio of the number of ozone molecules to other molecules in a given volume) rarely exceeds 10 parts per million by volume. Certain chemical and meteorological processes control the ozone content of the stratosphere. The release of chlorine atoms from chlorofluorocarbons (CFCs) has led to the increasing depletion of ozone observed since 1979. Other halogen compounds (principally those containing bromine) also contribute to the problem, but the CFCsare the major cause of current ozone loss.

The interacting processes that contribute to the stratospheric ozone budget are illustrated in Fig. 4.12. Ozone is formed when solar radiation of wavelength shorter than 240 nm breaks up molecular oxygen and the resulting oxygen atoms combine with other oxygen molecules. Ozone is destroyed when it reacts with oxygen atoms to give two oxygen molecules. It can also be lost through reactions which belong to several chemical "families" that include oxides of hydrogen (HO_x), nitrogen (NO_x), chlorine

(ClO$_x$) and bromine (BrO$_x$). These families are found in the stratosphere in mixing ratios of no more than a few parts per billion by volume (a few parts per trillion for BrO$_x$), but they are of crucial importance because they can catalyze ozone destruction (Garcia 1994).

Nitrous oxide (N$_2$O) is the main source of NO$_x$ in the stratosphere; methane and water vapor supply the hydrogen atoms that maintain HO$_x$ concentrations. Stratospheric ClO$_x$ comes partly from methyl chloride (CH$_3$Cl). The largest sources of stratospheric chlorine are CFCs, and over three quarters of the chlorine in the stratosphere today has originated from the breakdown of CFCs.

The importance of CFCs for the stratospheric chlorine budget is surprising because natural sources (e.g. volcanoes and oceans) give off much more chlorine than CFCs. However, most of the chlorine naturally released into the atmosphere is highly soluble and quickly removed by precipitation in the lower atmosphere. CFCs, however, were designed to be unreactive at sea level to be safe in industrial applications. Because of this stability, CFCs are transported by atmospheric motions to the stratosphere, where solar radiation is strong and intense enough to break them down, releasing chlorine. The fate of these chlorine atoms and their role in ozone destruction depends upon complex interactions between photochemistry and transport in the stratosphere. This transport is very important: if atmospheric transport played no role, ozone would be expected to be most abundant in the tropics, but this is not the case. Ozone content is highest at high latitudes, in winter and spring, the opposite of what would be expected from photochemistry alone. To explain this behavior, transport by stratospheric circulation has to be considered.

4.11.1
The HO$_x$ Family of Reactions

The involvement of HO$_x$ species in the catalytic destruction of stratospheric ozone was first suggested by Bates and Nicolet (1950). The key active HO$_x$ species in this respect is HO, which initiates the catalytic cycle

$$HO + O_3 \rightarrow HO_2 + O_2; \tag{4.26}$$

$$\underline{HO_2 + O \rightarrow HO + O_2} \tag{4.27}$$

Net: $O_3 + O = 2 O_2$.

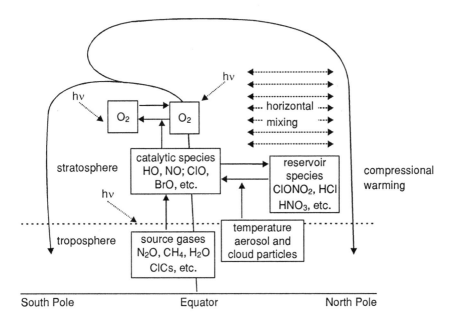

Fig. 4.12. Ozone chemistry and circulation of the stratosphere. The *heavy lines* represent meridional circulation streamlines for Northern Hemisphere winter. The meridional circulation is strongest in the winter hemisphere where there is also efficient north-south mixing (*dashed lines*) due to planetary wave breaking. The meridional circulation transports ozone poleward and downward from the tropical production region. The species involved in catalytic cycles are produced from gases originating in the troposphere. Strong ozone destruction by chlorine occurs when stratospheric conditions (flow temperatures, presence of aerosol or cloud particles) shift the partitioning of chlorine from reservoir to active species. (After Garcia 1994)

and, in the lower stratosphere, is involved in the sequence

$$HO_2 + O_3 \rightarrow HO + 2\,O_2, \tag{4.28}$$

$$HO + O_3 \rightarrow HO_2 + O_2, \tag{4.29}$$

Net: $2\,O_3 = 3\,O_2$.

The HO radical is generated in the stratosphere by the reaction of photochemically produced $^1O_2$ (singlet oxygen) with H_2O

$$^1O + H_2O \rightarrow 2\,HO. \tag{4.30}$$

Reactions which terminate the cycle are

$$HO + HO_2 \rightarrow H_2O + O_2, \tag{4.31}$$

$$HO + HONO_2 \rightarrow H_2O + NO_3, \tag{4.32}$$

$$HO + HO_2NO_2 \rightarrow H_2O + NO_2 + O_2, \tag{4.33}$$

$$HO + H_2O_2 \rightarrow H_2O + HO_2, \tag{4.34}$$

$$HO_2 + HO_2 \rightarrow H_2O_2 + O_2. \tag{4.35}$$

The main reactions involved in the HO_x cycle are summarized in Fig. 4.13. At altitudes >30 km the $HO + HO_2$ reaction is dominant. At still higher altitudes in the upper stratosphere and mesosphere, an additional termination step plays a role:

$$H + HO_2 \rightarrow H_2 + O_2. \tag{4.36}$$

4.11.2
The NO_x Family of Reactions

The important atmospheric species in the NO_x chemical family include N, NO, NO_2, NO_3, N_2O_5, $ClONO_2$, $HONO_2$ and HO_2NO_2. In the NO_x chemical family the photostationary concentration of ozone is controlled primarily through the important catalytic cycle:

$$NO + O_3 \rightarrow NO_2 + O_2, \tag{4.37}$$

$$NO_2 + O \rightarrow NO + O_2, \tag{4.38}$$

Net: $O_3 + O = 2\ O_2$.

Much concern has arisen over the potential depletion of the stratospheric ozone layer through this catalytic cycle, initiated by the direct injection into the stratosphere of large amounts of nitrogen oxides by supersonic aircraft (Johnston 1971). Significant coupling exists between the NO_x catalytic cycle and the HO_x and ClO_x catalytic cycles. Part of this coupling occurs through reactions which directly affect the partitioning of the odd hydrogen and odd chlorine species into different active species, for example, $HO_2 + NO \rightarrow HO + NO_2$ and $ClO + NO \rightarrow Cl + NO_2$. Coupling is also shown in the sequestering of HO_x and ClO_x species (Fig. 4.14).

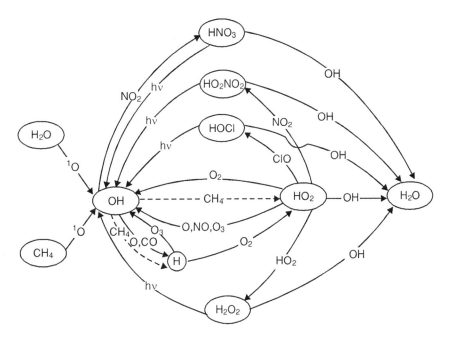

Fig. 4.13. Schematic diagram of the HO_x family of reactions. (After Atkinson et al. 1989)

4.11.3
The Halogen Family of Reactions

The reactions of the halogen family were central to the hypothesis that CFCs released as aerosol propellants could cause large depletions of stratospheric ozone (Fig. 4.15) because of their inert characteristics in the troposphere and their photodissociation to release Cl atoms in the stratosphere (Atkinson et al. 1989):

$$ClO + NO_2 + M \rightarrow ClONO_2 + M, \tag{4.39}$$

$$ClONO_2 + hv \rightarrow Cl + NO_3. \tag{4.40}$$

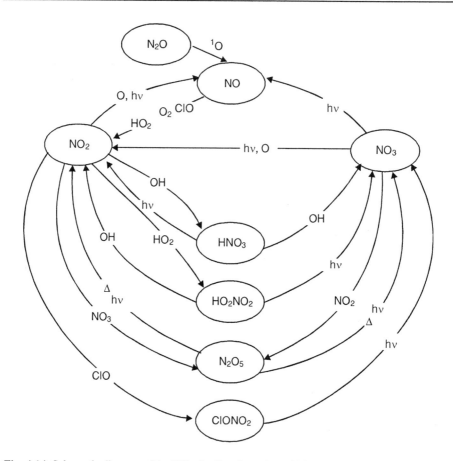

Fig. 4.14. Schematic diagram of the NO_x family of reactions. (After Atkinson et al. 1989)

the formation of $ClONO_2$ and its slow photolysis leads to a reduction in the concentrations of the active Cl and ClO species. This highlights the strong coupling between the different families (see Figs. 4.13 to 4.15 which also summarize the main features of the ClO_x cycle of reactions), and presented a challenge to experimentalists because of the large number of elementary reactions which required study.

Figure 4.16 illustrates the effect of CFCs on stratospheric ozone. When gases containing chlorine, e.g. CFCs, are broken down in the atmosphere (upper), each chlorine atom sets off a reaction that may destroy thousands of ozone molecules. The general trend of a depleting ozone layer since 1979 is shown in Fig. 4.17 and the mechanism by which catalysts (X) destroy atmospheric ozone is explained in Fig. 4.18.

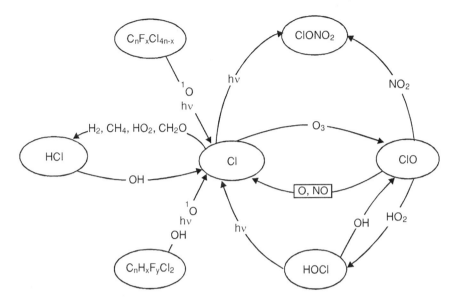

Fig. 4.15. Schematic diagram of the ClO_x family of reactions. (After Atkinson et al. 1989)

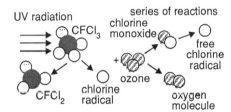

Fig. 4.16. Effect of CFCs on stratospheric ozone (UNEP 1992)

4.12
Ozone Layer Thickness

Averaged over Europe, the thickness of the ozone layer in the period from December 1991 to March 1992 was 10% lower than the long-term mean over the same four months and it was 13% lower for the same period in 1992/1993. The Royal Netherlands Meteorological Institute (KNMI) performs a daily analysis of ozone layer data which are collected by the American NOAA weather satellites.

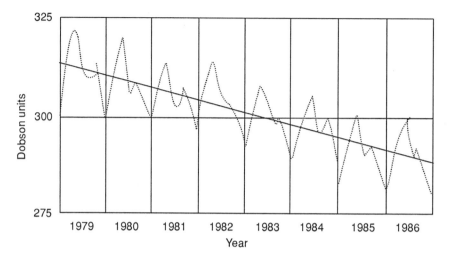

Fig. 4.17. Daily global mean total ozone in Dobson units (DU, 1 DU = 10^{-3} bar cm^{-1})

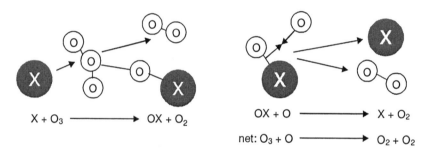

Fig. 4.18. Ozone destruction by catalysts. X catalyst

The satellite measurements (Fig. 4.19) reveal that the thickness of the ozone layer at mid-latitudes varies with the seasons: the highest ozone values are normally measured in the spring (April), and the lowest in the autumn (October). The figure shows that the ozone layer in recent times, and especially in the spring (day numbers 0 to 150), is thinner than normal. The differences between the 1991 and the 1992 values are smaller in the autumn. The thin patches in the ozone layer, especially above Western Europe, can be partially explained on the basis of meteorological circumstances: during two winters there has been a remarkable recurrence of a weather situation in which sub-tropical, ozone-poor air has been transported to Western Europe.

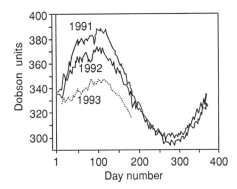

Fig. 4.19. The diurnal value of the mean thickness of the ozone layer above the temperate latitudes of the Northern hemisphere. From *top* to *bottom* are the years 1991, 1992 and 1993. (No data were obtained for a number of days in 1993, hence the break in the line)

Why this situation occurred with greater frequency is not clear but a possible factor may well be the pollution of the stratosphere with aerosols emitted by the Mt. Pinatubo volcano (Philippines) during its eruption in June 1991. These dust particles absorb or reflect part of the solar radiation, so that the warming of the Earth's surface is reduced. Another possible cause may be the chemical breakdown of ozone by reactions with anthropogenic chlorine and bromine compounds. High concentrations of chlorine oxides have been detected above the Northern latitudes, which suggests that ozone is depleting. It is not known how far this cause has contributed to the depletion.

The large number of dust particles in the stratosphere may contribute to an increased depletion of ozone. In the course of the coming years, the amount of dust in the stratosphere will reduce gradually, which may gradually reduce the influence of the Pinatubo emissions. On the other hand, the concentration of chlorine and bromine compounds in the atmosphere will remain high until well into the next century, which may lead to long-term ozone depletion.

Aerosol cloud condensation nuclei (CCN) may increase cloud droplet concentration and cloud reflectance (albedo) of incoming solar radiation. Atmospheric aerosol particles are both biogenically derived from dimethylsulfide (DMS) oxidation and by anthropogenic activities, particularly sulfates from SO_2 emissions, organic condensates and soot from biomass combustion.

Recent applications of coupled atmospheric chemical/radiative transfer models, by utilizing empirical aerosol scattering properties, have demonstrated that anthropogenically derived sulfate aerosols cause clear-sky climatic forcing which, when averaged over the globe, is comparable in magnitude, but opposite in sign, to forcing by CO_2.

Unlike CO_2, anthropogenically derived aerosol particles are not uniformly distributed over the globe but are mainly found over industrialized areas in the Northern Hemisphere. In fact, the present day aerosol forcing in many regions of the Northern Hemisphere, as an annual average, can offset the combined greenhouse effect of CO_2, CH_4, N_2O, CFCs and O_3 (Fig. 4.20; Guicherit 1993).

Besides changes in the ozone column density distribution, the direct and indirect effects of tropospheric aerosol currently pose the largest uncertainty in model calculations of the climate forcing due to anthropogenic induced changes in the composition of the atmosphere. Changes in the ozone column density distribution form an uncertain factor in model calculations. The reason for this is that although CO_2 and water vapor are the main greenhouse gases, they already cover most of the region of maximum infrared absorption; apart from a "window" which is occupied by N_2O, CH_4, CFCs and O_3. This means that the concentration of CO_2 and water vapor can change with little direct effect on radiation, and that other greenhouse gases will have more influence per molecule on radiative forcing. A complicating factor is, however, that the magnitude of the O_3-forcing effect is height dependent, with a maximum around the tropopause and an "opposite effect" above 30 km. The implications for climate forcing are at the moment very uncertain.

Sulfate aerosol particles play two potential roles in the radiative climate of the Earth: scattering of sunlight in cloud-free air, thereby reducing solar irradiance at the ground, and their cloudy sky forcing potential by increasing cloud albedo. The main sources of sulfur emissions are anthropogenically derived SO_2, estimated to be about 71 Tg S yr^{-1} and DMS release from the oceans, estimated to be about 40 Tg S yr^{-1}. DMS has played a central role in climate regulation throughout the late geological history of the Earth. A 30% increase in DMS emissions may have a cooling effect of about 1 K, which is twice as much as, but opposite in sign to, the anthropogenically induced greenhouse forcing by the major greenhouse gases at the moment (Guicherit 1993).

Globally, dimethylsulphoniopropionate (DMSP), which is mainly produced by phytoplankton and microalgae in the marine environment, is probably the most important source of DMS. Although the global oceanic flux of DMS is now quite well established, the biological processes leading to this flux are poorly understood. These processes, however, are of crucial importance because it is believed that only less than 10% of the DMS produced in the aquatic environment enters the air compartment and more than 90% is being recycled in the oceans. Future trends of DMS emissions are essential for climate modeling of the feedback mechanism caused by atmospheric sulfate aerosols due to DMS release from the oceans and hence for the appraisal of future climate change. Algae differ greatly in their DMS production potential. Changing environmental factors, as a consequence of global change (such as solar (UV) irradiance, temperature and nutrients), influence species composition, total phytoplankton population, biological processes and hence DMS production.

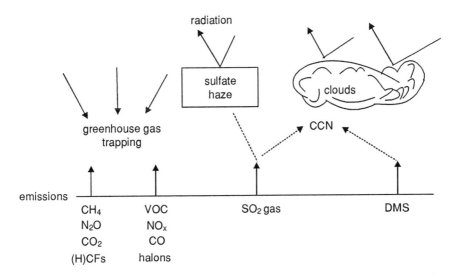

Fig. 4.20. Gases, aerosols and clouds have an important influence on the radiation budget of the atmosphere. Gases may trap infrared radiation leading to temperature increase. Aerosols and clouds determine the Earth's albedo to a great degree by reflection of solar radiation, which leads to cooling. Aerosols play a significant role in cloud formation. Clouds also provide an infrared blanket for the surface of the Earth by absorption and downward emission of infrared radiation, leading to temperature increase. Globally averaged, there is a prevalent net cooling effect due to clouds. (After Guicherit 1993)

4.13
Total Column Ozone

The Nimbus-7 TOMS (Total Ozone Mapping Spectrometer) has recorded the atmospheric total column ozone since November 1978. From 1979 to 1991 the amount of total column ozone decreased over most of the globe. Small (3 to 5%) losses at mid-latitudes, larger (6 to 8%) losses at high latitudes and no losses near the equator were reported (Stolarski et al. 1992; Gleason et al. 1993). The most dramatic ozone decease has been observed each year in the spring time Antarctic ozone hole region; the 1992 ozone amounts there were about 50% of the 1979 amounts. For latitudes between 65°S and 65°N, the average area-weighted ozone loss rate for all seasons (1979 to 1991), after correction for solar cycle and quasi-biennial oscillation (QBO) effects has been estimated to be 2.7 + 1.4% per decade (Stolarski et al. 1992; Gleason et al. 1993).

Analysis of the 13-year ozone data shows that most of the ozone depletion occurred at mid and high latitudes (Stolarski et al. 1992; Gleason et al. 1993).

Gleason et al. (1993) have shown that the 1992 global average total ozone was 2 to 3% lower than any earlier year. Ozone amounts were low in a wide range of latitudes in both the Northern and Southern Hemispheres, and the largest decreases were in the regions from 10°S to 20°S and 10°N to 60°N. Global ozone in 1992 was at least 1.5% lower than predicted by a statistical model.

Significantly, 1992 was the first time that ozone amounts showed a simultaneous sustained decrease over a wide latitude range in both hemispheres. Although the mechanism for this ozone decrease is unknown, the first guess would be that it was related to the continuing presence of aerosol from the Mount Pinatubo eruption. There are three possibilities related to the presence of the aerosol:

1. Direct chemical loss through increased heterogeneous processing,
2. An aerosol-induced change in radiative heating which can directly affect ozone transport, or
3. Changes in photochemical production or loss rates caused by the temperature changes resulting from the aerosol heating.

The Antarctic springtime minimum observed by TOMS has shown similar low total ozone amounts in 5 out of the last 6 years (up to 1993). The spring of 1992 was one of the lowest years in this record. The areal extent of the Antarctic ozone hole has essentially remained constant. In 1992 the area was 15% larger for about 2 weeks compared with previous years. The total ozone observations are consistent with the removal of 90% or more of the ozone in the region affected by PSCs. The decline in ozone is clearly driven by the increase in stratospheric chlorine released from industrial halocarbons.

The previously determined trends in respect of Northern mid and high latitude ozone have continued. These trends can be explained by increasing chlorine. The mechanism likely involves heterogeneous reactions on particulate surfaces, but the relative roles of reactions on background sulfate aerosols versus reaction on PSCs in and near the vortex are uncertain. Heterogeneous reactions convert chlorine from its inert reservoirs to the catalytically active ClO, thus enhancing the potential for ozone depletion. At the same time they convert nitrogen oxides to the inert reservoir, nitric acid, decreasing the potential impact of NO_x on ozone.

In late 1991, after the Mt. Pinatubo eruption, the equatorial ozone measured by TOMS decreased by 4–5%. This decrease appears to be consistent with enhanced upward velocities driven by the heating associated with absorption of solar radiation by the volcanic aerosol. There have been new record lows in 1992 and early 1993. TOMS data reached a record 2–3% below any previous year in the 14-year record. Ground-based stations reached record lows for over 3 decades of records. The low total ozone amounts in 1992 occurred over a broad range of latitudes in both hemispheres, but not in the equatorial region. The continued low total ozone in January to March of 1993 was most pronounced at mid- to high northern latitudes. It appears that the low observed total ozone was caused, in part, by chemical reactions involving chlorine and bromine.

Satellite data did not show any significant trends in the equatorial region during the 1980s and early 1990s. There is a great paucity of reliable ground-based measurements in the tropics to determine trends independently.

4.14
Vertical Distribution of Ozone

The climatic effects of O_3 changes are strongest when they occur in the lower stratosphere and upper troposphere. From October 1984 through March 1993, there was a significant decrease in ozone in the upper stratosphere. The trend was strongest at mid-high latitudes (-5 to -7% per decade) and indistinguishable from zero in the tropics. The trend derived for 1984–1993 was larger than that in 1979–1981. In the middle stratosphere (25–35 km), there was no significant trend in any data set. For the lower stratosphere (16–25 km), the trend from October 1984 – June 1991 was about 10% per decade and occurred about 2 km lower in altitude than that from 1979 – 1981. The trends in the lower stratosphere mainly determined the trend in the ozone column.

For the troposphere (up to about 16 km), long-term data are available for only ten stations at northern high and mid-latitudes and at one Southern Hemisphere mid-latitude station. Trends in northern stations range from 0 to +2–3% per year for 1970 – 1990. There is a crossover from a decrease in ozone in the lower stratosphere to an increase in the troposphere that occurs within a few km of the tropopause. It is quite difficult to obtain any reliable trends in ozone in the upper troposphere where O_3 changes affect climate more than changes in other regions, because of high variability.

4.15
Mount Pinatubo

In June 1991, Mt. Pinatubo (Philippines) erupted and its debris that spread through the stratosphere contributed to ozone losses, as the particles changed the chemistry of the stratosphere. Outside the polar regions, global ozone plummeted to a record low: 4% below levels typical of the past dozen years and 3% below even the very lowest level seen in those years. The low was measured in late 1992 and early 1993 by the total ozone mapping spectrometer (TOMS) on the Nimbus-7 satellite.

Some parts of the global ozone layer suffered more than others. TOMS measured below-normal levels of ozone everywhere outside the poles (between 65°N and 65°S) except near the equator. Between 10°S and 20°S and north of 10°N, the levels were lower than anything the satellite had ever seen. Late in 1992, between 30° and 60°N, levels were 9% below normal and at 60°N ozone was down 14% (see Kerr 1992).

After the eruption of Mt. Pinatubo, it was predicted that ozone in temperate latitudes might fall as much as 10% to 20%; the actual losses reached only 2% to 4%. It was thought that the volcanic debris, like the ice particles in the polar clouds, would catalyze the destruction of nitrogen oxide that normally locks up chlorine and thus protects ozone. The disparity between predictions and reality has created doubt as to whether volcanic debris makes much difference in stratospheric chemistry. Instead, the Pinatubo debris may have worked much of its effect in some other way, such as by interfering with the transport of ozone around the stratosphere. By absorbing solar radiation and warming parts of the stratosphere, the cloud may have changed the high altitude winds that sweep ozone out of the tropical stratosphere, where much of it forms, and around the rest of the globe (see Kerr 1992).

4.16
Ozone Holes

The occurrence of an ozone hole over the Antarctic in spring time was announced by Farman et al. (1985). Observations showed that the ozone loss occurs rapidly, with most of the ozone in the polar lower stratosphere, between approximately 15 and 20 km, being removed in about 6 weeks in the Antarctic spring. In winter, the circulation in the polar stratosphere of both hemispheres is dominated by a low pressure vortex of strong westerly winds. In the Antarctic, these winds are mostly circumpolar. At the center of this vortex temperatures can remain low during winter and early spring. In contrast, the Northern Hemisphere winter circulation is more disturbed and variable. The center of the vortex does not usually correspond to the pole, and the minimum temperatures tend to be several degrees higher than in the south (Pyle et al. 1992).

In the south, temperatures drop low enough for the formation of polar stratospheric clouds (PSCs). PSCs form at temperatures a few degrees above the frost point, below about 195 K, and appear to be composed initially of nitric acid trihydrate. With a fall in temperature, water ice forms. The PSCs probably act as surfaces for heterogeneous reactions which convert chlorine from reservoir forms, such as HCl and $ClONO_2$, into radicals which can destroy ozone.

Halons are similar to CFCs but contain bromine. They were developed towards the end of the Second World War as fire extinguishing agents for use on tanks. Their commercial use in fire fighting equipment has accelerated since the 1970s. CFCs and halons are very valuable substances because of their low toxicity, non flammability, ability to produce a fine spray, compressibility, long-term stability and low cost.

Early studies on the ozone layer soon showed up the danger of chlorine in the upper atmosphere. The link between CFCs and ozone depletion had long been suspected. The United States and several Nordic countries banned CFC propellants in spray cans in the 1970s. It is now scientifically established that ozone is being depleted and that CFCs and halons are responsible. Because of their chemical stability CFCs and halons survive in the atmosphere a very long time, long enough to be transported into the stratosphere. It is only at this level that sufficient UV radiation is found to break the CFC molecules down. Consequently, chlorine and bromine atoms are freed to become involved in the catalytic destruction of ozone. There is less certainty about the rates of depletion. Recent studies indicate that a decrease of approximately 2 to 3% has taken place over the Northern Hemisphere since 1970. There is less data for the Southern hemisphere but it seems likely that its depletion rates are slightly higher. The conditions in the stratosphere above the Antarctica during spring allow other catalytic processes to occur. In essence ice particles form a reaction surface which enables the free chlorine from the CFCs to act more effectively.

Scientists expect a major volcanic eruption to emit sulfuric acid into the stratosphere. This too could act as a reaction surface with consequent drastic reductions in the ozone layer. Some predictions have indicated depletion rates of up to 50% in the first year after an eruption. This could occur because free chlorine and bromine atoms are now present in the stratosphere.

4.16.1
The Antarctic Ozone Hole

Over the past 15 years, there has been a rapid and unexpectedly high loss of ozone in the stratosphere each spring above Antarctica (Fig. 4.21). Every spring, up to 95% of stratospheric ozone is destroyed at a height of 12–24 km above the Earth's surface – the heart of the polar ozone layer. The ozone hole is sometimes larger than the area of the continental USA. The latest measurements show that from 12 to 50 km above the Earth's surface, half the ozone layer above Antarctica is destroyed each spring. The rate at which ozone in the stratosphere above mid to high latitudes is now disappearing is much faster than was predicted.

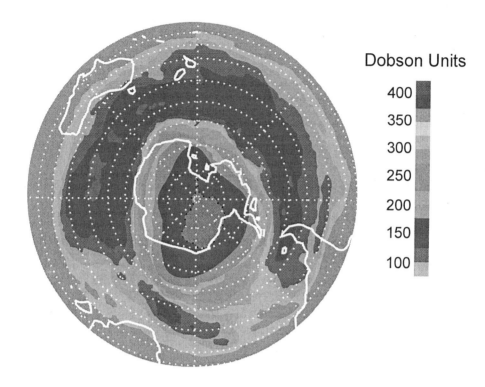

Fig. 4.21. Satellite mapping of the ozone hole over Antarctica on 6 October 1996

There is strong scientific evidence that man-made chlorine and bromine compounds are responsible for these losses. In the stratosphere above Antarctica during the spring, when the ozone hole is at its largest, amounts of reactive chlorine compounds, which cause rapid and large-scale destruction of ozone, are up to 100 times greater than at other times of the year. The process of ozone destruction above Antarctica actually begins in the winter, when a vortex of extremely cold air blows around the pole. No sunlight enters the vortex in the Antarctic winter night, and no air from warmer latitudes moves into the vortex for ozone destruction by allowing clouds of nitric acid and water ice to form in the lower polar stratosphere (UNEP 1992).

Reactions between particles in these polar stratospheric clouds and stable chlorine and bromine compounds play a strong role in large-scale ozone depletion above Antarctica. Once the stable compounds of chlorine and possibly bromine become unstable and change into highly reactive forms, sunlight can tear apart ozone molecules and destroy them rapidly. Chlorine chemistry accounts for most of the ozone destruction that occurred in the Antarctic spring of 1987. The process begins with the formation of polar stratospheric clouds. These trigger the breakdown of stable chlorine molecules into highly reactive forms. Chemical reactions at the surface of PSCs prevent nitrogen oxides from slowing down the rate of ozone destruction – as they normally do when they capture reactive forms of chlorine and change them back into more stable molecules. In PSCs, by contrast, nitrogen oxides are tied up in the long-lived species HNO_3 and hence cannot convert chlorine monoxide into reservoirs of more stable chlorine compounds. Consequently, highly reactive chlorine compounds stay in the stratosphere in larger quantities and for longer than normal, increasing rates of ozone destruction (UNEP 1992).

The seasonal formation of the Antarctic ozone hole remains the most dramatic example of stratospheric ozone depletion. The hole is a sharp decline in the total ozone content over Antarctica, from September through November, when the breakdown of the southern winter circulation regime mixes polar and mid-latitude air. In September 1993, the ozone column at the South Pole had dipped below 100 DU for the first time since records have been kept, representing almost a two-thirds loss from the pre-ozone-hole values.

The springtime decrease in total ozone over Antarctica is due to marked reductions in ozone abundance at altitudes between 12 and 22 km. It was realized during the 1970s that continued CFC emissions would eventually lead to stratospheric ozone depletion; however, the calculations predicted significant losses only above 30–35 km. At these altitudes, where the ozone density is much smaller than in the lower stratosphere, ozone reductions have only a small impact on the total ozone column.

Nobody expected ozone depletion by ClO_x catalysis in the lower stratosphere because most of the atmospheric chlorine at these altitudes was believed to be tied up in the "reservoir" species, chlorine nitrate ($ClONO_2$) and hydrogen chloride (HCl), which do not react with ozone. This mystery was solved when it was realized that "heterogeneous reactions" played a significant role in stratospheric chemistry. Heterogeneous reactions occur when a gas collides with solid or liquid particles, reacting with another species present on the surface or in the interior of the particle. In 1986, it was thought that such processes free stratospheric chlorine from its reservoir species. Further, the heterogeneous process

$$N_2O_5 + H_2O \rightarrow 2\,HNO_3, \tag{4.41}$$

which produces the relatively long-lived HNO_3, is also important for efficient ozone catalysis. In the absence of this last reaction, N_2O_5 is quickly photolyzed to yield NO_2, which can recombine with ClO to again sequester chlorine in $ClONO_2$, thereby counteracting the effect of reactions (4.14) and (4.24). The chlorine released from its reservoir species by reactions (4.14) and (4.24) destroys ozone rapidly through a catalytic cycle, depending on the formation and splitting of the ClO dimer, Cl_2O_2. As solar radiation is required to split Cl_2O_2 at polar latitudes, the catalytic cycle can only occur when sunlight returns to the polar cap in the spring, consistent with the temporal behavior of Antarctic ozone depletion (Garcia 1994).

The heterogeneous chemistry mechanism also explains the formation at high latitudes during winter of PSCs, whose droplet surfaces provide sites for heterogeneous processes (Hamill and Toon 1991). These PSCs contain two types of particles. Type I particles are frozen mixtures of water and nitric acid, which form when the ambient temperature drops below about 193 K. Type II particles are much larger, and are made chiefly of water ice – they grow only when the temperature falls below 187 K. The very cold environment required for the formation of these clouds explains why they occur with much greater frequency in the Southern Hemisphere, where wintertime temperatures are usually 15 K colder than in the Arctic (Garcia 1994). In Antarctica, most of the ozone loss occurs in September, when the southern polar cap is sunlit, but the air is still very cold and isolated from mid-latitude air. In the Northern Hemisphere, in contrast, planetary waves tend to upset the wintertime circulation, raising temperatures above 210 K before the middle of March. Thus, when sunlight returns to the Arctic the conditions necessary to keep chlorine activated by reactions (4.23, 4.25) and (4.41) are no longer present.

Balloon and satellite measurements during the 1993 Southern Hemisphere spring revealed 15% less ozone over Antarctica than during 1992 thinning, leaving the protective ozone shield at less than one third its normal thickness. These additional losses are surprising because in recent years the depletion of ozone has been almost complete at the altitudes where conditions normally favor its destruction by chlorine from man-made CFCs. Only if favorable conditions, including extreme cold and the presence of the fine particles that trigger the ozone depleting reactions, had spread to other altitudes could more ozone disappear. This may have happened, thanks to lingering debris from the 1991 eruption of Mount Pinatubo and perhaps unusual cold at high altitudes (Kerr 1992).

In 1993, the erosion of Antarctic ozone appeared on schedule, but by early October it had extended well beyond the region where ozone loss had tended to be complete: a 1 to 2 km thick layer centered at an altitude of 17 km. Instead, balloon-borne instruments showed total depletion from 14 to 19 km. Just 90 DU of ozone were left in the Antarctic stratosphere this spring, compared with 105 DU in the year 1992, which in turn was 5 to 10 DU below preceding years. (At other seasons, Antarctic ozone levels are about 280 to 300 DU).

Something similar also happened in 1992, when heavy ozone depletion occurred for the first time at altitudes below 14 km. Since the high-altitude ice clouds that activate ozone destroying chlorine are centered at an altitude of around 17 km, something else had to be taking their place at lower altitudes. The best candidate seemed to be the haze of sulfuric acid particles lofted by Pinatubo's eruption. And that same, lingering haze may probably be responsible for the low-altitude losses in this year. However, volcanic haze does not explain the enhanced losses seen between 18 and 23 km.

The Antarctic ozone hole has become a serious problem. Average amounts of ozone measured each October, and also daily maximum and minimum amounts have decreased sharply during the past several years. The ozone hole has also spread into southern mid-latitudes, possibly because of the export of ozone-poor air from the Antarctic stratosphere late each Austral spring. Increased intensities of ultraviolet-B radiation have been measured beneath the hole, with intensities more typical of the summer occurring in early spring. Chlorine from CFCs is the main ozone-destroying agent and the activity of chlorine is maximized by chemical reactions on the surfaces of PSC particles. Since CFCs survive for many years in the atmosphere it means that the Antarctic ozone hole will continue to develop each year, and the amount of ozone in the Southern Hemisphere will continue to fall.

Cicerone (1994) suggested injecting thousands of tons of alkanes (such as ethane or propane) into the polar stratosphere over Antarctica for about a month each year. Alkanes are relatively short-lived. This means they do not spread globally. There are many uncertainties underlying this idea but it needs critical debate and consideration of pros and cons, especially of any likely side effects, etc. It seems that though ozone depletion might be slowed through annual alkane injections, there may be certain other plausible outcomes in which small injections could worsen ozone depletion. There is a direct reaction of HCl and HOCl on the surfaces of PSCs: this reaction may well alter the response to alkanes.

4.16.2
Sulfate Aerosols and Ozone Depletion

During winter in polar regions, ozone is depleted mainly through the catalytic action of chlorine liberated by reactions on polar stratospheric cloud (PSC) particles. At low to mid-latitudes, greater solar illumination cause the destruction rate of ozone to be determined by a combination of catalytic cycles. The dominant destruction cycle in these latitudes involves nitrogen oxides. Laboratory studies have shown, however, that the nitrogen oxides N_2O_5 and $ClONO_2$ react on the surface of sulfuric acid solutions with compositions similar to those of stratospheric aerosol particles.

$$\text{aerosol}$$
$$N_2O_5 + H_2O \rightarrow 2\ HNO_3, \tag{4.42}$$

$$ClONO_2 + H_2O \rightarrow HOCl + HNO_3. \tag{4.43}$$

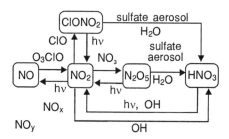

Fig. 4.22. Diagram denoting the principal species of the NO_y reservoir (= $NO + NO_2 + 2\ N_2O_5 +$ $HNO_3 + ClONO_2 + HO_2NO_2 + ...$) and reaction pathways. The NO_x/NO_y ratio within the reservoir is set by the balance of many reactions involving O_x, HO_x, Cl_y and Br_x species. The *thickness of the arrows* is nominally proportional to the conversion rate between component species. The effect of the N_2O_5 and $ClONO_2$ reaction with H_2O on sulfate aerosol is to reduce the steady-state abundance of NO_x within NO_y. N_2O_5 is produced mainly at night in the recombination of NO_2 and NO_3 and destroyed during the day by photodissociation, reforming NO_2. NO_3 is formed mainly at night in the reaction of NO_2 with O_3. $ClONO_2$ is formed in the recombination of ClO and NO_2. HNO_3, the reaction product of (4.42) and (4.43) and principal NO_y species, reforms NO_x during the day through photolysis. In addition, increases in reactive chlorine in the form of ClO and in the hydroxyl radical, $OH^\bullet$, are expected for sulfate aerosol at or above background values

Sulfate aerosol particles, formed from biological and volcanic sources of sulfur, occur throughout the lower stratosphere. The reaction probability of (4.42) is nearly independent of temperature, but the reaction probability of (4.43) increases exponentially with decreasing temperature, becoming equal to that of reaction (4.42) below 200 K. Both reactions decrease NO_x (= $NO + NO_2$) and increase HNO_3 within the reactive nitrogen reservoir, NO_y (see Fig. 4.22). Reducing NO_x changes the balance among the various photochemical cycles that produce and destroy ozone. Thus production of the inactive chlorine reservoir species $ClONO_2$ and HCl is reduced, thereby increasing ClO and its associated ozone loss cycles.

In situ measurements of stratospheric sulfate aerosol, reactive nitrogen and chlorine concentrations at mid-latitudes confirm the importance of aerosol surface reactions that convert active nitrogen to a less active, reservoir form. This makes mid-latitude stratospheric ozone less vulnerable to active nitrogen and more vulnerable to chlorine species. The effect of aerosol reactions on active nitrogen depends on gas phase reaction rates, so that increases in aerosol concentration following volcanic eruptions can have only a limited effect on ozone depletion at these latitudes. Aircraft emissions are a significant source of NO_x in the upper troposphere and lower stratosphere and hence affect photochemical ozone production and loss in both regions (WMO/UNEP 1992). In

1990 and 1992 the United Nations' Intergovernmental Panel on Climate Change prepared a report on the knowledge then available on the natural greenhouse effect and its enhancement by human activities. It also formulated the following conclusions about the consequences of that enhancement for our climate:

1. In the past century the temperature has increased on a global scale. However, the temperature trends in the past century cannot be ascribed with certainty to the enhanced greenhouse effect because of our still limited knowledge and the natural variability of the climate.
2. There is every possibility/chance of further global warming. The disruption of the climate system by the greenhouse effect continues steadily. It is well established that this leads to global warming. From the increase of concentrations of greenhouse gases in the atmosphere an increase in radiative forcing is evident. For the beginning of the next century, this is likely to lead to an increase in the Earth's temperature by at least 1 °C. This increase can be counteracted by cooling effects, for example by an increase of the amount of aerosols in the atmosphere. Such effects have a much shorter time scale than the greenhouse effect, therefore they cannot form a structural counterbalance to it.
3. The ozone layer is affected by chlorine and bromine from CFCs and halons, especially above the Antarctic. The proposed measures, laid down in the Montreal protocol and its modifications, however, hold some prospects of a recovery of the ozone layer. A new matter of concern is that the mutual influence of the greenhouse effect and the ozone budget is insufficiently known.
4. The thickness of the ozone layer varies from place to place. Short periods of extra ozone reduction have been observed in some areas. These could immediately be related to exceptional weather conditions. There is also a long-term trend. Over the period 1978–1990 a statistically significant decrease of several percent of the total amount of ozone has been recorded above the Northern hemisphere. However, an increase in the amount of UV-B radiation at sea level has not been observed yet.

4.16.3
Stratospheric Ozone Depletion by $ClONO_2$ Photolysis

Springtime ozone depletion over Antarctica is believed to be due to catalytic cycles involving chlorine monoxide. The latter is formed by reactions on the surface of polar stratospheric clouds (PSCs). When the PSCs evaporate, ClO in the polar air can react with NO_2 to form the reservoir species $ClONO_2$. High concentrations of $ClONO_2$ can also be found at lower latitudes because of direct transport of polar air or mixing of ClO and NO_2 at the edges of the polar vortex. $ClONO_2$ can take part in an ozone-depleting catalytic cycle, whose significance is not clear.

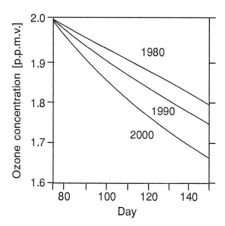

Fig. 4.23. The concentration of ozone at an altitude of 19 km, 65°N, as a function of time (day 75 = 16 March). Initial (day 74) conditions (in volume mixing ratio) for the years 1980, 1990 and 2000 $ClONO_2 = 2, 3, 4 \times 10^{-9}$; $BrO = 10, 15, 23 \times 10^{-12}$; $HNO_3 = 10, 9, 8 \times 10^{-9}$; for all years: $CH_4 = 0.5 \times 10^{-6}$, $H_2O = 5 \times 10^{-6}$, $H_2 = 0.5 \times 10^{-6}$, $CO = 2 \times 10^{-8}$, $H_2O_2 = 1 \times 10^{-12}$, $ClO = HCl = NO = NO_2 = N_2O_5 = BrONO_2 = CH_3O_2 = 0$, $T = 210$ K, $p = 64.5$ mbar. (From Toumi et al. 1993)

Toumi et al. (1993) presented model simulations of ozone concentrations from March to May both within the Arctic vortex and at a mid-latitude Northern Hemisphere site. They found increasing ozone loss from March to May. The $ClONO_2$ cycle seems to be responsible for a significant proportion of the simulated ozone loss. This cycle is not as limited as the other chlorine cycles to the timing and location of PSCs; it may therefore play an important role in ozone depletion at warm mid-latitudes (Toumi et al. 1993).

High levels of $ClONO_2$ have been recorded for both hemispheres at the edges of the polar winter vortex and inside the Arctic spring vortex. Toumi et al. showed that ozone depletion does not cease once $ClONO_2$ is formed. Figure 4.23 shows ozone depletion at an altitude of 19 km (near the peak of the ozone layer) from March to May, assuming different total chlorine and total bromine concentrations that are representative of the years 1980, 1990 and 2000. After 75 days there is a depletion of ozone of 9.5, 12.5 and 16% for these respective years. This is a significant additional change as models suggest about 20% chemical depletion of ozone from December to March for a 1990 atmosphere. It may therefore be concluded that there is significant additional ozone depletion months after heterogeneous reactions have ceased. Comparing the 1990 run with the 1980 run shows that ozone was 3.3% lower in 1990, which is about half of the observed column ozone trend for these years.

It is also noteworthy that ozone depletion later in the season is more damaging as the UV-B flux is largest in the summer (Toumi et al. 1993). Ozone is naturally produced and destroyed through photochemistry that occurs in the stratosphere. Before about 1980, the minimum of sunlight that occurs during the polar winter was insufficient to trigger substantial O_3 production or destruction but with our present elevated concentrations of stratospheric chlorine, rapid O_3 loss can occur.

During the Antarctic late winter and early spring, the decrease in ozone that is a result of anthropogenic chemical loss can overwhelm the naturally occurring increase due to transport of O_3-rich air from lower latitudes. It is this scarcity of O_3 in column measurements that is called an O_3 hole. The seasonal onset of the chemical O_3 loss is not as easily observed because it is masked by increases that are the result of air transport. In the Northern Hemisphere vortex, the period of significant loss is relatively short because of the warmer, less stable Arctic vortex dissipating by late winter. In this case, a weaker loss is masked by a stronger seasonal increase that is due to transport. Although an O_3 hole does not form, the Arctic vortex contains less O_3 than normal (Proffitt et al. 1993). Concern over O_3 loss has intensified with recent reports of significant column O_3 decrease over the heavily populated mid-latitudes in all seasons and both hemispheres. It appears that the decreases are due to chemistry occurring at mid-latitudes and transport of O_3-poor air from polar areas to mid-latitudes.

Measurements made in the outer ring of the Northern polar vortex from October 1991 through March 1992 have shown an altitude-dependent change in ozone, with a decrease at the bottom of the vortex and a substantial increase at the highest altitudes accessible to measurement (Proffitt et al. 1993). The increase is the result of ozone-rich air entering the vortex, and the decrease reflects ozone loss accumulated after the descent of the air through high concentrations of reactive chlorine. The depleted air that is released from the bottom of the vortex significantly reduces column ozone at mid-latitudes (Proffitt et al. 1993).

Browell et al. (1993) measured stratospheric ozone and aerosol distributions across the wintertime Arctic vortex from January to March 1992 with an airborne lidar system as part of the 1992 AASE II measurements. Aerosols from the Mt. Pinatubo eruption were found outside and inside the vortex with distinctly different distributions that clearly identified the dynamics of the vortex. Changes in aerosols inside the vortex indicated advection of air from outside to inside the vortex below 16 km. No PSCs were observed and no evidence was found for frozen volcanic aerosols inside the vortex. Between January and March, ozone depletion was observed inside the vortex from 14 to 20 km with a maximum average loss of about 23% near 18 km (Browell et al. 1993).

4.17
Ozone Depletion at Mid-Latitudes

The ozone layer is not only diminishing above Antarctica but also at mid-latitudes, e.g. Southern America, Australia and New Zealand. Reduced ozone levels have been recorded throughout the year at latitudes much nearer the equator, in both hemispheres, and these decreases are largely due to chlorine and bromine. Decreases in winter in the Northern Hemisphere and also in the spring and summer in both hemispheres at both middle and high latitudes have been measured.

The rate of loss appears to be accelerating. The loss per decade was about 2% higher during the 1980s than during the 1970s (UNEP 1992). Ozone depletion at low latitudes is partly due to the spread of ozone-poor air when the polar vortex breaks up in late spring. The vortex may also process a larger volume of air, changing stable chlorine into more reactive forms and priming this air to destroy ozone as it moves to lower latitudes.

Sulfuric acid particles may also trigger reactions, causing rapid destruction of stratospheric ozone. This may be particularly important after a volcanic eruption. It could also allow a higher rate of ozone destruction above heavily polluted areas (UNEP 1992).

Ozone has also been depleted in the Northern Hemisphere. In northern mid-latitudes, the ozone layer in the 38–43 km slice of the stratosphere thinned by 5–13% between 1979 and 1986, and continues to thin during the winter at altitudes of up to 25 km. During the past few years an Arctic hole has formed, though it has not been as deep or durable as the Antarctic hole. While ozone depletion over Antarctica happens over largely unpopulated regions, the ozone hole over the Arctic could seriously affect the populations of Northern Europe, Canada, Greenland and Siberia.

How do mid-latitude ozone changes originate? This has been studied by observations of ozone depletion following volcanic eruptions. Enhanced ozone loss was observed after the 1991 eruption of Mount Pinatubo (Philippines). Aerosol particles are present in the stratosphere; the largest, with radii over 0.1 mm, are most abundant at altitudes between 15 and 25 km where their concentrations reach 1 cm^{-3}. This aerosol layer has mainly sulfuric acid droplets derived from gases of tropospheric origin, mainly carbonyl sulfide (COS) and sulfur dioxide (SO_2). COS is produced by natural and industrial sources which are constant functions of time. In contrast, SO_2 is injected into the stratosphere by volcanic eruptions, so it is not constant over time after major volcanic eruptions; aerosol concentrations in the lower stratosphere often rise as much as 100 times. Volcanic eruptions in the tropics are far more significant than those occurring at higher latitudes, since meridional circulation favors the global spread of material injected in the tropics. After the eruption of Mount Pinatubo the volcanic material spread throughout the whole stratosphere, essentially as expected from the equator-to-pole direction of the meridional circulation (Garcia 1994).

Reactions occurring on stratospheric aerosols can also free chlorine from reservoir species, in the same way that PSCs do at high latitudes in winter. Reactions (4.42) and (4.43) are both possible on sulfate aerosols. The reaction rate for N_2O_5 hydrolysis (4.42) is mostly independent of temperature, but the rates of these reactions are strongly dependent on the extent of hydration of the sulfuric acid droplets, which itself is a strong function of temperature. Reactions (4.42) and (4.43) become important even at temperatures of 210 K, i.e. warmer than the 193 K necessary for PSC formation and implying that sulfate aerosols might activate chlorine at latitudes outside the polar caps, where the temperature seldom drops below 210 K. Activation of chlorine on sulfate aerosols can thus explain the ozone depletion in mid-latitudes.

Heterogeneous processes enhance the abundance of active chlorine in the lower stratosphere, increasing its impact as an ozone sink. The impact of these processes before the emergence of the ozone hole remained small due to the much smaller amount of chlorine in the stratosphere before the advent of CFCs. The chlorine released from the breakdown of CFCs has quadrupled the chlorine load of the stratosphere, bringing about a major change in the ozone budget. A doubling of CO_2 is expected to lead to colder stratospheric temperatures. In the troposphere, CO_2 traps outgoing infrared emission from the warm surface layers, causing the "greenhouse warming". In the stratosphere, however, its effect is to enhance infrared emission to space, so increases in atmospheric CO_2 would actually induce cooling. Lower polar stratospheric temperatures during northern winter would produce a more stable circulation in the Arctic, similar to that which now prevails over Antarctica. Under these conditions, the development of an Arctic ozone hole remains a distinct possibility (Garcia 1994; Austin et al. 1992). The scientific evidence linking CFCs to stratospheric ozone depletion rests on the establishment of three facts:

1. CFCs get into the stratosphere where they are destroyed, becoming the major source of stratospheric chlorine,
2. Chlorine destroys ozone, and
3. The amount of chlorine in the stratosphere is large enough to cause the observed ozone depletion.

CFCs are undoubtedly the chief source of chlorine in the stratosphere. At the 55 km level, model calculations show that 90% of the chlorine is in the form of HCl and 80% of the fluorine is in HF. Then, from HCl and HF measurements, the increases in stratospheric chlorine and fluorine between 1991 and 1995 can be deduced. These are in excellent agreement with the average increases in total chlorine and fluorine concentration in CFCs and other compounds measured at many ground stations worldwide from 1985 to 1992. The growth in CFC emissions was fairly constant during the 1980s and only began to fall off in 1992.

The average time for CFCs to get to 55 km is estimated to be 5.9 years. By applying this time lag to anthropogenic chlorine containing compounds and natural methyl chloride produced at the surface (but not to surface HCl, because it is too water soluble to reach the stratosphere in any quantity), a chlorine concentration at 55 km of 3.44 ± 0.23 parts per billion by volume (ppbv) has been predicted which is very similar to the observed value of 3.3 ± 0.33 ppbv. Thus, both the observed increases and absolute values confirm that CFCs and other man-made compounds (e.g. carbon tetrachloride, methyl chloroform and HCFCs) are the source of about 80% of stratospheric chlorine. HCl from volcanoes and sea spray can contribute at most only 5%. HCl is the main component of stratospheric chlorine, but another product of CFC photochemistry, chlorine monoxide (ClO), is the main ozone destroyer. The mid-latitudes, where ClO exerts its greatest effect on ozone depletion are also those where ozone decreases are largest below 25 km and near 40 km.

The best support for these links comes from the dramatic loss associated with the Antarctic ozone hole and the 15–20% loss seen in the Arctic winter stratosphere. These cold polar regions have clouds that initiate a powerful chlorine-dominated chemistry. Aircraft, balloons, satellite and instruments on the ground have all recorded large amounts of ClO and a rapid decline in ozone. The ozone loss matches that calculated by using the observed amount of ClO and other reactive compounds (see Brune, 1996).

4.18
The Ozone Depletion Controversy

Some critics have stated that the ozone depletion theory is a sham. So much chlorine is getting into the stratosphere from sea salt, volcanoes, and burning biomass, that CFCs could not possibly have a noticeable effect on the ozone layer (see Taubes 1993). Rush Limbaugh, best-selling author of *The Way Things Ought to Be*, insists that the theory of ozone depletion by CFCs is a hoax, zoologist Dixy Lee Ray, makes the same argument in her book, *Trashing the Planet*.

R. Maduro and R. Schauerhammer have argued that natural sources of chlorine in the stratosphere dwarf any contributions from CFCs. It is felt that in one eruption, Mount Pinatubo spewed forth more than a thousand times the amount of ozone-depleting chemicals than all the fluorocarbons manufactured in history. And the result was only a minor depletion of ozone. Meanwhile, volcanoes have been spewing chlorine for billions of years, and yet the ozone has not gone. Atmospheric scientists counter that these claims have been intensively studied and found incorrect (see Taubes 1993).

Much of the acrimonious debate over ozone depletion has centered on the claim that chlorofluorocarbons pale into insignificance alongside natural sources of chlorine in the stratosphere. If so, chlorine could not be depleting ozone as atmospheric scientists claim, because the natural sources have existed since time immemorial, and the ozone layer has not disappeared.

Maduro and Schauerhammer calculate that 600 million tons of chlorine enters the atmosphere annually from seawater, 36 million tons from volcanoes, 8.4 million tons from biomass burning and 5 million tons from ocean biota. It contrast, CFCs account for a mere 750 000 tons of atmospheric chlorine a year. Besides disputing the numbers, scientists have both theoretical and observational bases for doubting that much of this chlorine is getting into the stratosphere, where it could affect the ozone layer.

However, there is one crucial problem with the argument: chlorine from natural sources is soluble, and so it gets rained out of the lower atmosphere. CFCs, in contrast, are insoluble and inert and thus make it to the stratosphere to release their chlorine. Further, work on stratospheric chemistry does not support the idea that natural sources contribute much to the chlorine there. If sea salt were making it up to the stratosphere, then there should be evidence of sodium from the salt in the lower stratosphere, which is not there. Chlorine from biomass burning should also have a distinctive signature: the chlorine-containing compound methyl chloride. However, the contribution of methyl chloride appears to be much less than that from CFCs.

Even if seawater and biomass do not hold up as major sources of stratospheric chlorine, Limbaugh, Ray, Maduro, and Schauerhammer point to volcanoes as a source that may suffice on its own to render CFCs irrelevant. Mount Erebus volcano in Antarctica has been erupting large quantities of chlorine constantly since 1973. Atmospheric researchers believe that Erebus, although 14 000 feet high, is still several kilometers below the base of the stratosphere in Antarctica, and it does not erupt explosively, which is a necessary condition to lift chlorine from volcanoes into the stratosphere. Thus it appears that volcanoes play a relatively minor role.

For the global picture, atmospheric researchers point to measurements from the ATMOS instrument, which flew on a space shuttle in 1985. The instrument precisely determined the total chlorine budget in the stratosphere by making measurements of 30 molecular signatures, including the major CFCs, as well as their sinks and sources. The measurements showed that chlorine is bound up in CFCs at lower levels of the stratosphere and in the predicted by-products of CFC breakdown, HCl and hydrogen fluoride (HF), at higher levels – just as the ozone theory predicts (see Taubes 1993). Thus the balance of evidence suggests that CFCs make the major contribution to stratospheric chlorine.

S. Fred Singer has countered Maduro and Schauerhammer's arguments about natural sources of chlorine, calling them "red herrings and completely false". Singer also once believed that natural sources of stratospheric chlorine overwhelm any man-made contribution, but the data have convinced him that CFCs are the major source. It is well established that levels of CFCs are increasing in the stratosphere and that chlorine levels are rising in tandem. And the evidence that the Antarctic ozone hole is caused by chemical reactions, in which chlorine plays a key role, is robust.

Yet current understanding of global ozone behavior is burdened with some uncertainty. Among these uncertainties are whether ozone depletion in the Northern hemisphere is due to natural variation and changes in atmospheric circulation, chlorine from CFCs, or some combination of both. Another crucial unknown is whether ozone depletion has led to a measurable increase in the flux of ultraviolet radiation at the Earth's surface. One study showed no increase in ultraviolet radiation in eight locations in the United States, and perhaps a slight decrease. However, the data were obtained from instruments that were not built for measuring yearly trends (Taubes 1993). More recent studies have indicated a significant increase of solar UV at mid and high latitudes linked with stratospheric ozone depletion.

On the one hand, recent evaluations of stratospheric and global tropospheric ozone trends indicate substantial anthropogenic impacts that, if allowed to continue, can result in widespread and unacceptable damage. On the other hand, current and proposed remediation efforts have resulted and will result in severe and potentially unacceptable socioeconomic impacts.

4.18.1
Stratosphere-Troposphere Exchange Process

A need is being felt to elaborate a global measurement strategy to help plan future campaigns of studying stratosphere-troposphere exchange (STE) processes. The traditional local approach has limitations when trying to quantify the global STE; making spot measurements to study STE is like looking at each eddy to study turbulence. The need for chemical airmass characterization to document the transport implies the use of chemical tracers of different lifetimes, in addition to physical quantities. Availability of global measurements of these parameters from satellite platforms is not likely to materialize in the near future.

4.18.2
Stratospheric Temperature Trends Assessment (STTA)

In the lower stratosphere, the model predictions have turned out to be somewhat uncertain because of the incomplete knowledge of the vertical profile of the O_3 loss, the forcing of the surface-troposphere system and the role of changes of O_3 versus changes in well-mixed situations; it is still not settled whether in the upper/middle stratosphere, the CO_2 signal is visible or not and, if not, if the absence of signal is due to the O_3 decrease; there is large dynamic variability on the time scale of interest, and the understanding of the sources of the variability needs to advance in parallel.

4.19
Ozone Trends

Many people would like to have a better understanding of the O_3 trends in the lower stratosphere. Some aim to look at the likely quality of the O_3 measurements to be made in the coming years, to help develop a validation and intercomparison program, and to keep an eye on the developments of total O_3 data quality or trends, especially in the tropics. The available data sets include the following key long-term data sets: ozone sondes, SAGE (stratospheric aerosol and gas experiment), SBUV and SBUV/2 and the Umkehr data set.

Future work is being planned to include a comparison of O_3 profiles, a full description of the SAGE algorithm, a study of the August 1995 Mauna Loa/SAGE O_3 profile intercomparison and analysis of the SME/correlative measurements with the European Brewer–Mast O_3 sondes. The subject of stratosphere-troposphere exchange (STE) is crucial to many areas of atmospheric science and has aroused present concerns about the impact of aircraft emissions on the ozone layer. Previous research on STE had focused virtually entirely on the behavior of the tropopause itself and on the mesoscale phenomenology of strong mixing events such as mid-latitude tropopause folds and deep tropical convection. It is now clear that STE is only one aspect of a global scene of transport and mixing of chemical species and involves the dynamics of the entire stratosphere. Rather than focusing on STE per se, it would be more sensible to consider the broader question of transport and mixing in what is now referred to as the lowermost stratosphere, viz., that part of the stratosphere, whose isentropic surfaces are connected to the troposphere (it is also sometimes called the stratospheric part of the atmospheric middleworld, see Shepherd 1996).

There is an intimate linkage between global and local aspects of exchange, as evidenced from some observed tropical water vapor profiles. There is a crucial distinction between the lowermost stratosphere and the "overworld" (the part of the stratosphere not connected to the troposphere along isentropic surfaces). The overworld is partitioned by a subtropical transport barrier which maintains a contrast between ascending and descending air. Exchange between the overworld and the lower-most stratosphere tends to be dominated by the zonal mean vertical (diabatic) mass flux, which is significantly controlled by an extratropical wave driving the "extratropical pump". In contrast, exchange across the tropopause itself is strongly affected by eddy fluxes.

Although there is a general connection between tropical ascent and extratropical wave driving, a detailed quantitative understanding of tropical upwelling is lacking. The subtropical transport barrier is highly important for STE. It is analogous to the polar-vortex transport barrier, though rather more porous; both are, to some extent, a consequence of potential vorticity (PV) mixing in mid-latitudes, creating edges on each side.

The tropical "tropopause layer" extends from about 15 to 18 km altitude, representing the transition region between clearly tropospheric and clearly stratospheric dynamics. There is a strong seasonal cycle of tropospheric CO_2 which means that CO_2:N_2O correlations in the lowermost stratosphere can be used to infer constraints on tropical-to-mid-latitude quasi-isentropic transport, suggesting transport time scales in the order of 1–2 months. Corroborative evidence using a combination of CO_2, H_2O and N_2O measurements suggests that the main route by which air reaches the lowermost stratosphere from the troposphere is indeed this quasi-isentropic transport across the subtropical edge of the tropical tropopause layer, rather than across the mid-latitude tropopause itself or "up-and-over" in the overworld via the Brewer–Dobson circulation (both of which require diabatic transport). This possible short-circuiting of the Brewer–Dobson circulation is an important effect to quantify (see Shepherd 1996). What do we know about tropopause folds? Turbulent exchange processes are very important. For example, the air in a tropopause fold is typically a 50/50 mixture of tropospheric and stratospheric air.

4.20
Unresolved Issues

The following appear to be some important outstanding issues (see Shepherd 1996) that need to be critically studied:

1. Understanding of how the tropopause is maintained (and evolves seasonally),
2. Quantitative theory of the details of tropical upwelling,

3. Quantitative theory of dispersive mixing and of transport barriers,
4. Relation between the residual circulation and the transport circulation,
5. Determination of the most appropriate measures of STE,
6. Understanding of the tropical "tropopause layer" (roughly 360–380 K), including a quantitative determination of water vapor mixing ratios,
7. Observational constraints on lateral mixing in the tropical stratosphere, especially the lowest part,
8. Determination of mix-down time of filaments in the lowermost stratosphere and mid-latitude upper troposphere,
9. Observational constraint on mid-latitude mixing,
10.Possible use of radars to determine vertical fluxes of tracers, and
11.Relation, if any, between turbulent intensity and vertical diffusivity.

The problem of STE is fundamentally a fluid dynamical problem of transport and mixing. A comprehensive understanding of STE therefore requires a comprehensive theoretical framework to understand transport and mixing.

4.21
Ozone Changes and Climate

Changes in the concentrations of atmospheric gases which interact with the incoming solar and/or the outgoing terrestrial infrared radiation lead to a perturbation in the radiative balance of the Earth's surface and the atmosphere. This constitutes a radiative forcing of the climate system. Thus, because of their infrared absorption features, increases in the concentrations of the well-mixed greenhouse gases are potentially significant for global climate change. The ozone molecule also can absorb infrared; in addition, ozone has absorption bands in the ultraviolet and visible spectrum. Therefore, changes in the concentration of ozone imply complex changes in the Earth's radiative balance.

Ramaswamy (1993) has considered the observed losses of stratospheric ozone in the tropical, middle and high latitudes over the past decade, as well as the increases in tropospheric ozone in some areas of the Northern hemisphere. He has investigated the resultant radiative forcing and the response of the climate system by analyzing the simulated results from a globally – and annually averaged – radiative convective model (RCM), a latitudinally-dependent fixed dynamical heating (FDH) model and a three-dimensional general circulation model (GCM). By making comparisons between the simulated and the observed temperature changes in the stratosphere, the atmospheric effects caused by the resulting changes in ozone were assessed.

Satellite and ground-based platforms indicate that there have been significant losses of ozone during the past decade in the mid to high latitudes of both hemispheres. The TOMS (total ozone mapping spectrometer) satellite data point to a reduction in the column ozone, and the ground-based and SAGE (stratospheric aerosol and gas experiment) satellite data indicate that these losses have occurred mainly below 25 km in the lower stratosphere. There is now considerable evidence that a significant portion of the losses is due to heterogenous chemical reactions involving the halocarbons (WMO 1992). These changes in ozone substantially perturb both visible and long wavelength radiation (Ramanathan and Dickinson 1979; Ramaswamy et al. 1992). While the solar effects due to ozone losses are determined solely by the total column ozone amounts, the long wavelength effects are determined both by the amount and its vertical location. The solar radiative process produces a warming tendency for the surface-troposphere system while the long wavelength radiative processes have a cooling effect. There appears to be an enhancement of the long wavelength effect over the solar, giving rise to a negative radiative forcing, the magnitude of which depends on the season and the geographical region. This can be contrasted with the positive radiative forcing (greenhouse effect) by the non-ozone gases which constitutes the basis for the global warming phenomenon. At higher latitudes, the ozone forcing can be a significant fraction of the sum of the (positive) non-ozone gases forcing over the same time period in both hemispheres (Ramaswamy 1993).

Imposed middle and high latitude decadal ozone losses tend to cause a cooling of the lower stratosphere. This does not only occur at mid and high latitudes, but extends to the tropical lower stratospheric region as well. It appears to be a dynamic feedback to the radiative forcing. The GCM response also indicates that the global lower stratosphere response is essentially one of radiative adjustment. A cooling of the upper troposphere also occurs in the GCMs. In the troposphere, ozone is a greenhouse gas, exerting a radiative effect much like CO_2. Tropospheric concentrations of ozone (see Fig. 4.24) are influenced by the distribution of CH_4, CO, NO_x and non-methane hydrocarbons. Although tropospheric ozone makes up only about 10% of all the ozone in the atmospheric column, its presence is central to the problem of the oxidizing efficiency of the troposphere. Ozone sonde measurements indicate that ozone has increased by 1–1.5% per year in the free troposphere over Europe during the last 25 years. A similar trend has previously been reported for stations influenced by regional air pollution.

Even though tropospheric ozone is only a small fraction of the total ozone in the atmospheric column, the pressure broadening effect makes the effective long wavelength optical path length in the troposphere comparable to that for stratospheric ozone. A 10% increase at all altitudes in the mid-latitude troposphere yields a positive radiative forcing of 0.1 W m^{-2} in that region. The forcing increases approximately linearly with increase in tropospheric ozone, with the largest sensitivity due to changes around the tropopause region (Lacis et al. 1990).

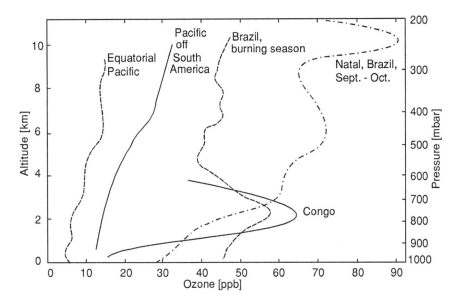

Fig. 4.24. Vertical profiles of O_3 in the tropical troposphere

Although the radiative effects due to increases in tropospheric ozone are extremely significant for the "greenhouse" climate forcing, there is insufficient evidence that such increases are taking place globally, especially in the radiatively significant upper tropospheric regions. One problem is the lack of sufficient number of monitoring stations outside of the land area in the populated regions of the globe. Thus, the actual global forcing due to tropospheric ozone over the past decade can only be speculated (Ramaswamy 1993). Recent changes in atmospheric ozone appear to have perturbed the global radiative energy budget, culminating in the following climatic implications:

1. More ultraviolet radiation is incident on the surface-troposphere system which has ramifications for the chemistry of the troposphere and biological activity on Earth.
2. There is a steepening of the equator-to-pole gradient of the trace gases (including all the "greenhouse" gases and ozone) radiative forcing.
3. If the stratospheric ozone losses are indeed due to heterogeneous chemical reactions involving the CFCs, then it constitutes an offset to the direct "greenhouse" forcing by CFCs.
4. Because ozone plays a leading role in the thermal balance of the stratosphere, there may be some significant impact on the stratospheric (and, possibly, the tropo-

spheric) circulation, with a concurrent effect on temperature changes and specific humidities in the lower stratosphere.

5. The reported increases in tropospheric ozone suggest a significant contribution to the "greenhouse" effect. This yields an effect that is opposite to the negative radiative forcing caused by the loss of stratospheric ozone (Ramaswamy 1993).

The effect of the decadal ozone change in the atmosphere may be examined from two different perspectives. In the first, stratospheric ozone loss and tropospheric ozone increase arise due to entirely different chemical reasons. Thus, one can attribute separate radiative forcing values to the changes, with the stratospheric loss acquiring a negative sign while the tropospheric gain constitutes a positive forcing. In the second, the total effect of stratospheric loss and tropospheric increase may be examined as a net atmospheric effect, irrespective of the cause. Then, the tropospheric gain and the stratospheric loss might generate a small residual surface-troposphere radiative forcing.

According to Ramaswamy, any analyses of atmospheric ozone must consider three important issues. First, the loss of the lower stratospheric ozone exerts a substantial effect on the temperatures there. This would occur in spite of a significant tropospheric increase. Indeed, observed long-term trends in some regions do suggest a cooling trend in the lower stratosphere which would be in accord with expectations of ozone loss. Considerable difficulties remain, though, in attributing a part or all of the trends to the ozone losses. Second, tropospheric ozone increases have not yet been confirmed on a hemispheric or global basis whereas there is evidence of a global stratospheric loss. Third, ozone radiative forcing is extremely sensitive to the altitude and the spatial extent of the losses (Lacis et al. 1990).

4.22
Tropospheric Ozone

Tropospheric ozone significantly influences the radiative forcing of the troposphere-surface climate system. Absolute ozone density between 8 – 12 km in altitude exerts the strongest effect on the troposphere-surface climate system. The photo-dissociation of O_3 determines the oxidation efficiency of the troposphere and, consequently ozone has indirectly modified the concentration and lifetime of other gases, especially carbon monoxide, methane and non-methane hydrocarbons (Mohnen et al. 1994).

By absorbing outgoing long wavelength radiation, O_3 acts as a greenhouse gas. It also absorbs solar radiation, in particular, the UV-B radiation. Any changes in ozone vertical distribution tend to disturb the solar and long wavelength radiative forcing of the troposphere-surface climate system. Ozone density in the 8 – 12 km altitude range strongly modifies the rate of radiative forcing.

Ozone differs markedly from other greenhouse gases. It is a product of photochemical reactions in the atmosphere and triggers the formation of the hydroxyl radical (OH$^{\bullet}$) in atmospheric background air. This radical not only controls the fate of carbon monoxide and methane but also produces the peroxy radical HO_2. The conversion of NO to NO_2 by HO_2 leads to further O_3 formation due to the rapid photolysis of NO_2 which yields the oxygen atom required to produce O_3. However, at very low NO concentrations (less than 10 parts per trillion) HO_2 will actually destroy ozone. Nitric oxide is conserved in the complex process of ozone production and catalyzes ozone formation. Thus, if nitric oxide concentrations change, they will likely modulate O_3 production and, indirectly, OH formation. Thus, there is strong coupling between ozone and methane, and the tropospheric oxidation efficiency. Changes in the upper tropospheric ozone are attributed to:

1. Downward transports of ozone-rich stratospheric air through the tropopause during folding events that are associated with mid-latitude cyclones, and
2. An in situ photochemical production from precursor gases with nitric oxide acting as a catalyser. The residence time of O_3 in the troposphere is a few days or weeks, but its lifetime in the lower stratosphere is much longer.

Some important chemical processes/reactions in the free troposphere (8 – 12 km range) are the following:

$$NO + HO_2 \rightarrow NO_2 + OH, \tag{4.44}$$

$$NO_2 + hv \ (\lambda < 400 \ nm) \rightarrow {}^3O + NO, \tag{4.45}$$

$${}^3O + O_2 + M \rightarrow O_3 + M, \tag{4.46}$$

$$O_3 + hv \ (\lambda < 320 \ nm) \rightarrow {}^1O + O_2, \tag{4.47}$$

$${}^1O + H_2O \rightarrow 2 \ OH, \tag{4.48}$$

$$OH + CO \rightarrow HO_2 + CO_2. \tag{4.49}$$

A critical study of tropospheric ozone in the tropical and subtropical regions is particularly important for the following reasons:

1. Industrial and agricultural growth will be very large in these areas, accompanied by great increases in the emissions of industrial and agricultural atmospheric pollutants, including the chemical precursors of O_3: carbon monoxide, hydrocarbons and NO_x.
2. Because of the large fluxes of solar UV and the higher temperatures and water vapor content, atmospheric photochemistry and the self-cleansing processes are very

strong in the tropical troposphere. Anthropogenic perturbations could affect the atmospheric levels of several important greenhouse gases (e.g. CH_4, HCFCs, HFCs) with potential consequences for the global climate.

3. Because of the high input of solar radiation at the Earth's surface, convection processes maximize in the tropics, resulting in fast and efficient vertical transport of pollutants from the surface to high altitudes, thus affecting the chemistry of the upper troposphere. Because the atmospheric lifetime of NO_x increases strongly with altitude, the O_3 production efficiency also increases, implying a likelihood of a substantial increase in upper tropospheric O_3 concentrations. It is also in these altitude regions that O_3 is most effective as a GHG especially in the tropics.

When the sun shines brightest, it creates the greatest risk of high levels of ozone and other photochemical oxidants in the atmosphere. In the presence of nitrogen oxides and volatile organic compounds, the sunlight sets off a long chain of reactions. But it is not that oxidants only pose a danger during periods of high summer: equally disquieting is the fact that background levels of ozone are steadily increasing and have doubled over the last century in central and northwestern Europe. The concentrations that are now being recorded are high enough to affect people's health, corrode materials, damage vegetation, and contribute to global warming.

Ozone affects mucous membranes, the lungs and the eyes. As a strong oxidant, it induces irritation in nose and the throat. According to the WHO guideline, the average eight-hour O_3 concentration must not exceed 120 µg m^{-3} of air. At that level, however, there is no safety margin. For weak and sensitive persons (about a tenth of the population), the limit is lower. The lung function of children and young people can be impaired after exposure to 160 µg m^{-3} during only 6 hours. A limit of 100 µg m^{-3} has recently been proposed for ozone in Great Britain. In order to come down to that level, the emissions of nitrogen oxides and VOCs (volatile organic compounds) will have to be reduced by up to 95%.

Urban air contains a mixture of several pollutants. It becomes difficult to distinguish the effect of any one from that of the others. Ozone is suspected to join with other pollutants to cause allergies, asthma and lung dysfunction. The photochemical smog formed during periods of high pressure in summer can also include particles containing organic substances and heavy metals with a highly adverse effect on health.

High levels of sulfur dioxide pose a serious threat to cultural objects and buildings. But the breakdown of materials goes unchecked despite markedly reduced levels of sulfur dioxide – a contributing cause probably being the increased concentrations of oxidants and nitrogen compounds, especially nitrogen dioxide and nitric acid. The latter is a highly corrosive substance formed through photochemical reaction.

Nitrogen dioxide, ozone and sulfur dioxide usually corrode stone more strongly in combination than separately. To protect materials, ozone concentrations should not be allowed to exceed 50 µg m^{-3} as long-term average. Ozone and other oxidants (such as peroxyacetyl nitrate, PAN) also damage vegetation.

Plants that are most sensitive to ozone are those with short-lived leaves, such as spinach and clover. Visible damage may appear on the leaves after only a few hours exposure to concentrations of 120 µg m^{-3}. The damage is not so easily discernible in plants having long-lived leaves, but their life span is nevertheless shortened by high concentrations. While the knowledge of the manner in which ozone affects plants is scanty, there is considerable evidence of damage being caused to various cell structures as a result of ozone penetrating into the stomata. O_3 also upsets the functioning of the stomata, thereby affecting the moisture balance of the plants. In coniferous forests, the leaves age more quickly as a result of exposure to ozone. Increasing concentrations of ozone also enhance global warming, since O_3 absorbs radiation that would otherwise have escaped from the atmosphere.

4.22.1
The Meteorological Environment of the Tropospheric Ozone Maximum over the Tropical South Atlantic

Global distributions of total column ozone have been obtained from the total ozone mapping spectrometer (TOMS) for about a decade. TOMS data usually indicate strong gradients of total ozone at mid and high latitudes. These horizontal distributions and their evolution are governed by large-scale atmospheric circulations which vary greatly from day to day. In particular, the depth of the troposphere and the position of the jet stream are closely related to column ozone. Ozone concentrations above the tropopause have been measured by the stratospheric aerosol and gas experiment (SAGE) and can be obtained by subtracting SAGE stratospheric values from TOMS total column amounts. These tropospheric values agree closely with those from ozone sondes (Fishman et al. 1990). From these maps of tropospheric residuals, the largest concentration (45 DU) is located off the west coast of Africa during the Austral spring (September – November).

Biomass burning may be a significant source of atmospheric trace gases such as ozone. Oxidation of CO, along with methane and nonmethane hydrocarbons, is an important source of tropospheric ozone when sufficient NO_x is present. A substantial portion of the production of these precursor gases is attributed to tropical biomass burning. The tropospheric ozone maxima in the southern tropical Atlantic and Indian Oceans may be due to the transport of byproducts from biomass burning over Africa and South America (Krishnamurti et al. 1993). The lifetime of ozone is an important factor when considering its transport. Since both water vapor and ultraviolet radiation are abundant in the tropical boundary layer, the lifetime there is only 2 to 5 days

(Fishman et al. 1991). Thus, ozone in the low levels of the atmosphere will not survive long enough to be transported over large distances. Lifetimes in the middle troposphere are much longer, up to 90 days. This means ozone can be transported much farther if it can be lifted out of the boundary layer, to where its lifetime is longer and winds are stronger. Cumulus convection, especially in the dry season, acts very efficiently in vertically lifting the ozone to higher levels (Pickering et al. 1992).

4.22.2
Climatic Impact of Ozone Change and Aerosols

There are three issues of some importance to the radiative balance of the Earth. One relates to the effect of sulfate aerosols which are derived mainly from SO_2 produced by fossil-fuel burning, though volcanoes can periodically emit amounts that for some time dominate all other sources. The second issue concerns the effect of reduced concentrations of ozone in the lower stratosphere caused by the use of CFCs (here also volcanic emissions may play an important episodic role). The third issue involves the long-term increase in tropospheric ozone resulting from increased emissions of nitrogen oxides and reactive hydrocarbons. The first two factors tend to exert a net cooling influence on the climate, the third causes a net heating. For example, unlike the radiative forcing by the globally distributed GHGs such as carbon dioxide and methane, the forcing by sulfate aerosol and ozone changes is strongly regional and seasonal. Again, in contrast to the historic record of atmospheric CO_2 and CH_4, the amounts of atmospheric aerosol and ozone present in the near and distant past, whether from natural or human sources, are not well described.

4.22.3
Tropospheric Aerosols

Anthropogenic aerosols influence atmospheric radiative transfer by direct light scattering and by enhancing the reflectivity of clouds. They may also modify the rate of precipitation from clouds and thereby alter the hydrological cycle. Radiative forcing by tropospheric aerosols is quite different from the trace gas forcing historically considered by climatologists; the short residence time of the aerosol (~ 1 week) makes it difficult to model the spatial distribution of aerosol loading and its interaction with the meteorology and radiation. This short residence time also has implications for possible future changes in emissions since, in contrast to the longer-lived GHGs, aerosol forcing would closely follow emissions.

Key anthropogenic influences are sulfate, organic aerosols, soot, and possibly mineral dust. Their properties depend strongly on particle size and morphology and on whether the aerosol is internally or externally mixed. For spherical or near-spherical particles characteristic of inorganic salt aerosols and organic aerosols, the particle radius is critical. Particle size of single component inorganic salts depends on relative humidity, increasing with increasing relative humidity above the deliquescence point.

For externally mixed aerosols, one may treat the influence of the multiple components (scattering or absorbing) as additive. However, concern remains regarding internally mixed aerosols (two or more components in individual particles), for which the optical effects are not simply additive. Absorption by soot may be strongly enhanced depending on its state of dispersion in an internally mixed aerosol. Relative humidity influence on the radius of mixed salt aerosols is fairly well described, but confidence weakens as the organic component increases (see IPCC/WMO 1993). Sulfate comes from SO_2 emitted from fossil fuel combustion. It is present mainly in the Northern Hemisphere, resulting in a major hemispheric asymmetry in forcing. Estimates of its global forcing range from -0.6 to -2.0 W m^{-2} (see IPCC/WMO 1993). The total current direct forcing due to the biomass burning aerosols has been estimated to be about -0.8 W m^{-2} (see IPCC/WMO 1993). The magnitude of the forcing may have increased by 0.35 W m^{-2} (from -0.45 to -0.8 W m^{-2}) since 1850.

4.22.4
Mineral Dust

Soil dust from deserts and arid farmlands makes up a large fraction of the mass of atmospheric aerosol, but it may not be a dominant factor in anthropogenic climate forcing, firstly because the particle sizes are large enough that the scattering efficiency per unit mass is low ($\sim$0.5 m^2 g^{-1}), and most of the scatter is in the forward direction. Also, a large fraction of soil dust is natural. The total forcing by soil dust is likely to be of the order of -1 W m^{-2}; however, the anthropogenic fraction of this is quite small.

The expanding human habitations in the semi-arid regions located mainly on the outskirts of the major deserts have modified the land use of these highly sensitive areas. More intense agricultural and pastoral activities have allowed aeolian erosion to act more efficiently on larger surfaces previously protected during a longer time by the natural vegetation. The effects of soil-derived dust on climate mostly depend on the size distribution of such particles. The submicron component of these particles is greatly enhanced for intense dust events as compared to moderate rising conditions (IPCC/WMO 1993). The more intense the dust storm, the smaller are the dust particles. Biomass burning has the potential to act as a significant source of soil-derived dust by the re-mobilization of dust which has been deposited previously (IPCC/WMO 1993).

4.22.5
Indirect Effects of Aerosols

Cloud albedo is sensitive to droplet concentration if other cloud properties are unchanged. Increasing aerosol concentrations in the size range of 0.1 to 1.0 μm can increase the mean droplet concentrations and hence cloud albedo. The relationship between these parameters depends on the cloud type, the source and age of the aerosol and the unperturbed aerosol concentration. Though the global forcing due to the indirect effect of anthropogenic aerosols has not been critically estimated, the effects appear to be significant. An estimate of the potential effect is that a droplet concentration change of 15% in marine stratus could probably decrease the global heat budget around 1 °C m^{-1}, so long as the other cloud properties remain unchanged.

Industrial emissions influence the CCN concentrations and cloud droplets on scales of hundreds of kilometers, but the geographical extent of the influence and the magnitude of the hemispheric or global concentrations are not well understood. Increases in droplet concentration appear to reduce the rate of conversion of cloud water to precipitation, if other cloud properties remain unchanged, but the effect cannot be quantified. Precipitation forms a significant part of the water budget of layer clouds. Any reduction in precipitation efficiency will naturally lead to increased cloud lifetimes, increased cloud cover and increased atmospheric water content. This will in turn increase the albedo and also increase cloud optical thickness.

The microphysical and scattering properties of cirrus clouds depend on the numbers of ice-forming nuclei present in the upper troposphere. Aerosol may be formed in the upper troposphere by gas-particle conversion, and this region may also be influenced by intrusions of stratospheric aerosol.

4.22.6
Stratospheric Aerosols

While the stratospheric aerosol load shows massive changes due to explosive volcanic eruptions leading to a mass increase by a factor of 100 or more, no such variations in the overall aerosol mass load in the troposphere are seen. On the contrary, the size distribution and composition of stratospheric aerosols are rather uniform compared to the strong variability both in size and chemical composition of tropospheric aerosol particles, which are increasingly due to man's activities. Furthermore, whereas the spatial distribution of tropospheric aerosol particles is rather patchy, in the stratosphere it is quite uniform. The long-term stratospheric aerosol record shows at least three trends:

1. Episodic volcanic enhancements,
2. Polar stratospheric clouds (PSCs) and clouds just above the tropical tropopause, and
3. A background level. At normal stratospheric temperatures, aerosols are most likely supercooled solution droplets of $H_2SO_4 \cdot H_2O$. The primary source of stratospheric aerosols is volcanic eruptions that are strong enough to inject SO_2 buoyantly into the stratosphere. Aerosol sizes range from hundredths of a micrometer to several micrometers.

Sometimes after an eruption, and following SO_2 conversion to H_2SO_4 and subsequently sulfuric acid aerosols, aerosol loading decreases at rates varying with altitude and latitude. The decay appears to be due to sedimentation, subsidence and exchange through tropopause folds. The net effect of this postvolcanic dispersion and natural cleansing is a greatly enhanced aerosol concentration in the upper troposphere after a major eruption. Except immediately after an eruption, stratospheric aerosol droplets tend to be concentrated into three distinct latitudinal bands, one over the equatorial region ($\pm 30°$) and the others over each high latitude region, 50 to 90°N and S. Following a low latitude eruption, aerosol is dispersed into both hemispheres within about 6 months, whereas following a mid-to-high latitude eruption, aerosols tend to stay primarily in that hemisphere where the eruption has occurred. Potential sources of a background aerosol component include carbonyl sulfide (COS) from the oceans, low level SO_2 emissions from some volcanoes and various anthropogenic sources, e.g. industrial and aircraft emissions.

Polar stratospheric clouds (PSCs) are those formed at cold stratospheric temperatures primarily in the polar regions and at altitudes of less than 25 km. They occur in two general classes: type 1, HNO_3/H_2O clouds which form at temperatures slightly above the frost point, and type 2, H_2O ice clouds which form below the frost point. Both classes efficiently catalyze direct chlorine activation through surface heterogeneous chemical reactions. PSCs are responsible for the stratospheric dehydration and denitrification frequently observed in the Antarctic and, to some extent, also for denitrification in some Arctic winters. Satellite sightings of Antarctic PSCs occur in the May–October period mostly in August. Sightings of Arctic PSCs have so far been confined to the December–February period.

4.23
Impacts of Ozone and Carbon Dioxide on Crop Yields

Compared to preindustrial levels, anthropogenic emissions lead to considerable increases in the concentrations of several trace gases such as CO_2, CH_4, N_2O, CFCs and tropospheric O_3 in the atmosphere. Though major concerns are actually related to future consequences of the radiative forcing of these trace gases for the global climate system, CO_2 and surface O_3 have become important in directly affecting plant growth and crop production (e.g. CO_2 fertilization, yield reduction due to O_3).

O_3-induced crop losses in Germany have been estimated for the period from 1984–1992 by calculating exposure statistics from statewide records of surface O_3 concentrations in forested and rural as well as suburban and urban sites. Ambient ozone concentrations were higher at rural or forested locations, and showed a less pronounced diurnal variation and an increasing overall trend compared with urban sites. Crop losses to O_3 are usually expressed as a relative yield change compared to an assumed "background level" of O_3, which roughly reflects the surface O_3 concentration at the end of the last century. Average potential yield losses from 1984–1992 ranged between 0.1 (barley, relatively O_3-tolerant) to 32% (O_3-sensitive spring wheat).

However, the above responses were estimated under the then prevailing levels of CO_2 and do not account for the potential CO_2-fertilization effect resulting from the increased level of CO_2 in the atmosphere from roughly 290 ppm at the end of the last century to the present values of over 350 ppm. On the other hand, most CO_2 enrichment studies of crops under ambient or near-ambient conditions were not controlled for O_3 or other air pollutants. Hence, a realistic baseline yield, i.e. the yield potential one century ago without CO_2 fertilization and O_3 pollution, is hard to ascertain for most crops. There are some indications that potential crop losses to O_3 concentrations, while disregarding a concomitant increase of atmospheric CO_2, are overestimated. Similarly, future prospects of regional crop production in a high CO_2 world strongly depend on the estimated trend of tropospheric O_3.

4.24
The Montreal Protocol

An important protocol was signed by 27 countries in Montreal in September 1987 (the Montreal protocol on Substances that Deplete the Ozone Layer). This committed every signatory state to reduce its use of certain CFCs by 50% of their level of use in 1986 by 1999. In 1990 the Montreal protocol was further strengthened. The use of many CFCs and halons was to be eliminated by the turn of the century. More than 80 countries agreed to these changes. The protocol regulates a wide range of substances, including five CFCs and three halons. It specifies heavy cuts in the consumption of CFCs, and it provides tough trade sanctions against countries that do not join the treaty.

The atmospheric concentrations of man-made chlorine are increasing. The abundance of stratospheric chlorine is now about 3.4 ppb, but chlorine will peak at the turn of the century at about 4.1 ppb even if all nations adhere to the Montreal protocol by eliminating CFCs by 2000 (Kerr 1992). All that chlorine will do more damage than has been assumed hitherto. In January 1992 ozone losses of up to 20% were found in the Northern Hemisphere and a maximum depletion over Russia of 40 to 45% below normal for a few days (Rowlands 1993). These findings prompted the World Meteorological Organization (WMO) to issue the following statement: "On the whole, the 1991–92 winter can be classified among those with the most negative deviation of systematic ozone observations, which started in the mid-1950s" (WMO/UNEP 1992). Over 70 countries agreed to move the phase-out deadline for CFCs and carbon tetrachloride (used in dry cleaning) forward from the year 2000 to 1996. Other additions to the protocol defined an end to the use of halons (used in fire extinguishers) by 1994 instead of the year 2000 and also of methyl chloroform by 1996 instead of 2005. There is, however, little agreement but continuing debate with respect to the listing and reduction of methyl bromide, the establishment of a phase-out timetable for the less dangerous hydrochlorofluorocarbons (HCFCs, substitutes for CFCs) and finally, establishing the multilateral fund to assist developing nations.

The final agreement contains clauses to cover the special circumstances of several groups of countries, and it is quite flexible; it can be tightened as the scientific evidence strengthens, without having to be completely renegotiated. Indeed, it sets the "elimination" of ozone-depleting substances as its "final objectives". The Protocol came into force on 1 January 1989, by when 29 countries and the EEC, representing approximately 82% of world consumption had ratified it. Since then many more countries have joined. Table 4.2 shows the estimated end-of-use dates world-wide for controlled CFCs. The heart of the treaty is a series of ever more stringent limits on CFC use. Each party was required to at least freeze its consumption of the five controlled CFCs (CFC-11, CFC-12, CFC-113, CFC-114 and CFC-115) at 1986 levels by 1 July 1989. By mid-1993 consumption had to be no more than 80% of 1986 levels, and by mid-1996 it had to be down to half of this. Meanwhile, consumption of the three halons (Halon-1211, Halon-1301 and Halon-2402) had to be at least frozen at 1986 levels by 1 February 1992. These chemicals are much less widely used than the CFCs, but global consumption has been growing even more rapidly.

Table 4.2. Estimated end-of-use dates for controlled CFCs (UNEP 1992)

Use/application	Sector	Date
Refrigeration		1989–2015
	Domestic	1995–1999
	Commercial/retail	1989–1999
	Transport	1989–2010
	Cold storage	1989–2005
	House air conditioning	1991–2015
	Industrial	1989–2010
	Heat pumps	1989–2005
	Mobile air conditioning	1994–2010
Flexible foams		1989–1993
Rigid foams		1989–1995
Solvents		1989–1995
	Electronic	1995–1997
	Metal cleaning	1993–1996
	Dry cleaning	1993–1995
Miscellaneous		
	Aerosols (non-medical)	1990–1995
	Aerosols (medical)	1995–2000
	Sterilization	1990–1995

Consumption is calculated by weighting the chemicals according to their power to destroy the ozone layer. They are all assigned figures corresponding to their relative "ozone depleting potential" (ODP): CFC-11, CFC-12 and CFC-114 are weighted at 1 in the Protocol and the other values range from 0.6 for CFC-115 to 10 for Halon-1301. So, for example, each tonne consumed of CFC-11 is equivalent to 600 kg of CFC-115. The actual amounts of each chemical that a country consumes (that is, its production plus its imports, minus what it exports) are multiplied by their ODPs and the resulting figures added together to give the total "consumption", which has first to be frozen and then reduced. This gives countries flexibility, while protecting the ozone; a country can decide to cut its use of any combination of the substances in meeting the targets.

The protocol also requires countries to make similar cuts in the production of the chemicals, but gives them wider options than over consumption levels. Production levels need to be frozen and cut in parallel with consumption. But it is recognized that some countries might need some extra allowance, firstly to produce chemicals for developing countries (which are treated rather differently under the treaty), and secondly because the international industry will rationalize itself as it shrinks (as a factory in one country closes, part of its production may be transferred to a plant in another country).

So any country is allowed 10% extra production on the CFC and halon freezes and on the first CFC cut and a 15% excess at the second and larger cut (UNEP 1989).

Several groups of countries secured concessions to help them over particular difficulties and to respect international equity. Countries with small CFC industries – producing less than 25 000 tonnes a year – are allowed to trade production excesses greater than those listed above with other parties, provided that their total combined production does not exceed the limits.

A major concession was granted to developing countries. Provided its yearly consumption of the chemicals is less than 300 g a head (i.e., less than a quarter of the US level), a developing country could delay implementation of the phase-down schedule for 10 years so as to meet its "basic domestic needs". Developing countries may need to increase their consumption as they develop, but still be brought under long-term control. Developing countries that joined the treaty were guaranteed access to alternative substances and technology and have been offered subsidies and other incentive.

All parties were required to ban imports of the bulk chemicals from non-parties by the beginning of 1990, and effectively do the same for imports of products containing the chemicals within another 3 years. Developing countries enjoying the 10-year delay in implementation were not to export the chemicals to non-parties from the start of 1993. Such stringent restrictions in trade provided an incentive to countries to join the treaty or lose their markets and supplies. Equally important, the treaty bound the parties to monitor the effectiveness of the control measures in 1990 and every 4 years afterwards on the basis of expert assessments.

Computer models predicted at the time that if both developed and developing countries observed the treaty, it would stabilize depletion of the ozone layer at a few percent during the first decade of the next century; without it, massive depletion would develop. In human terms, the treaty would prevent 1 860 000 deaths from skin cancer and 38 million cataracts among people born before the year 2075. It would also cut the contribution of CFCs to global warming by about a third (UNEP 1989).

Since the protocol, events have developed rapidly. New evidence has showed that much tighter controls are needed than agreed to under the protocol. In 1987 in Chile, the airborne Antarctic ozone experiment was conducted with two aircraft, packed with scientific instrumentation (a converted DC8 airliner with a team of scientists aboard, and a one-man single-engined ER-2, an upgraded version of the U2 spy plane) which could fly 20 km up into the stratosphere. In August and September 1987, both planes flew 12 missions – a total of 175 000 km – into the area of the ozone hole. Until then, the suspicion that CFCs were primarily responsible for the hole was just the most likely theory among many. But the direct measurements taken by the airborne experiment proved the chemicals to be responsible beyond reasonable doubt. Measurements from these flights showed that as they traveled south, levels of chlorine monoxide increased and ozone levels fell correspondingly (Fig. 4.25).

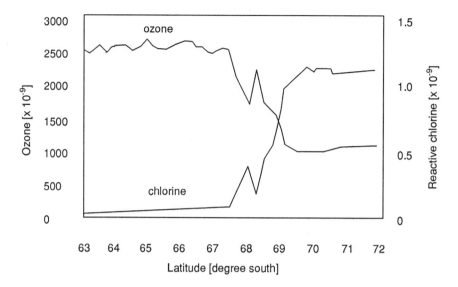

Fig. 4.25. Comparison of stratospheric concentrations of ozone and chlorine measured during an ER-2 flight from Chile to the Antarctic

Within 3 years of the publication of the first paper reporting the hole, the world's scientists had unraveled the complex meteorology and chemistry of the Antarctic stratosphere and knew what was causing it. In March 1988, another study convincingly demonstrated that the ozone layer was being destroyed on a global scale; the ozone layer had decreased by about 1% in summer and 4% in winter between 64 and 30°N, the area of the world (outside Antarctica) where the most reliable data were available. The same study also reported a year-round depletion of 5% at all latitudes south of 60°S (UNEP 1989). Over the North Pole, it was found that the crucial ozone-depleting chemical formed by the breakdown of CFCs was over 50 times more abundant than it should be. The conditions were similar to those found in Antarctica immediately before the opening of the ozone hole.

The Montreal protocol was ahead of the scientific knowledge of its time, but science has since overtaken it. It has been demonstrated that it does not go far enough. It will allow levels of chlorine in the atmosphere to double or more over the next 50 years and rise to about five times the 1986 level by the end of the next century. At the time it was agreed, the best computer models suggested that depletion of the ozone layer would stabilize within 30 years, under its provisions, and then possibly recover. But the work of the ozone trends panel strongly suggests that the models underestimated the danger (UNEP 1989). Furthermore, the Protocol was never designed to heal the Antarctic ozone hole; now that we know that CFCs are to blame, we also know that stricter controls will be necessary if the hole is not to deepen any further.

In meetings in May 1989 held in Helsinki and London, 81 countries unanimously agreed to phase out the production and consumption of the CFCs controlled by the Montreal protocol as soon as possible but not later than the year 2000, to phase out halons and to control and reduce other significant ozone-depleting substances as soon as feasible. It was also agreed to hasten the development of environmentally acceptable substitutes.

Industry has also moved fast. By the end of 1988, voluntary 90% cutbacks in the use of CFCs in propellants had been announced by industry in Belgium, the Federal Republic of Germany, the Netherlands, Switzerland and the United Kingdom. The United States food packaging industry comprising some 100 companies, voluntarily abandoned CFCs in December 1988. Several American multinational companies engaged in CFC production announced, even before the Montreal protocol came into force, that they planned to phase out manufacture of all CFCs controlled by its provisions (UNEP 1989). Serious efforts are being made to find substitute chemicals that do not damage the ozone layer. Also, many companies are concentrating on recovering and recycling CFCs so as to re-use them instead of releasing them to the atmosphere. Recycling and conservation also offer one of the best short-term prospects for controlling the use of halons.

The world's response to the first global environmental threat of ozone depletion has been encouraging. In a short time, more than 150 countries have agreed to phase out chlorofluorocarbons and other ozone-destroying chemicals. Together we have built a means of dealing with a common problem that can be useful when tackling other environmental threats.

Sweden has played a key role in bringing different nations together to solve the problem of ozone depletion. A ban on the use of chlorofluorocarbons (CFCs) in sprays was adopted in Sweden as far back as 1979. In 1988 the Swedish parliament banned the use of CFCs after 1995. One of the goals of this plan was to demonstrate that a phase out was technically and politically possible. By January 1995, 93% of ODS use had been phased out in Sweden.

According to the World Meteorological Organization, for a few days in early 1996, ozone values fell to unprecedented lows for the northern latitudes – 45% ozone deficiencies were reported over the sub-polar region from Greenland to Scandinavia to the western part of the Russian Arctic. Ozone values as low as 250 mbar cm were recorded for many days. This caused the monthly mean ozone deficiency to exceed 20%. Fortunately, in contrast to Antarctica, the extremely low ozone values in the Arctic lasted only for weeks and not months. During the Antarctic spring of 1995, the surface covered by the ozone hole exceeded 20 million km^2 for more than 40 days, and from its appearance in August to its last days in early December, was the longest on record.

Current models predict that ozone depletion will peak sometime around the year 2000, and ozone levels are expected to return to normal (pre 1979) by around 2065.

Some countries have made significant advances in their environmental policy implementation during the last decade but have failed on a number of important issues such as curbing CO_2 and NO_x emissions. In contrast, reductions in SO_2, CFC and VOC are less closely linked to the growth in manufacturing which is the crux of the environmental problem.

The Montreal protocol on curbing of CFC emissions, besides obliging countries to reduce their own emissions, also provides for grants to help developing countries. However successful the CFC approach may be, it does not necessarily lead to a reduction in the amount of raw materials and energy consumed. Environmentally damaging substances are being replaced by less damaging or benign substances. However, the agreements reached in the EC on the stabilization of CO_2 emissions go further towards reducing consumption. The EC has agreed to establish a ceiling for itself. New industrial developments will only be permitted if the resulting rise in CO_2 emissions is compensated for elsewhere. However, some are skeptical about the measures being taken by industrialized countries in an attempt to adapt their patterns of production and consumption. We must ensure that means are redistributed. This can be done in some countries, for example, by investing some of the money available for wage rises in the environment, at both national and international level, rather than paying it out in increased wages.

The redistribution and adjustment of the Western lifestyle are important issues which the new environment policy must address. The slow progress made in this may has largely been due to the growth in car ownership. Something must be done to curb car use. Emphasis needs to be placed on clean production for the domestic market. The environment will have to be evenly distributed among over 7 billion people.

"England used half of the world's resources to reach its present prosperity. How many worlds would India need to achieve the same?" It was with these prescient words that, over half a century ago, Mahatma Gandhi unwittingly laid the basis for Agenda 21. Countries which work together to increase their economic growth also increase their environmental problems. With the growth of the world economy, local and regional environmental problems have assumed global proportions. International cooperation has acquired a new dimension: industrialized and developing countries are in it together, ecologically as well as economically.

For centuries there has been international trade, but the enormous growth in production and the goods we produce have begun to pollute the global environment. Feed for Dutch cattle is grown in South East Asia, leading to a manure surplus in Netherlands and soil erosion in Asia. To provide the Dutch with aluminum for their beer cans, large areas of the Brazilian rainforest are cleared, while the waste mountain grows ever larger in the Netherlands. It has been clear since the 1970s that the countries of the world are working in tandem to pollute the environment we all rely on. Similarly, we have to work together to clean it up. Although we are nowhere near a solution, many forms of cooperation are beginning to emerge.

In the Netherlands, a ban on the use of CFCs in cooling systems, insulation, spray cans, solvents and detergents has come into force since January 1993. The ozone-depleting gas halon will also no longer be used in fire extinguishers. The legislation which brought about this ban – the CFC Decree – is in line with the agreement reached internationally in Copenhagen in November 1992. It was decided that CFC production in industrialized countries should cease by 1996, 4 years earlier than agreed in 1990. Developing countries will have a further 10 years to introduce the ban. Western countries agreed to deposit 240 million dollars in a fund which will be used to help developing countries to wind down their CFC production. But the use of HCFCs will be permitted until 2030 which means that the ozone layer is anything but safe, even after the promises made in Copenhagen.

Imperial Chemical Industries Ltd. (ICI) recently developed KLEA-135a, a product containing HFCs, which has an ozone depletion potential (ODP) of zero. HFCs are hydrocarbons which contain no chlorine, but hydrogen atoms instead. Industry is investing in them as an alternative to HCFCs, as substitutes for CFCs. According to ICI, KLEA-134a offers the most effective and energy-efficient solution. Dupont de Nemours also recently launched an HFC-based product, SUV-A, in the Netherlands. However, even if the ODP might be zero, chlorine is still released during production. It is also a greenhouse gas, so HFCs may not really be the ultimate solution.

4.24.1
Effects of the Protocol

A halving of the concentration of ozone above a 75 ppb limit will be the main beneficial effect, at the close of the 1990s, of the international agreement to reduce emissions of volatile organic compounds (VOCs). Yet the agreement may on the whole, fail to bring the concentrations down to below the critical levels (Agren 1992). Most of the 22 signatories to the VOC Protocol to the UN ECE Convention on long range transboundary air pollution (November 1991) have agreed to reduce their emissions of VOCs by 30% between 1988 and 1999.

A major difficulty in assessing the effects of the protocol was the totally inadequate reporting of emissions from most countries. By agreement within the ECE, all countries have to report their national emissions of VOCs, man-made and excluding methane. But many have not sent in any reports at all, others have included methane, and some have failed to distinguish between anthropogenic and natural emissions. It appeared that a reduction of VOC emissions according to the protocol would not have any great effect on the long-term mean levels of ozone. They would be 4–8% lower in northwestern Europe, but only 1–4% lower elsewhere. The principal aim of the protocol was however to bring down the occurrence of episodes with peak levels of ozone. The greatest improvement would be a reduction of the number of ppb-hours in excess of 75 ppb - amounting to 40–60% over the greater part of Europe.

The number of ppb-hours in excess of 40 ppb is calculated to drop by 15–20% in northwestern Europe, by 5–15% in the rest of western Europe and 5% or less in the east. Particular interest attaches to the excess of ozone above 40 ppb, because this has been proposed as the base level for determining the critical levels for ozone (Agren 1992). As for the critical exposure level for ozone, the UN ECE task force on Mapping has suggested a cumulative exposure during daylight in the course of a growing season of 300 ppb-hours above the 40-ppb baseline. Some regard this as being too high. It appears from the study that the proposed critical level of 300 ppb-hours above 40 ppb is already being greatly exceeded over practically the whole of Europe, and that it will continue to be exceeded even after implementation of the VOC protocol. Bringing down concentrations to below the critical level will require a very large reduction of the emissions both of NO_x and VOCs in Europe (Agren 1992).

Nothing can be confidently predicted about the future political aspects and implications of the ozone layer, but some general trends are discernible. It is highly likely that the phaseout schedules for each of the controlled chemicals will be brought forward. A consensus is emerging that HCFCs will be more tightly regulated. The future of methyl bromide as a controlled chemical is less clear: in future, changes to its control schedule may not require amendments to the protocol but only adjustments. It may also be safely predicted that there may be pressures in the future to amend the protocol so that other chemicals containing either chlorine or bromine would be brought under the agreement. The number of possible ozone depleting substances is enormous.

The participation of countries, especially developing nations, in the protocol will continue to grow. Scientific findings have encouraged politicians to accept more rapid phaseout schedules. But the power of industrialized interest in the political process is equally undeniable. Indeed, science, economics and politics will all be crucial in the future progress of the international community to protect the ozone layer.

4.25
Recent Ozone Developments

Recent findings have indicated that statistically significant ozone reductions are continuing everywhere except in equatorial regions. The intensity and size of the Antarctic ozone hole has increased to record high levels in the last few years. Unprecedented ozone decreases have been recorded over mid and high latitudes of the Northern Hemisphere. Such changes are generally consistent with our understanding of ozone depletion processes related to increasing levels of halocarbons and temperature changes. More serious ozone depletion can be expected over the next few decades until stabilization occurs as a result of actions taken under the Montreal protocol.

There has been much progress in the understanding of stratospheric processes, and the ability to simulate and predict ozone changes has improved with new three-dimensional modeling capabilities that have become available. But uncertainties remain over ozone loss at mid and high latitudes, sources and sinks of some ozone depleting substances (notably methyl bromide) and the microphysical and chemical processes, particularly the role of iodine compounds associated with ozone depletion. Not all the recorded loss of ozone in the Northern Hemisphere is attributable to known ozone depleting substances or processes. It seems probable that increasing greenhouse gas concentrations are also promoting ozone depletion by cooling the stratosphere.

Although there has been a significant increase in the number of UV-B monitoring sites, only minor improvement has occurred in the quantification of the environmental effects of increased UV-B radiation. This is because most of the recommendations that were made in the 1991 UNEP Panel Report on Environmental Effects are still unfulfilled. In the light of the above situation, the following recommendations were made by the WMO/UNEP in 1992:

1. Maintenance and further development of monitoring and archiving of measurements of stratospheric and tropospheric ozone, including vertical profiles and other trace species and aerosols, are essential, and the development and implementation of new observational capabilities such as aircraft and satellite-based measurements should be pursued,
2. Increased understanding and quantification of stratospheric and tropospheric processes through routine monitoring and experimental campaigns are necessary to understand current changes and to further develop and implement predictions of stratospheric change both for the short and long terms,
3. The interactions between ozone and climate and the impact of aircraft emissions need to be given higher priority in research than up to now, and
4. Research on the effects of UV-B, and efforts to monitor such effects, need to be intensified substantially. Base line data on biological systems should be established. International co-ordination and co-operation across all areas of impacts should be encouraged to establish the interactions with other environmental factors such as climate change.

WMO is giving particular emphasis to improvements of the global ozone observing system (GOOS) which has provided the essential data in the study and detection of the depletion of the ozone layer. WMO has also started issuing near-real-time information on the state of the ozone layer during the Antarctic spring and Northern Hemisphere winter-spring seasons as operational activity.

There has been a considerable increase in the number of UV-B monitoring sites both for broadband and spectral measurements and progress is being made towards better coordination and standardization of these measurements.

The atmospheric observations, laboratory investigations and modeling studies of the past 5 years have allowed better understanding of the anthropogenic and natural chemical changes in the atmosphere and their relation to the decline in the Earth's ozone and possible effects on the radiative balance of the climate system. These studies indicate that the byproducts of human-made CFCs and halons are the culprit for the Antarctic ozone loss and support preventive actions agreed by the Montreal protocol aimed at recovery to the natural state of the ozone layer sometime in the second half of the 21st century.

The total ozone decline, which started in the 1970s continues. It is statistically significant all year round except over the 20°N-20°S tropical belt. The GOOS quality controlled data from more than 40 stations with long-term observations supplemented by another 100 stations and satellites operating during the last 2 decades show that the ozone decline over the mid and polar latitudes is about 10% relative to the ozone levels of 1950s and 1960s. Taking into account known natural variability (e.g. annual and solar cycles, the quasi-biennial oscillation), the decline in both hemispheres is particularly strong during the winter and spring (over 6–7% per decade) and is only half of that during the summer and autumn seasons. Detailed studies show a statistically significant increase in the rates of the negative ozone trends by about 1.5–2.0% during 1981–1991 compared with 1970–1980. As the ozone loss is latitudinally and logitudinally non-symmetrical, this may have important consequences in the long run.

The main ozone losses occur in the lower stratosphere, where over mid-latitudes, especially during November–May, the partial ozone decline has exceeded 20% during the last two decades. At the same time, all European and Japanese ozone sounding stations have recorded increases of the tropospheric ozone from the 1960s through the 1980s, and these can have a significant radiative impact.

Ozone concentrations in the boundary layer over populated regions of the Northern Hemisphere have increased by more than 50% during the past three decades due to the photochemical production from anthropogenic precursors (e.g. CO, NO_x, hydrocarbons). Export of ozone from North America is a significant source for the North Atlantic region during summer, and biomass burning is a source of ozone (and carbon monoxide) in the tropics during the dry season. An increase in UV-B radiation may change the concentrations of some chemically and climatically important tropospheric constituents such as OH, CO, and CH_4. This can cause a decrease in tropospheric ozone in the background-clean troposphere but, in some cases, it will increase production of ozone in heavily polluted places.

An accurate assessment of the radiative effect of ozone changes is not possible because of lack of detailed information on the variation in vertical distribution of ozone with latitude and longitude. However, calculations by WMO (1994) support earlier conclusions that lower stratospheric ozone depletion in recent decades has resulted in a negative radiative forcing (i.e., a cooling effect on the climate) and has offset by about 15–20% the positive greenhouse forcing due to increases in other gases. The increase of tropospheric ozone since pre-industrial times may have enhanced the total greenhouse forcing by as much as 20%. The tropospheric ozone increase from the 1970s to the 1980s has caused positive radiative forcing equal to the forcing due to the changes of all other greenhouse gases during the same period. Such changes can probably have an impact on the radiative balance of the Earth-atmosphere system as well as the thermal structure of the atmosphere and hence cause some unpredictable changes to atmospheric circulation patterns.

The drastic decline of total ozone over the Antarctic appearing during the Austral spring since the early 1980s when the total ozone decreases to less than 200 mbar cm (commonly called the ozone hole) continues with new strength. Ozone values in August 1995 were generally 25–30% below the pre-ozone-hole averages. The ozone layer over the Antarctic during the Austral spring of 1995 underwent massive destruction especially during the second half of September or October when the deficiency was about 50% from the pre-ozone hole 1957–1979 average and for a few days reached 70%. Starting at the end of September and lasting for a period of six consecutive weeks, the balloon soundings at Marambio, Neumayer and Syowa pointed to virtual complete loss of the ozone at altitudes between 14 and 20 km. During this period, the spread of the ozone hole (area covered with less than 220–200 mbar cm of total ozone) exceeded 20 million km^2 and lasted as long as 40 days: Comparing the number of days when the ozone-hole-covered area exceeded 15 million km^2 during the past 5 years, the 1995 event was the longest lasting. In 1991 there were 32 days, in 1992 49, in 1993 63, in 1994 55 and in 1995 71. The ozone hole event in 1995 did not disappear until early December and thus was the longest lasting on record so far (WMO 1996).

In the mid and polar latitudes of the Northern Hemisphere during the first 3 months of the winter spring season (December 1995 to February 1996) the average ozone amount was about 5 to 10% below the long-term (1957–1979) average. In contrast with the Antarctic, the extremely low ozone values in the Arctic lasted for weeks but not for months.

References

Agren C (Dec. 1992) VOCs/ozone: effects of the protocol. Acid News 5:6–7

AtkinsonR, Baulch DL, Cox RA, Hampson RF Jr, Kerr JA, Troe J (1989) Towards a quantitative understanding of atmospheric ozone. Planet Space Sci 37:1605–20

Austin J, Butchart N, Shine KP (1992) Possibility of an Arctic ozone hole from a double-CO_2 climate. Nature 360:221–223

Banks HJ (1995) Agriculture production without methyl bromide – four case studies. CSIRO Division of Entomology, Canberra

Barrie L-A, Bottenheim JW, Schnell RC, Crutzen PJ, Rasmussen RA (1988) Ozone destruction and photochemical reactions at polar sunrise in the lower Arctic atmosphere. Nature 334:138–41

Bates DR, Nicolet M (1950) The photochemistry of water vapor. J Geophys Res 55:301–306

Boden et al (March 1992) Trends-91, highlights. ORNL-CDIAC-49 Oak Ridge

Browell EV, Butler CF, Fenn MA, Grant WB, Ismail S, Schoeberl MR, Toon OB, Loewenstein M, Podolske JR (1993) Ozone and aerosol changes during the 1991–1992 airborne Arctic stratospheric expedition. Science 261:1155–58

Chapman S(1930a) A theory of upper atmospheric ozone. Mem Roy. Meteor Soc 3:103–107

Chapman S (1930b) Ozone and atomic oxygen in the upper atmosphere. Philos Mag 10:369–75

Cicerone RJ (1994) Fires, atmospheric chemistry, and the ozone layer. Science 263:1243–44

Crutzen PJ (1970) The influence of nitrogen oxides on the atmospheric ozone content. Quart J Roy Meteor Soc 96:320–30

Dijkstra E (1994) Small enterprise development and cooling technology. GATE 2/94:27–30

Farman JC, Gardiner BG, Shanklin JD (1985) Large losses of total ozone in Antarctica reveal seasonal ClO_x/NO_x interactions. Nature 315:207–210

Fishman J, Fashruzzaman K, Cros B, Naganga D (1991)Identification of widespread pollution in the Southern Hemisphere deduced from statellite analysis. Science 252:1963–96

Fishman J, Watson CE, Larsen JC, Logan JA (1990) The distribution of tropospheric ozone determined from satellite data. J Geophys Res 95:3599–3617

Garcia R (April, 1994) Causes of ozone depletion. Physics World :49–55

Gleason JF, Bhartia PK, Herman JR, McPeters R, Newman P, Stolarski RS, Flynn L, Labow G, Larko D, Seftor C, Wellemeyer C, Komhyr WD, Miller AJ, Planet W (1993) Record low global ozone in 1992. Science 260:523–526

Guicherit R (6 Oct. 1993) Oceanic dimethylsulfide (DMS) and climate. Change (Newsletter) :7–9

Hamill P, Toon OB (1991) Polar stratospheric clouds and the ozone hole. Physics Today 44:34–36

Hunt BG (1966) Photochemistry of ozone in a moist atmosphere. J Geophys Res 71:1385–90

IPCC/WMO (17–19 May, 1993) Workshop report "the impact on climate of ozone change and aerosols". A joint workshop of IPCC and the International Ozone Assessment Panel. Max Planck Institute for Meterology, Hamburg.

Johnston HS (1971) Reduction of stratospheric ozone by nitrogen oxide catalysts from supersonic transport exhaust. Science 173:517–519

Kerr RA(1992) New assaults seen on Earth's ozone shield. Science 255:797–98

Koop T, Carslaw KS.(1996) Melting of H_2SO_4·4 H_2O particles upon cooling: implications for polar stratospheric clouds. Science 272:1638–1640

Krishnamurti TN, Fuelberg HE, Bensman EL, Sinha MC, Oosterhof D, Kumar VB (1993) The meteorological environment of the tropospheric ozone maximum over the tropical South Atlantic. Proc

(Abst) Internat Conf 25–30 Jan. 1993, Sustainable Development Strategies and Global/Regional/Local Impacts on Atmospheric Composition and Climate. Ind Inst Technol, New Delhi, pp 114–16

Lacis AA, Wuebbles DJ, Logan JA (1990) Radiative forcing by changes in the vertical distribution of ozone. J Geophys Res 95:9971–9981

Logan JA (1985) Tropospheric ozone: seasonal behavior, trends, and anthropogenic influence. J Geophys Res 90:10463–69

Mano, S., Andreae, M.O. (1994) Emission of methyl bromide from biomass burning. Science 263:1255–57

McConnel JC, Henderson GS, Barrie L, Bottenheim J, Niki H, Lanfford CH, Templeton EMJ (1992) Photochemical bromine production implicated in Arctic boundary-layer ozone depletion. Nature 355:150–152

Mohnen VA, Goldstein W, Wang W-C (Sept. 1994) Assessing tropospheric ozone as a climate gas. Global Change Newsletter No 19:1–3

Molina LT, Molina MJ (1987) Production of Cl_2O_2 from the self reaction of the ClO radical. J Phys Chem 91:433–39

Molina MJ, Rowland FS (1974) Stratospheric sink for chlorofluoromethanes: Chlorine atom-catalysed destruction of ozone. Nature 249:810–812

Newman P, Lait LR, Schoeberl M, Nash ER, Kelly K, Fahey DW, Nagatani R, Toohey D, Avallone L, Anderson J (1993) Stratospheric meteorological conditions in the Arctic polar vortex, 1991 to 1992. Science 261:1143–46

Pickering KE, Thompson AM, Scala JR, Tao WK, Simpson J (1992) Ozone production potential following connective redistribution of biomass burning emission. J Atmos Chem 14:297–313

Proffitt MH, Aikin K, Margitan JJ, Loewenstein M, Podolske JR, Weaver A, Chan KR, Fast H, Elkins JW (1993) Ozone loss inside the northern polar vortex during the 1991–1992 winter. Science 261:1150–54

Pyle JA, Carver G, Grenfell JL, Kettleborough JS, Lary DJ (1992) Ozone loss in Antarctica and the implications for global change. Phil Trans R Soc Lond B 338:219–26

Ramanathan V, Dickinson RE (1979) The role of stratospheric ozone in the zonal and seasonal radiative energy balance of the Earth-troposphere system. J Atmos Sci 36:1084–1104

Ramaswamy V (1993) Perturbation of global climate due to atmospheric ozone changes: significant chemistry climate interactions. Abst Internat Conf on Sustainable Development Strategies and Global/Regional/Local Impacts on Atmospheric Composition and Climate, 25–30 Jan., 1993. Ind Inst Technol New Delhi, pp 152–154

Ramaswamy V, Schwarzkopf MD, Shine KP (1992) Radiative forcing of climate from halocarbon induced global stratospheric ozone loss. Nature 355:810–812

Ravishankara AR, Solomon S, Turnipseed AA, Warren RF (1993) Atmospheric lifetimes of long-lived halogenated species. Science 259:194–196

Rodriguez JM (1993) Probing stratospheric ozone. Science 261:1128–29

Rowlands IH (1993) The fourth meeting of the parties to the Montreal protocol: Report and reflection. Environ. 35:25–34

Salawitch RJ, Wofsy SC, Gottlieb EW, Lait LR, Newman PA, Schoeberl MR, Loewenstein M, Podolske JR, Strahan SE, Proffitt MH, Webster CR, May RD, Fahey DW, Baumgardner D, Dye JE, Wilson JC, Kelly KK, Elkins JW, Chan KR, Anderson JC (1993) Chemical loss of ozone in the Arctic polar vortex in the winter of 1991–1992. Science 261:1146–49

Shepherd, T.G. (Jan. 1996) SPARC workshop on stratosphere-troposphere exchange. SPARC Newsletter (CNRS, Verrieres-le-Buisson (France) No 6 :6–13

Shorter JH, Kolb CE, Crill PM, Kerwin RA, Talbot HW, Hines ME. Harriss RC (1995) Rapid degradation of atmospheric methyl bromide in soils. Nature 377:717–19

Stolarski R, Bojkov R, Bishop L, Zerefos C, Staehelin J, Zawodny J (1992) Measured trends in stratospheric ozone. Science 256:342–49

Taubes G (1993) The ozone backlash. Science 260:1580–83

Thompson AM (1992) The oxidizing capacity of the Earth's atmosphere: probable past and future changes. Science 256:1157–61

Tolbert MA (1996) Polar clouds and sulfate aerosols. Science 272:1597

Toohey DW, Avallone LM, Lait LR, Newman PA, Schoeberl MR, Fahey DW, Woodbridge EL, Anderson JG (1993) The seasonal evolution of reactive chlorine in the Northern Hemisphere stratosphere. Science 261: 1134–36

Toon O, Browell E, Gary B, Lait L, Livingston J, Newman P, Pueschel R, Russell P, Schoeberl M, Toon G, Traub W, Valero FPJ, Selkirk H, Jordan J (1993) Heterogeneous reaction probabilities, solubilities, and the physical state of cold volcanic aerosols. Science 261:1136–40

Toumi R, Jones RL, Pyle JA (1993) Stratospheric ozone depletion by $ClONO_2$ photolysis. Nature 365:37–39

UNEP (1989) Action on ozone. United Nations Environment Programme, Nairobi

UNEP (1992) Methyl bromide: its atmospheric science, technology and economics. United Nations Environment Program, Nairobi

UNEP (1992) The impact of ozone-layer depletion. UNEP/GEMS Library No 7. UNEP, Nairobi

Van Haasteren J (1994) Methyl bromide damages the ozone layer. Change 16–17

Van Zijst P (Oct 1993) Thinner ozone layer leads to more UV radiation. Change 16:1–3

Visscher PT, Culbertson CW, Oremland RS (1994) Degradation of trifluoroacetate in oxic and anoxic sediments. Nature 369:729–31

Wallington TJ, Schneider WF, Worsnop DR, Nielsen OJ, Schested J, Debryn WJ, Shorter JA (1994) Environmental impact of CFC replacements – HFCs and HCFCs. Environ Sci Technol 28:320–26

Wania F, Mackay D (1993) Organochlorine compounds in polar regions. Ambio 22:10–18

Webster CR, May RD, Toohey DW, Avallone LM, Anderson JG, Newman P, Lait L, Schoeberl MR, Elkins JW, Chan KR (1993) Chlorine chemistry on polar stratospheric cloud particles in the Arctic winter. Science 261:1130–1134

Wilson JC, Jonsson HH, Brock CA, Toohey DW, Avallone LM, Baumgardner D, Dye JE, Poole LR, Woods DC, Decoursey RJ, Osborn M, Pitts MC, Kelly KK, Chan KR, Ferry GV, Loewenstein M, Podolske JR, Weaver A (1993) In situ observation of aerosol and chlorine monoxide after the 1991 eruption of Mount Pinatubo; effect of reactions on sulfate aerosol. Science 261:1140–43

WMO (1989) Scientific assessment of stratospheric ozone. World Meteorological Organization. Global Ozone Research and Monitoring Project, Report No. 20, Vols I, II. Geneva

WMO (1992) Scientific assessment of ozone depletion: (Co-Chairpersons Watson RT and Albritton D), WMO Report No 25 (Chapter 7), Geneva

WMO/UNEP (1996) Report of the third meeting of the ozone research managers of the parties to the Vienna convention for the protection of the ozone layer. 19–21 March 1996. WMO Report No 41, Geneva

WMO/UNEP (1992) Global ozone research and monitoring project Rep N. 25. World Meteorological organization. On the State of the Ozone Layer in 1992. WMO, Geneva

5 Solar Ultraviolet Radiation

5.1
Introduction

The human eye responds to light with wavelengths from about 700 nm (red) to 400 nm (violet). Radiation with wavelengths shorter than the human eye "sees" is called "ultra-violet" (UV; Fig. 5.1). UV radiation is further subdivided into three wavelength bands: UV-A (315–400 nm), UV-B (280–315 nm) and UV-C (220–280 nm).

UV-B and ozone are closely interrelated in the atmosphere. Though the total O_3 (stratosphere + troposphere) is important, the larger part in the reduction of the surface flux density of UV-B is played by the stratospheric O_3. UV-B contributes about 1.5% of solar irradiance, but its share at the Earth's surface drops to about 0.5% after attenuation by passage through the atmosphere (Table 5.1). With overhead sun and typical stratospheric O_3 levels, a 10% O_3 decrease will cause a 20% increase in flux density at 305 nm, a 250% increase at 290 nm, and a 500% increase at 287 nm. Frederick et al. (1989) have shown the dependence of surface UV-B on total O_3 column levels.

Table 5.1. Relative solar radiation flux densities at surface level

Wavelength [nm]	Description	Percent
<280	UV-C	0
280–315	UV-B	0.5
315–400	UV-A	5.6
400–800	Visible	51.18
>800	Infrared	42.1

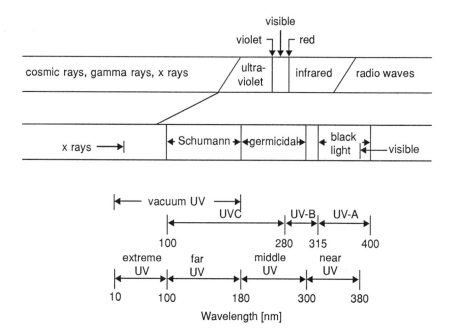

Fig. 5.1. Electromagnetic spectrum with enlargement of the ultraviolet (UV) region

Sunlight is received both as direct rays and as diffuse light. The latter is scattered by particles in the atmosphere. The sky is blue because air molecules scatter blue light more than red light (Raleigh scattering). UV radiation is scattered even more than blue light. Being shaded from the sun's direct rays therefore provides only partial protection from UV exposure because of the high level of diffuse UV rays.

Typical window glass transmits less than 10% of sunburning UV. Sunblock creams work by absorbing or reflecting UV rays. The SPF (sun protection factor) rating of a sunblock cream indicates its effectiveness. For example, an SPF of 15 means that skin damage will occur after 15 times as long as it would on untreated skin (i.e., if it is correctly applied, the cream should block about 93% of the damaging radiation). Figure 5.2 shows how much more solar radiation reaches a unit area of the Earth's surface in the tropics than at higher latitudes.

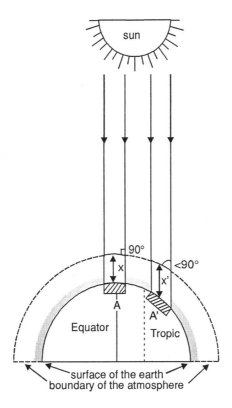

Fig. 5.2. Incidence of solar radiation per unit area of the Earth's surface in the tropics and at higher latitudes. Note the difference in the angle of incidence, the spread of rays on the ground area (*area A < A'*), and distance traveled through the atmosphere (*length x < x'*)

5.2
Ozone depletion and solar UV

The amount of ozone found in the stratosphere has a bearing only on the UV-B radiation. The sun emits a very broad band of light with significant levels down to 100 nm and below. Oxygen efficiently absorbs radiation below 200 nm, producing atomic oxygen that in turn produces ozone. About 10% of the total ozone is in the troposphere, but most is in the altitude range of 12–40 km, with a peak concentration at about 20 km. Ozone absorbs strongly in the mid UV (Fig. 5.3). It is the tail of the absorbance that is responsible for the effective 295 nm cutoff of terrestrial UV.

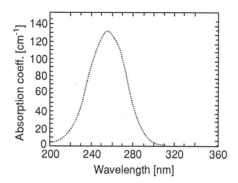

Fig. 5.3. Absorption spectrum of gaseous ozone at standard temperature and pressure. (Koller 1965)

Any loss of stratospheric ozone means that more of the damaging UV-B would reach the Earth's surface. UV-B has adverse effects on human health, vegetation, aquatic organisms, and even some non-living materials. Enhanced incidence of UV-B has the potential to intensify global warming. This effect is due to the fact that ozone acts as a greenhouse gas in the upper troposphere and lower stratosphere.

Ozone forms a fragile shield that is highly effective. It is scattered quite thinly through the 35 km-deep stratosphere. Its concentration in the stratosphere varies with height; but it never makes up as much as one hundred thousandth of the atmosphere around it. Yet this filter effectively screens out almost all the harmful ultraviolet rays of the sun. The shorter the wavelength of ultraviolet radiation, the greater the harm it can do to life – but the better it is absorbed by the ozone layer. Relatively short wavelength ultraviolet radiation, known as UV-C, is lethal to living things – and is almost totally screened out. Longer wavelength ultraviolet, UV-A, is relatively harmless and is almost transmitted through the atmosphere. In the middle lies UV-B (wavelength range about 280 to 315 nm), which is less lethal than shorter wavelength radiation but still dangerous; it is most of this UV-B that is absorbed by the ozone layer.

According to Frederick et al. (1989), the biologically effective UV-B irradiance at the Earth's surface varies with the elevation of the sun, the amount of atmospheric ozone and with the abundance of atmospheric matter generated by natural and anthropogenic processes which have some scattering and absorbing properties. Figure 5.4 A shows the solar global spectral irradiance in the UV-A and UV-B regions, computed for normal ozone concentrations and for a 16% ozone-depleted condition. Figure 5.4 B illustrates the relative enhancement at different wavelengths resulting from ozone depletion.

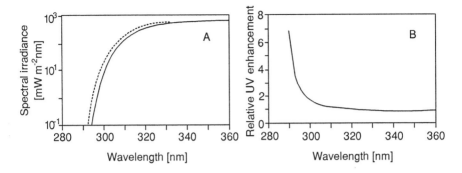

Fig. 5.4. A Solar global spectral irradiance over the UV-A and UV-B portions of the spectrum, computed for normal ozone concentrations and solar zenith angles appropriate for midday in the summer at temperate latitudes. The *dashed line* represents the irradiance for the same conditions, but with a 16% ozone reduction. **B** The relative enhancement at different wavelengths resulting from ozone reduction

The UV-B spectral region is defined as wavelengths from 280 to 315 nm but for practical purposes no solar radiation penetrates to the ground at wavelengths between 280 and 290 nm, even when there is very little ozone left in the stratosphere.

Much attention has focused on the detection and interpretation of changes in the atmospheric ozone layer, both the total column abundance and its vertical distribution. Photobiologically, the concern is with the UV irradiation, including an action spectrum appropriate to the response under study (Frederick 1993). The total amount of ozone in a column of the atmosphere and the UV-B irradiance are intrinsically related, but the UV spectral irradiance at the Earth's surface depends on several variables that determine the transmission of the atmosphere; these include vertical distribution of ozone as well as the influence of gaseous and particulate air pollutants, clouds and haze. These factors complicate the passage of radiation through the atmosphere and cause differences in surface UV-B radiation between regions located at the same latitude. However, notwithstanding the various complications, the dominant variables that determine the 24-h integrated UV irradiance at the ground are the elevation of the sun above the horizon at a fixed local time and the duration of daylight. These factors are related to the nature of the Earth's orbit around the sun and lead to latitudinal and seasonal variations in surface UV irradiance that far exceed the changes predicted at any mid-latitude location as a consequence of the decline in column ozone since 1970 (Frederick 1993).

The total ozone mapping spectrometer (TOMS) on the Nimbus 7 satellite (Stolarski et al. 1991) has conducted measurements over the entire period from 1979 through 1991; statistically significant downward trends in column ozone outside the tropics have been revealed (Niu et al. 1992) during all seasons of the year. A decline in column ozone has been identified during summer, the time of year when surface UV irradiance is greatest at any fixed local time or when integrated over 24 h. The magnitude of the downward trend depends on season. The zonal mean trend reported for latitudes between 30 and 50°N is in the range of -6.6 to -6.7% per decade during winter. The analogous values during summer are in the range of -2.6 to -3.0% per decade (Frederick 1993).

UV radiation measurements are made in both Belgium and the Netherlands. Earlier measurements revealed no clear trend in the UV-B radiation. In recent years, new equipment has been developed in Europe which allows more accurate UV-B radiation measurements to be made. Recent measurements (Fig. 5.5) show that the current mean intensity, as well as the maximum intensity of UV-B radiation is greater than in 1992. A preliminary analysis of the measurement data in April 1993 reveals that there is a clear (inverse) relationship between the thickness of the ozone layer and the UV-B intensity.

Kerr and McElroy (1993) have shown that at a carefully monitored site in Toronto, wintertime levels of UV-B radiation increased more than 5% every year from 1989 to 1993, as ozone levels dropped. This finding is important since skeptics claim that air pollution and clouds, by absorbing ultraviolet light, can negate the effects of ozone loss. It has also been shown that the UV-B increase was due to ozone loss and not to clearer skies or less low-level air pollution. This is because clouds, haze and sulfate particles from power plants block radiation across the entire UV-B band while ozone leaves its mark only at the shortest wavelengths. At 324 nm, where ozone should have little effect, Kerr and McElroy found no changes in intensity. But at 300 nm – closer to the peak of ozone absorption – UV-B intensity shot up, with summertime intensities increasing by 7% each year and wintertime intensities jumping by 35%.

5.2.1
UV Variability

Ozone loss has been greatest near the polar regions but other factors have caused high UV intensities to occur at lower altitudes (i.e. nearer the equator). The major factor which controls UV is the angle of the sun's rays (solar elevation). When the sun is low in the sky, sunlight has to pass through more atmosphere before reaching the ground than when the sun is high, so damaging UV radiation reaches a maximum at around the solar noon.

Thick, extensive cloudiness and the concentrations of particles and molecules (other than ozone) capable of reflecting/absorbing UV in the atmosphere also reduce UV-B radiation by reflecting light rays back into space.

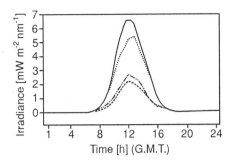

Fig. 5.5. Measurements of UV-B intensity as a function of the hour of the day. From *top* to *bottom*; the maximum value for February 1993, the maximum value for February 1992, the mean value for February 1993 and the mean value for February 1992. (KNMI; van Zijst 1993)

Sunlight reaching the Earth is also attenuated by scattering, which is more intense at shorter wavelengths. At wavelengths around 300 nm, more than half of the radiation is scattered (Koller 1965). This means that significant amounts of UV-B come from whole-sky radiation and less as direct radiation from the direction of the sun (Pickett 1994).

Ozone levels vary naturally. At high latitudes the ozone level increases through the winter with a maximum early in the spring. It then decreases through the summer reaching a minimum in the autumn. There is relatively more ozone at high latitudes with greater seasonal variation.

5.2.2
North-South Differences

For equivalent latitudes, locations in the Southern Hemisphere are expected to receive approximately 15% more UV than in the Northern Hemisphere because of differences in ozone and because the Earth is slightly closer to the sun during the Southern Hemisphere summer. Actual measurements have shown that sunburning UV in the Southern Hemisphere can exceed that at comparable latitudes in Europe by over 50%. This stronger difference is due to the build-up of pollution (tropospheric ozone and emissions from industry and vehicles) in the atmosphere over Europe, which reflects or absorbs incoming UV-B.

5.3
Tropospheric Transmission in the Ultraviolet

Ozone near the ground, in the sense of an air pollutant, has probably increased in many areas over the past few decades, with the exception of urban centers that already violated air quality standards. In these centers, efforts to control pollution levels have tended to reduce surface ozone amounts. A change in the ozone amount in the lower atmosphere has some bearing on the UV-B irradiance reaching the Earth's surface. An increase in ground ozone can combine with its decrease in the upper atmosphere to produce a net decline in total column ozone. However, under certain conditions it is possible for this reduction in column ozone to be accompanied by a reduction in the surface UV-B irradiance. This seeming contradiction comes from the fact that the total UV-B irradiance consists of two components, a direct component coming from the discrete position of the sun and a scattered, diffuse component which may roughly be taken as isotropic over the hemisphere. The ratio of diffuse irradiance to direct irradiance increases as altitude decreases as a consequence of efficient scattering in the lower atmosphere (Frederick 1990, 1993).

Absorption by gaseous air pollutants in major urban areas can reduce the broadband UV irradiance by several percent, and scattering by clouds and haze provides a large and highly variable attenuation of sunlight (Frederick and Snell 1990; Frederick et al. 1993; Madronich 1991). Frederick et al. examined the effects of absorption and scattering in the troposphere on solar ultraviolet radiation reaching the ground. They noted that the attenuation provided by clouds and haze undergoes an annual cycle. The monthly mean ultraviolet irradiance ranged from about 84% of the clear-sky value for June 1991 to about 49% for January 1992. Average ultraviolet irradiances for June and July of 1992 were about 11% and 22% lower than in corresponding months of 1991, due to differences in local cloudiness. The attenuation of total sunlight provided by local clouds and haze was the same as their attenuation of ultraviolet radiation. Frederick et al. also found a statistically significant negative correlation between the UV values and ground-level ozone when the atmosphere was relatively free of clouds and haze. This work shows that gaseous air pollution can have a detectable effect on ultraviolet radiation reaching the ground.

Scattering by clouds reduces the annually integrated surface UV irradiance to levels between about 60 and 75% of the values prevailing under perpetually clear skies (see Madronich 1991; Frederick and Snell 1990; Frederick and Weatherhead 1992). Furthermore, sulfur-based aerosols in the lower atmosphere scatter UV radiation back into space. The decline in UV-B irradiance related to increased aerosol loading has been estimated (Liu et al. 1991) to be in the range of 5 to 18%. This decrease would have more than offset the change brought about by the downward trend in stratospheric ozone over the past several years, irrespective of the increase in ozone near the ground. However, sulfur emissions are no longer increasing, whereas the decline in stratospheric ozone continues. According to Frederick (1993), it is unlikely that changes in the opacity of the troposphere will continue to offset the UV radiation effect of the ongoing decline in stratosphere ozone (Frederick 1993).

Brühl and Crutzen (1989) showed that tropospheric O_3 causes a disproportionately strong attenuation because a large fraction of the UV-B reaching the surface is diffuse radiation and hence has a relatively longer path through the troposphere than through the stratosphere.

5.4
Measurement of UV

It is quite difficult to make accurate measurements of UV because of the steeply sloping spectrum in the UV region and because calibration standards in this spectral region are imperfect. Trend detection is also difficult because there are several other factors besides ozone which affect the UV. Nevertheless, a clear relationship between ozone and UV has been demonstrated.

UV can be measured in two ways. One uses broad-band instruments to measure the total energy received over a broad range of wavelengths in the UV band. The Robertson Berger meter, for example, mimics the response of human skin to UV radiation to provide a measure of the erythemal (sunburn) energy. With broad-band instruments one can obtain a continuous series of readings over time. However, their applicability is fairly limited because they are difficult to calibrate, and different biological processes have different wavelength sensitivities.

Recently, in Europe, a new monitoring network (ELDONET) for solar radiation using three-channel dosimeters (UV-A, UV-B, PAR = photosynthetic active radiation) has been installed from Abisko (northern Sweden, 68°N, 19°E) to Tenerife (Canary Islands, 27°N, 17°W). Some of the instruments have been placed in the water column (North Sea, Baltic Sea, Kattegat, eastern and western Mediterranean, North Atlantic), establishing the first network of underwater dosimeters for continuous monitoring (Fig. 5.6). The second approach is spectroradiometry. It involves scanning the whole UV band, at, say, 1-nm intervals to build up a detailed spectrum. Because each scan takes several minutes to complete, continuous time series are not practical. Spectral data can be weighted for any biological process and can be used to understand the causes of changes in UV.

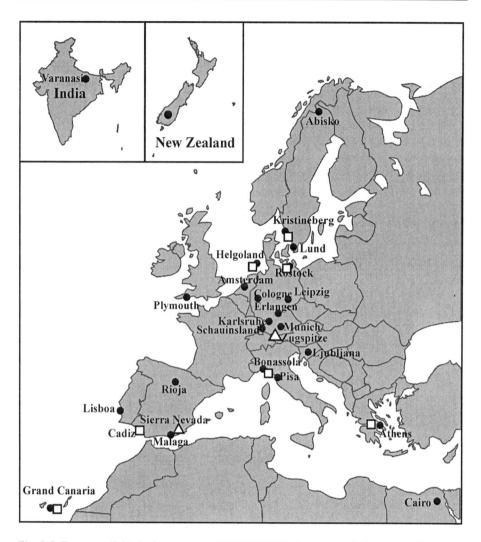

Fig. 5.6. European light dosimeter network (ELDONET). Locations of the terrestrial, aquatic and high latitude instruments. Closed circles terrestrial instruments, open squares aquatic instruments, open triangles high altitude stations

It is also possible to understand UV changes by calculating the amount of UV reaching the ground in clear-sky conditions. These calculations, which can also be useful in quality control of measurements, are based on knowledge of the sun angle, time of year and ozone amount (which can be measured by satellites or ground-based instruments). Lubin and Jensen (1995) combined satellite-based measurements of ozone and cloudiness to estimate the ultraviolet radiation (UVR). They concluded that if reductions of stratospheric ozone continue at current rates, many temperate and high-latitude locations will experience rises of monthly averaged UVR levels.

The geographical and seasonal distributions are largely consistent with the increases in UVR estimated for cloud-free skies and support the view that UVR increases will be most pronounced at mid and high latitudes of both hemispheres. Summertime levels of UVR are perhaps already exceeding the background cloud variability in some parts of Europe, central Asia, North America and the Southern Hemisphere poleward of 30° (Madronich 1995).

In the tropics, where reported ozone trends are small, and in those regions of mid and high latitudes that normally experience highly variable cloudiness, the ozone-induced increases in UVR seem to be less meaningful. In particular, the rapid attainment of the threshold in tropical regions results mostly from low cloud variability rather than large reductions in ozone levels.

Measurements of terrestrial UV so far have shown significant increases in UV-B in the Arctic and Antarctic regions and slowly increasing levels at mid-latitudes. Changes in the amount of cloud cover also affect the terrestrial UV-B levels. Cloudy, rainy summers result in much less UV while drought conditions are accompanied by increased UV. From this, however, it should not be inferred that these effects will continue to offset the loss of ozone for much longer. The UV levels are likely to begin reflecting the ozone loss in the near future (Pickett 1994).

5.4.1
The UV Index

As some countries are now providing UV information to the public through media, an international agreement was reached in 1994 by the World Meteorological Organization and the World Health Organization on a way to standardize the reporting of UV so as to avoid confusion. A new standard, called the "UV index", is based on the UV irradiance, so that when the UV is more intense, the index is higher. The index is defined as the UV spectral irradiance weighted by the erythemal (sunburn) reference spectrum, and is given in units of W m^{-2}, multiplied by 40. The scaling by a factor of 40 was chosen to agree with a scale which had been developed and used in Canada for several years, so that the maximum values there were approximately 10. The USA had also adopted a similar scale. The scale is open-ended, but a UV index of greater than 10 is extreme and a UV index less than 1 is low. At the high-altitude Mauna Loa site in Hawaii, the UV index sometimes exceeds 18.

5.5
Biological Hazards of Ultraviolet Exposure

Ozone is an effective filter for solar short wavelength radiation. However, a small amount of UV-B radiation does manage to penetrate the shield and causes considerable harm. It damages the genetic material and is the main cause of skin cancer, which is already increasing rapidly around the world. It is the dominant cause of non-melanoma skin cancers and causes the rarer but virulent cutaneous malignant melanomas. A 1% loss of ozone may lead to a 2% increase in UV radiation on the ground and a 4% increase in skin cancers among fair skinned people.

UV-B also impairs the immune system, making it easier for tumors to take hold and spread. And the same process also makes people more vulnerable to infectious diseases that enter the body through the skin, such as herpes and the parasitic disease, leishmaniasis (UNEP 1989). It also produces cataracts, which blind some 15 million people and seriously impairs the vision of another 25 million worldwide.

Other forms of life also suffer. Two thirds of some 300 crops and other plant species tested for their tolerance to ultraviolet radiation have been found to be sensitive to it. Among the most vulnerable are crops related to peas and beans, melons, mustard and cabbage; increasing UV-B radiation has also been found to reduce the quality of certain types of tomato, potato, sugar beet and soybean (UNEP 1989). Forests are also vulnerable; several conifer seedlings are adversely affected by UV-B. Even its present levels are limiting the growth of some plants, and it is bound to cut the productivity of agriculture and forestry if it increases.

Similarly, UV-B radiation strikes beneath the surface of the sea, causing damage up to several dozen meters down in clear waters. It is particularly harmful to small creatures such as plankton (Häder et al. 1995), the larvae of fish, shrimp and crab and plants, essential to the food web of the sea. Even small increases could bring about important changes in underwater life, damaging fisheries. Many countries depend on fish to provide much of the protein in their people's diet; many developing countries are even more dependent on food from the sea (UNEP 1989, 1994). Merely the exclusion of UV-B under a "normal" ozone column (350–400 DU) has been shown to result in better growth and reproduction in several natural communities (Dey et al. 1988). Biological effects depend strongly on wavelength even within a few nanometers (Cullen et al. 1992), and small changes in ozone concentration result in disproportionate increases in the biological harmfulness of incident UV-B radiation (Karentz 1994).

5.6
UV Damage to Biological Molecules

Most biological species evolved after the initial formation of the stratospheric ozone layer. Indeed, this layer provided for them a strong protective umbrella against the adverse effects of UV by absorbing strongly the wavelengths shorter than about 320 nm (see Fig. 5.7). Undoubtedly, the most important biomolecule is the DNA which makes the genetic material of the vast majority of living cells. The absorption spectrum of DNA peaks at 260 nm, well below 315 nm, and drops by some three orders of magnitude at 320 nm (Fig. 5.7). For green plants, plastoquinone and plastoquinol are extremely important as they have important roles in photosynthesis; both of these molecules also absorb strongly at wavelengths shorter than 310 nm (Fig. 5.7). The percentage of UV-B in the solar output reaching the top of the Earth's atmosphere is less than 1.5% (20 W m^{-2}); the amount reaching the Earth's surface is less than 0.3% (2 W m^{-2}) due to the filtering effect exerted by various atmospheric chemicals. Even then, enough ambient UV-B does reach the biosphere to cause some damage to cellular DNA (Coohill 1991).

Vincent and Roy (1993) have reviewed the mutagenic and lethal effects of UV-B on aquatic organisms. DNA is a primary lethal target. There are also many other UV-sensitive biological molecules. Secondary effects of UV exposure include absorption by RNA, proteins, pigments and plant hormones. UV also induces photochemical reactions inside cells and in their ambient environment. Radicals produced by such reactions are often toxic, cause oxidative stress (Mopper and Zhou 1990) and inactivate photosynthesis, membrane transport, nutrient uptake and other metabolic activities (Karentz 1994).

The extent of UV sensitivity shown by any species depends on the number and efficiency of repair systems and the existence of avoidance strategies, e.g., behavior modification or production of sunscreens. Considerable variability in UV tolerance occurs at both intraspecific and interspecific levels (Bothwell et al. 1993; Karentz 1994).

In response to the environmental natural UV stress, cells have developed several mechanisms for minimizing and correcting UV-induced damage. These include (1) reduction of UV exposure by physical barriers or UV-absorbing compounds, (2) physiological mechanisms that can recognize and repair UV-induced damage and (3) the production of anti-oxidants that can neutralize the effects of radicals produced by UV photochemical reactions (Karentz 1994). The penetration of UV-B into the cells is largely a function of the cell diameter (see Table 5.2).

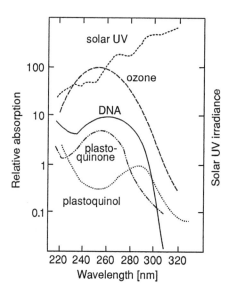

Fig. 5.7. The umbrella-like protection from solar UV afforded to several important biomolecules by the absorption characteristics of ozone. Solar UV is shown at the top of the atmosphere, and absorption spectra of ozone, DNA, plastoquinone and plastoquinol are shown (Coohill 1991)

Table 5.2. Estimate of the percent transmission of UV to the center of selected cells and viruses. All values are approximate, values at λ above 300 nm can vary widely due to the presence of endogenous chromophores

Biological sample	Diameter [μm]	Wavelength in nm		
		250	300	350
Bacteriophage (T$_2$)	0.1	86	100	100
Bacterial cell	1	78	98	100
Yeast cell	5	69	97	100
Mammalian cell (spherical)	20	20	91	96
Mammalian tissue (100 μm thick)	-	10^{-5}	39	66

5.7
Health Effects

As regards the effects of UV-B on human health, the skin and the eyes are the most important "target organs"; at the cellular and subcellular level, the cell membrane, proteins and DNA are most affected by UV radiation. Whilst there are a wide array of natural defenses, the high division rates of skin cells make them especially susceptible to genetic damage and the potentially lethal accumulation of mutations. Harmful effects come primarily from the production of high-energy oxygen species and other reactive radicals. These chemical species also occur as a result of a wide range of other biochemical processes. Under conditions of moderate UV-B exposure, their effects are counteracted by scavenging enzymes and vitamins with antiradical properties. Most attention has been focussed on UV-B, but it should be noted that UV-A, though less energetic, is also carcinogenic at high doses.

Exposed skin surface is irradiated differently depending on cultural and social behavior, clothing, the position of the sun in the sky and the relative position of the body. Exposure to UV-B of the more exposed skin surfaces (e.g. nose, ears) relative to that of the lesser exposed areas (such as underneath the chin) normally ranges over an order of magnitude. In cutaneous photobiology, radiant exposure is frequently expressed as "exposure dose" in units of $J\ cm^{-2}$ (or $J\ m^{-2}$). "Biologically effective dose", derived from radiant exposure weighted by an action spectrum, is expressed in units of $J\ cm^{-2}$ (effective) or as multiples of "minimal erythema dose" (MED).

The cumulative annual exposure dose of solar UVR varies widely among individuals in a given population, depending to a large extent on occupation and extent of outdoor activities. Indoor workers in mid-latitudes (40–60°N) receive an annual exposure dose of solar UVR to the face of about 40–160 MED, depending upon the extent of outdoor activities, whereas the annual solar exposure dose for outdoor workers is typically around 250 MED. These estimates are very approximate and subject to differences in cultural and social behavior, clothing, occupation and outdoor activities.

UVR is used in many different industries, yet there is a paucity of data concerning human exposure from these applications, probably because in normal practice sources are well contained and exposure doses are expected to be low. In recent years the use of tungsten-halogen lamps, which also emit UVR has been increasing for general lighting.

Studies of transmission in whole human and mouse epidermis and human *stratum corneum* in vitro show that these tissues attenuate radiation in the solar UVR range.

This attenuation, which is more pronounced for UV-B than for the UV-A wavebands, gives some protection from solar UVR to dividing cells in the basal layer.

UVR produces erythema, melanin pigmentation and acute and chronic cellular and histological changes in humans. Generally consistent changes are seen in experimental species, including the hairless mouse. The action spectra for erythema and tanning in humans and for oedema in hairless mice are similar. UV-B is three to four times more effective than UV-A in producing erythema. In humans, pigmentation protects against erythema and histopathological changes. People with a poor ability to tan, who burn easily and have light eye and hair color are at a higher risk of developing melanoma, basal cell and squamous cell carcinomas.

5.7.1
Molecular Targets of UV

Solar UVR induces a variety of photoproducts in DNA, including cyclobutane-type pyrimidine dimers, pyrimidine-pyrimidone photoproducts, thymine glycols, cytosine damage, purine damage, DNA strand breaks and DNA protein cross-links. Good information on biological consequences is available only for the first two lesions. Both are potentially cytotoxic and can lead to mutations in cultured cells. Cyclobutane-type pyrimidine dimers may be precarcinogenic lesions. The relative and absolute levels of each type of lesion vary with wavelength. Substantial levels of thymidine glycols, strand breaks and DNA-protein cross-links are induced by solar UV-A and UV-B radiation, but not by UV-C radiation. The ratio of strand breaks to cyclobutane-type dimer lesions increases as a function of increasing wavelength. In narrow band-width studies, the longest wavelength at which cyclobutane-type pyrimidine dimers have been observed is 365 nm, whereas the induction of strand breaks and DNA-protein cross-links has been found at wavelengths in the UV-B, UV-A and visible ranges. Non-DNA chromophores such as porphyrins, which absorb solar UVR, appear to be important in generating active intermediates that can lead to damage (Fig. 5.8). Solar UVR also induces membrane damage (Anonymous 1992).

DNA damage occurs in human skin cells in vivo after exposures to UV-A, UV-B and UV-C radiation. Most of the DNA damage after a single exposure is repaired within 24 h. The importance of these wavelength ranges depends on several factors. UV-C and UV-B are somewhat more effective than UV-A when compared on a per photon basis, probably due to a combination of the biological effectiveness of the different wavebands and of their absorption in the outer layers of the skin.

Non-mammalian systems				Mammalian systems			
Proka-ryotes	Lower eukaryotes	Plants	Insects	In vitro		In vivo	
				Animal cells	Human cells	Animals	Humans
D G	D R G A	D G C	R G C A	D G S M C A T I	D G S M C A T I	D G S M C DL A	D S M C A
+			+$^{\prime}$	+ + +$^{\prime}$	+ + +	+$^{\prime}$ +	+

Fig. 5.8. Summary table of genetic and related effects of ultraviolet B radiation. *A* aneuploidy, *C* chromosomal aberrations, *D* DNA damage, *DL* dominant lethal mutation, *G* gene mutation, *I* inhibition of intercellular communication, *M* micronuclei, *R* mitotic recombination and gene conversion, *S* sister chromatid exchange, *T* cell transformation, + considered to be positive for the specific endpoint and level of biological complexity, +$^{\prime}$ may be positive but data insufficient, - considered to be negative, ? considered to be equivocal or inconclusive (e.g. there were contradictory results from different laboratories, there were confounding exposures, the results were equivocal)

5.7.2
UV, Sunburn and Skin Cancer

Sunburn is proof that a person has been exposed to too much of UV-B radiation. The potential for developing skin cancer is related to the damage caused by exposure to UV. For non-melanoma cancers the relationship seems to be a linear increase in occurrence with increase in UV-B exposure. For melanoma, the situation is more complex, though UV-B exposure is probably an important factor.

Positive correlations occur between measures of solar skin damage and the prevalence of basal- and squamous-cell carcinomas. Several studies of white people in North America, Australia and some European countries show a positive association between incidence of and mortality from melanoma and residence at low latitudes. Studies of migrants suggest that the risk of melanoma is related to solar radiant exposure at the place of residence in early life. The body site distribution of melanoma shows lower rates per unit area on sites usually unexposed to the sun than on usually or regularly exposed sites (Anonymous 1992).

Solar radiation has been tested for carcinogenicity in several studies in mice and rats. Well-characterized benign and malignant skin tumors developed in most of the surviving animals. Sunlight is carcinogenic for the skin of animals. Broad-spectrum UVR (solar-simulated radiation and ultraviolet lamps emitting mainly UV-B) has been tested for carcinogenicity in many studies in mice, to a lesser extent in rats and in a few experiments in hamsters, guinea-pigs, opossums and fish. Benign and malignant skin tumors were induced in all of these species except guinea-pigs, and tumors of the cornea and conjunctiva were induced in rats, mice and hamsters. The predominant type of tumors induced by UVR in mice is squamous cell carcinoma. Basal-cell carcinomas have been observed occasionally in athymic nude mice and rats exposed to UVR. Melanocytic neoplasms of the skin were shown to develop following exposure of opossums and hybrid fish to broad-spectrum UVR (Anonymous 1992).

Studies in hairless mice demonstrated the carcinogenicity of exposures to UV-A, UV-B and UV-C. UV-B radiation is three to four orders of magnitude more effective than UV-A. Both short-wavelength UV-A (315–340 nm) and long-wavelength UV-A (340–400 nm) induced skin cancer in hairless mice. The carcinogenic effectiveness of the latter waveband is known only as an average value over the entire range. Solar and "solar-simulated" radiation from sunlamps (UV-A and UV-B) as well as UV-C are mutagenic (Fig. 5.9), induce sister chromatid exchange in mammalian, human and amphibian cells, induce micronucleus formation and transformation in mammalian cells, are mutagenic to and induce DNA damage and sister chromatid exchange in human cells in vitro and induce DNA damage in mammalian skin cells irradiated in vivo.

New findings with regard to the molecular events underlying UV-induced non-melanoma skin cancer have supported the earlier conclusion that these tumors often have mutations in the p53 tumor suppressor genes that are typical of UV-B (UV-B-signature alterations). Similar UV-B signature alterations in the so-called ptc tumor suppressor gene appear to be connected with both familial and spontaneous basal cell carcinomas (BCC). Recent epidemiologic studies have confirmed that BCC is not principally related to cumulative UV dose, but as with melanoma, is more related to childhood and intermittent (over-)exposures.

Several epidemiological studies have suggested that sunscreen use is not protective and may even be associated with increased risk. The hypothetical mechanism advanced for this effect is an increased importance of exposure to UV-A. Some new work on melanoma in animals has shown that short-term neonatal exposures to broadband UV-B/UV-A radiation (of approximately a week) result in highly aggressive (metastatic) melanomas.

The action spectrum of UV-B induced skin cancer induction seems to coincide with that for edema induction in mice and erythema induction in man. The action spectra for the latter skin reactions are very similar to those for DNA damage, if the transmission of the upper skin layers is taken into account (Zölzer and Kiefer 1993). The action spectrum for pyrimidine dimer formation agrees with that for inactivation and transformation in Syrian hamster embryo cells and with that for inactivation and mutation induction in Chinese hamster fibroblasts (Zölzer and Kiefer 1993). The above spectra do not differ significantly from those for inactivation and mutation induction in human cells, such as fibroblasts and keratinocytes. Calculations made by Zölzer and Kiefer (1993) have suggested that a 10% reduction in the ozone level would lead to about a 30% increase in the incidence of squamous cell carcinomas and a 22% increase in the incidence of basal cell carcinomas.

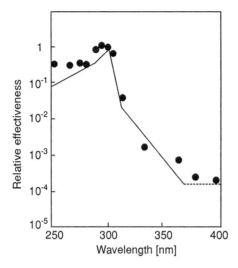

Fig. 5.9. Relative effectiveness of different forms of UV radiation in inducing DNA damage in mammalian skin cells irradiated in vivo

5.7.3
UV Effects on the Immune System

Relatively few investigations have been reported of the effects of UVR on immunity in humans, but changes do occur. Contact allergy is suppressed by exposure to UV-B and possibly to UV-A radiation. The number of Langerhans cells in the epidermis is decreased by exposure to UVR and sunlight, and the morphological loss of these cells is associated with changes in antigen-presenting cell function in the direction of suppression; this change could be due not only to simple loss of function but also to active migration of other antigen-presenting cells into the skin. A reduction in natural killer cell activity also occurs, which can be produced by UV-A radiation. These changes are short-lived. Pigmentation of the skin may not protect against some UVR-induced alterations of immune function (Fig. 5.8). There are at least three mechanisms by which UV-B exposures suppress cellular immunity: DNA damage, isomerization of urocanic acid and through the active metabolite of vitamin D.

5.7.4
UV Effects on the Eye

The different components of the human eye act as optical filters for the UVR range. Consequently, little or no UVR reaches the retina in the normal eye (Anonymous 1992). UV radiation damages the cornea and lens of the eye (Longstreth et al. 1995). Chronic, prolonged exposure to UV-B can sometimes cause cataracts. Although quantitative estimates of risk are available for skin cancer, no such estimates can be made for other risks of UV-B exposure.

5.7.5
Positive Effects of UV Radiation

Humans benefit from UV exposures through the formation of vitamin D_3. Even short-term exposure to UV generally suffices to gain adequate vitamin D_3. There is some evidence from tissue culture studies that vitamin D inhibits tumor cell growth; this may be the mechanism underlying an observed latitudinal gradient in breast and colon cancers, but the linkage is still conjectural. The production of the biologically active vitamin is self-limiting so that excessive exposures are not likely to be associated with any benefit.

UVR has been used for several decades to treat skin diseases, notably psoriasis. A variety of sources of UVR are employed, and nearly all emit a broad spectrum of radiation. A typical dose in a single course of UV-B phototherapy might lie between 200 and 300 MED. Humans with the inherited condition *Xeroderma pigmentosum* appear to have frequencies of non-melanoma skin cancer and melanoma that are much higher than expected. Some evidence suggests that the greatest excess occurs on the head and neck. Exposure to sunlight increases the risk of non-melanocytic skin cancer.

5.8
UV-B Effects on Plants and Terrestrial Ecosystems

Screening studies have now been carried out on around 300 terrestrial plant species, mostly of agricultural importance. There is considerable variability in their response: legumes are highly sensitive, with pea plants showing enhanced growth when "normal" natural sunlight is filtered to exclude UV-B. The number of studies on crop yields is much lower, with only one investigation covering several years which showed that additional UV-B (simulating 25% ozone depletion) caused significant changes in both crop quantity and quality for soybeans. Interactions with other environmental factors are also important, resulting in considerable inter-annual variation in the magnitude of these effects. Intra-specific differences in sensitivity may also be important. For soybeans, the genetic basis for such effects is being investigated, with the possibility that biotechnological manipulations could assist in ameliorating UV-B damage in this species and in other crop plants.

Very few studies have been carried out on trees; pine species are especially sensitive to UV-B. For Loblolly pines, there are cumulative effects during long-term experiments. Major uncertainties relate to the effects of UV-B on tropical forest species (not yet studied, yet responsible for about 50% of terrestrial primary production), indirect effects (e.g. on disease and competitive balance) and other whole ecosystem interactions under changing environmental conditions (e.g. enhanced CO_2 and climate change).

It seems likely that different plant species show different abilities to adapt to UV-B (Caldwell et al. 1995). Under increasing levels of UV-B, plant biodiversity in the natural environment is likely to change. We do not know if trees or perennial plants receiving chronic doses for several years will be affected more than herbs or annual plants. It seems certain that competitive balance between crops and their weeds will be upset. What we do not know is whether weeds will become more or less competitive under enhanced UV-B.

Targets of UV-B damage include photosynthesis, proteins and nucleic acids. When these targets become damaged by enhanced UV-B, the plants grow slower, take longer to reach flowering, give lower yields, and their competitive ability in the community is altered. The photosynthetic photosystem II is especially sensitive to UV-B. Some plants show leaf curling and wrinkling under UV-B stress, resulting in increased stomatal conductance and water loss.

Many pollutants and xenobiotics have enhanced negative impacts in the presence of light. Many aromatic xenobiotics absorb in the UV-B. Since plants can neither avoid UV-B nor airborne chemicals, they are a potential target of photoinduced chemical toxicity.

With respect to ecosystem-level effects of UV-B on plants, inter- and intra-specific differences exist in sensitivity and are partially related to geographical origin. Also, other environmental parameters can modify the UV-B effects. Increases in UV-B may not only alter plant susceptibility to disease but also modify plant-herbivore interactions and may change the rates of decomposition and nutrient cycling.

Many of the effects of UV-B may not involve damage per se. Instead the exposed plant may be using UV-B as a signal for altering growth form and some physiological processes. One example of such a change is the commonly observed increase in UV-absorbing protective compounds in plant leaves following exposure to UV-B. This partly explains many of the seemingly paradoxical results of previous studies suggesting stimulatory effects of UV-B in plants. The UV-B responsiveness is often very specific to species and depends on other environmental factors such as mineral nutrition, drought and air pollutants. This can lead to changes in species interactions and the properties of ecosystems. An indirect effect of elevated UV-B radiation is on decomposition of plant litter which may influence cycling of nutrients. Other such effects include influence on plant pathogen susceptibility and plant attractiveness to herbivores and pollinators. Many of these indirect effects may be UV-B action mediated through changes in plant structure, timing and secondary chemistry. Though indirect, these effects may ultimately be the most important ones for ecosystems. Recent studies indicate that detrimental effects of UV-B in evergreen woody plants may accumulate from year to year.

Much of our understanding of the impacts of increased UV-B and ozone levels on terrestrial vegetation (Tables 5.3) has come from studies of the effects of each individual factor. The interactions of environmental stresses on vegetation are rarely predictable. Also, much of the information has come from growth chamber, greenhouse or field studies using experimental protocols that make little or no provision for the stochastic nature of the changes in UV-B and ozone levels at the Earth's surface and hence excluded the roles of repair mechanisms (Runeckles and Krupa 1993). We know very little about dose-response relationships under field conditions. Though we have a good qualitative understanding of the adverse effects of increased levels of either factor on vegetation, the quantitative relationships are not known. In both cases, sensitivity varies with the stage of plant development. At the population and community levels, differential responses of species to either factor result in changes in competitiveness and community structure (Runeckles and Krupa 1993). Ozone generally inhibits photosynthetic gas exchange under both controlled and field conditions. UV-B is inhibitory in some species under controlled conditions but has no effect on others in the field.

A common metabolic response elicited by either of the two factors is increased secondary metabolism leading to the accumulation of phenolic compounds that, in the case of UV-B, offer the leaf cells some protection from radiation.

5.8.1
Effects of UV-B and Ozone on Terrestrial Vegetation

Little is known about the effects of simultaneous or sequential exposures to both UV-B and ozone. Since both increased surface UV-B exposures have spatial and temporal components, it is important to evaluate the different scenarios that may occur, bearing in mind that elevated daytime ozone levels will attenuate the UV-B reaching the surface to some extent (Runeckles and Krupa 1993).

Table 5.3. Effects of elevated UV-B and ozone on terrestrial vegetation. (After Krupa and Kickert 1989)

Plant characteristic	Effects	
	UV-B	Ozone
Photosynthesis	Reduced in many C_3 and C_4 species at low light intensities	Decreased in most species
Leaf conductance	Reduced (at low light intensities)	Decreased in sensitive species and cultivars
Water use efficiency	Reduced in most species	Decreased in sensitive species
Leaf area	Reduced in many species	Decreased in sensitive species
Specific leaf weight	Increased in many species	Increased in sensitive species
Crop maturation rate	Not affected	Decreased
Flowering	Inhibited or stimulated	Decreased floral yield, fruit set and yield, delayed fruit set
Dry matter production and yield	Reduced in many species	Decreased in most species
Sensitivity between species	Large variability in response among species	Large variability
Sensitivity between cultivars (within species)	Response differs between cultivars	Frequently large variability
Drought stress sensitivity	Plants become less sensitive to UV-B, but sensitive to lack of water	Plants become less sensitive to O_3, but more sensitive to drought
Mineral stress sensitivity	Some species become less and others more sensitive to UV-B	Plants become more susceptible to O_3 injury

5.9
Effects on Aquatic Ecosystems

Ultraviolet-B radiation appears to be a stress in aquatic and marine systems even for submerged organisms. Although UV-B penetration is reduced by plankton and dissolved organic material (Smith and Baker 1979), it still penetrates to a depth of approximately 10% of the euphotic zone before it is reduced to 1% of its surface irradiance level (Worrest 1983). This penetration can occur to depths of many meters in ocean waters, even up to 40–50 meters in some cases (Fleischmann 1989). The effects of UV-B on marine systems have often been studied primarily on plankton. Motility of phytoplankton, including the percentage of motile cells, the rate of movement of cells, and phototactic and gravitactic ability, are reduced specifically by UV-B (Ekelund

1990, 1991; Häder and Worrest 1991; Häder et al. 1994). Phytoplankton productivity can be reduced by UV-B through its disruption of photosystem II activity, DNA damage and inhibition of Rubisco activity (Lesser et al. 1994). The effects of reduced growth and productivity of phytoplankton on marine food webs may be far-reaching (Häder et al. 1994; Häder and Worrest 1997; Grobe and Murphy 1994).

High attenuation of UV-B transmission occurs in the presence of dissolved organic carbon (DOC) within a water column. For fairly clean lakes with only 2 mg DOC l^{-1}, 99% of the downwelling UV-B radiation tends to be absorbed in the top 1 m, with 90% being absorbed by the upper 0.5 m. Suspended particles and phytoplankton further attenuate the UV-B, often quite sharply, thereby limiting the impacts of UV-B to shallow waters.

In Canada, disturbing effects resulting from a combination of global warming, acid rain and ozone depletion have been observed. Normally, dissolved organic carbon (DOC) in freshwater lakes protects aquatic plants and animals from UV-B radiation. But global warming and acid rain now seem to be reducing DOC levels in lakes, thus exposing aquatic life to more UV-B. Measurements at several lakes in northwest Ontario over 20 years have shown that their DOC levels fell by 15–20%, allowing radiation to penetrate 2–6% deeper. In one lake, UV-B penetration increased from 30 to 275 cm.

In contrast to visible light, UV-B radiation reaching the Earth's surface is very diffuse. This high percentage of diffuse-to-direct UV-B radiation results in an essentially constant reflection of 6–8% from the aquatic surface, as well as the transmission of UV-B radiation across the air-water layer being mostly unaffected by all but the strongest surface turbulence (Bukata 1993). Aquatic ecosystems which are most vulnerable to UV-B effects are those with shallow water, lower DOC content and those at higher latitudes (Häder et al. 1994).

The precise impact of UV-B increases on ecosystems is often very difficult to estimate because of complex interactions between changes in water chemistry and the biology of most organisms. Both compounding and compensatory effects can be involved. For example, UV-B may decrease photosynthesis and growth rates of phytoplankton, but it can also increase the availability of iron which can sometimes limit algal production in marine waters. Increased iron availability compensates for the direct harmful effects of UV-B on algae. Alternatively, UV-B may also photo-oxidize certain vitamins which the algae need. This exacerbates any direct deleterious effect of UV-B on algae. The overall impact depends on the situation (e.g. freshwater versus marine) and the type of community affected (Bothwell 1993).

The knowledge of UV-B incident at the ocean surface and of the depth and magnitude of its penetration, as well as an estimate of the time spent by the phytoplankton at depth, is needed to determine the irradiance to which in situ phytoplankton communities are exposed to UV-B radiation. Such information is needed for quantification of the damage to phytoplankton as a function of the magnitude and duration of O_3-dependent UV-B exposure in underwater light fields.

There is concern that phytoplankton communities confined to near-surface waters of the marginal ice zone of the Southern Ocean will be harmed by increased UV-B irradiance penetrating the ocean surface, thereby altering the dynamics of Antarctic marine ecosystems. Results from a 6-week cruise (Icecolors) in the marginal ice zone of the Bellingshausen Sea during the Austral spring of 1990 indicated that as the O_3 layer thinned: (1) sea surface and depth-dependent ratios of UV-B irradiance to total irradiance (280 to 700 nm) increased, and (2) UV-B inhibition of photosynthesis increased. These findings suggest that O_3-dependent shifts of in-water spectral irradiances change the balance of spectrally dependent phytoplankton processes, including photoinhibition, photoreactivation, photoprotection and photosynthesis. A minimum of 6 to 12% reduction in primary production associated with O_3 depletion was estimated (Smith et al. 1992).

Effects of enhanced UV radiation on Antarctic marine ecosystems have been studied by Karentz (1994). The natural variability of marine production processes and the lack of baseline information (before ozone depletion events began in the 1970s) have made it difficult to show unequivocally that significant damage is occurring at the ecosystem level. Nevertheless, current estimates suggest that enhanced UV-B levels in the Antarctic spring reduce marine primary production, over very large areas, by around 10%.

Different phytoplankton species vary in their response to UV-B, with damage related to their surface area/volume ratio. Consequently, increased UV could change community structure in favor of larger-celled species, potentially affecting higher levels of the food chain (e.g. krill, penguins and marine mammals). Compounds that absorb UV-B (mycosporine-like amino acids, MAA) occur in phytoplankton, and these are also passed on to consuming organisms. However, they do not give complete protection. In sea urchins, the greatest concentration of UV-absorbers is found in the eggs, that are released into the (surface) water (hence with greater exposure to UV than the benthic adults). Recent experimental studies have shown that a period of particularly high natural UV-B radiation tends to cause a marked increase in embryo malformation at 1 m water depth (but not at 7 m).

There are some indications that ultraviolet-B radiation may be damaging populations of the foraminifera that live in coral reefs around tropical islands. Populations of some of these organisms have declined markedly, raising concerns about the effects on rates of coastal sedimentation since their shells produce up to 90% of the sand-sized sediments found on many tropical islands.

In near-shore and intertidal systems, macroalgae are major primary producers. They are a good source of food for diverse invertebrates and fish. Algal systems are highly productive. Kelp beds (*Laminaria* and *Macrocystis*) produce up to 2000 g C m^{-2} yr^{-1}

(Mann and Chapman 1975). These rates compare with those for the most productive of terrestrial ecosystems. Although they occupy a smaller total global area, benthic macrophytes in bays and estuaries have a productivity that can be up to three-fold greater than that of phytoplankton when compared on a per-area basis.

Wood (1987) studied the effect of solar UV on the kelp, *Ecklonia radiata*. Subcanopy sporophytes grew more rapidly in outdoor tanks to which solar UV-B was screened off than in unscreened tanks. Larkum and Wood (1993) measured the sensitivity of several benthic algae to artificial UV radiation from a xenon arc lamp. *Enteromorpha intestinalis* and *Porphyra* sp. showed very little sensitivity to UV. *Ecklonia radiata* and *Kallymenia cribrosa* suffered marked reduction in photosynthetic oxygen production. The above species are mostly tropical.

However, it is the temperate species that are more likely to receive increased UV-B doses due to ozone depletion. Accordingly, Grobe and Murphy (1994) studied the effects of UV-B on a specific, temperate-region intertidal macroalga, *Ulva expansa*. They measured the growth of segments of thallus under naturally occurring and supplemented levels of UV-B and showed that UV-B, even at current ambient levels, inhibits the growth of this alga (Grobe and Murphy 1994).

Segments of thallus collected from a natural population were grown in outdoor seawater tanks. Combinations of UV-B opaque screens, UV-B transparent screens, and UV-B lamps were used to study the effects of solar UV-B and solar plus supplemental UV-B on segment surface area, wet weight, and dry weight. Growth rates of segments were inhibited under both solar UV-B and solar plus supplemental UV-B treatments. Growth rates were also inhibited by high levels of photosynthetically active radiation, independent of the UV-B intensity. These results indicate that increases in UV-B resulting from further ozone depletion will have a negative impact on the growth of this alga.

Most macroalgae and seagrasses are restricted to their growth site (Lüning 1990). They show a distinct pattern of vertical distribution in their habitat. Some of these plants inhabit the supralittoral (coast above high water mark), exposed only to the surf, whereas others populate the eulittoral (intertidal zone), which is characterized by the regular temporal change in the tides (Häder 1997). Still others are restricted to the sublittoral zone. The range in exposure is remarkable, from over 1000 W m^{-2} (total solar radiation) at the surface, to less than 0.01% of that, which reaches the understory of a kelp habitat (Markager and Sand-Jensen 1994). Macroalgae have developed mechanisms to regulate their photosynthetic activity to adapt to the changing light regime and protect themselves from excessive radiation. They use the mechanism of photoinhibition to decrease the photosynthetic electron transport during periods of excessive radiation. This phenomenon facilitates thermal dissipation of excessive excitation. Different algal species occupy different depth niches and are adapted to different solar exposure (Häder and Figueroa 1997). They also differ in their ability to cope with enhanced UV radiation (Dring et al. 1996). A broad survey was carried out to understand photosynthesis in aquatic ecosystems and the different adaptation strategies to solar radiation of ecologically important species of green, red and brown algae from the North Sea, Baltic Sea, Mediterranean, Atlantic, polar and tropical oceans (Markager and Sand-Jensen 1994; Häder and Figueroa 1997; Porst et al. 1997).

Photoinhibition can be quantified by oxygen exchange (Häder and Schäfer 1994) or by PAM (pulse amplitude modulated) fluorescence measurements developed by Schreiber et al. (1986) and based on transient changes of chlorophyll fluorescence. Surface-adapted macroalgae, such as several brown (*Cystoseira*, *Padina*, *Fucus*) and green algae (*Ulva*, *Enteromorpha*), show a maximum of oxygen production at or close to the surface (Herrmann et al. 1995; Häder and Figueroa 1997); whereas algae adapted to lower irradiances usually thrive best when exposed deeper in the water column (the green algae *Cladophora*, *Caulerpa*, most red algae; Häder and Figueroa 1997). It is interesting to note that respiration is inhibited to a far smaller degree than photosynthesis.

PAM fluorescence allows the determination of the photochemical and non-photochemical quenching (Büchel and Wilhelm 1993). Even algae harvested from rock pools, where they are exposed to extreme irradiances, show signs of photoinhibition after extended periods of exposure (Fig. 5.10). Deep-water algae and those adapted to shaded conditions are inhibited even faster when exposed to direct solar radiation. Large differences were also found in the recovery between high light-adapted and protected species. Exclusion studies were carried out to determine the effects of solar UV-B and UV-A (Herrmann et al. 1995). Increasing exposure to solar radiation resulted in a shift of the light compensation point to higher irradiances. The compensation point defines the irradiance at which photosynthetic oxygen production and respiratory oxygen consumption balance each other. Exclusion of UV-B partially reduced the effects. This trend increased when about half or all of the UV-A radiation was excluded.

Chronic photoinhibition occurs when algae are exposed to excessive irradiance. The inhibition is characterized by photodamage of PS II reaction centers and subsequent proteolysis of the D1 protein (Critchley and Russell 1994). In contrast, dynamic photoinhibition is readily reversible and follows a diurnal pattern with the lowest quantum yield around or soon after noon (Hanelt et al. 1994; Häder and Figueroa 1997). The lowest light compensation point for photosynthesis has been reported in Arctic and Antarctic algae (Gómez et al. 1995; Gómez and Wiencke 1996). The long-term effects of solar UV on the primary productivity of macroalgae still need to be evaluated. Shallow water specimens in coral reefs undergo a 50% reduction in photosynthetic efficiency during the middle of the day and show a complete recovery by late afternoon.

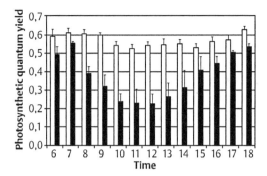

Fig. 5.10. Photosynthetic quantum yield measured on site using a PAM fluorimeter in the Mediterranean brown alga *Padina pavonica* harvested from 0 m (*closed bars*) and 6 m depth (*open bars*) at 1-h intervals. (From Häder 1997)

A number of cyanobacteria (see below), phytoplankton and macroalgae produce screening pigments which filter out some of the solar UV radiation. These sunscreen pigments are passed on to the primary and secondary consumers in aquatic ecosystems by means of the food chain. An interesting observation on the possible role of MAA in protecting against UV-B radiation has emerged from the field experiments conducted at San Salvador Island, Bahamas, in September 1991 by Gleason and Wellington (1993). Episodes of coral bleaching resulting from dissociation of endosymbiotic zo-oxanthellae from host coral tissues throughout the tropics were thought to result from increases in seawater temperature. However, the mass bleaching events that occurred throughout the Caribbean during 1987 and 1990 have not been explained by tempera-ture alone. Gleason and Wellington (1993) reported that irrespective of high water temperatures, short-term (three-week) increases in UV radiation of a magnitude possi-ble under calm, clear water column conditions can readily induce bleaching in reef-building corals. They observed that colonies of *Montastrea annularis* showed a grad-ual reduction in MAA concentration with increasing depth, indicating that deeper-water colonies (> 20 m) may be particularly vulnerable to sudden increases in UV ra-diation. Measurement of UV-absorbing pigments from colonies transplanted from 24 to 12 m depth also indicated that *M. annularis* is unable to counter rapid increases in UV irradiance through enhanced synthesis or accumulation of MAAs over a period of 21 days.

UV-B radiation brings about photochemical degradation of refractory macromole-cules into biologically labile organic compounds and it might influence the cycling of organic matter in the sea, which is believed to be largely mediated by bacterioplankton (Herndl et al. 1993). Herndl et al. reported that bacterioplankton activity in the surface layers of the oceans is suppressed by solar radiation by about 40% in the top 5 m of the water column in near shore waters, whereas in oligotrophic open oceans suppression might be detectable to a depth of >10 m. Bacterioplankton from near-surface (0.5 m depth) waters of a highly stratified water column were as sensitive to surface UV-B radiation as subpycnocline bacteria, indicating no adaptive mechanisms against surface solar radiation in near-surface bacterioplankton consortia. Surface solar radiation levels also photochemically degrade bacterial extracellular enzymes. Thus, enhanced UV-B radiation might lead to reduced bacterial activity and accompanying increased concentration of labile dissolved organic matter in the surface layers of the ocean, as bacterial uptake of this is retarded (Herndl et al. 1993).

5.9.1
Effects on Cyanobacteria

Cyanobacteria are cosmopolitan organisms inhabiting almost every terrestrial and aquatic ecosystem. They were the first organism to develop oxygenic photosynthesis which led to the evolution of an oxic atmosphere which resulted in the generation of a protective ozone layer in the stratosphere. These prokaryotic organisms are major biomass producers both in the sea and in terrestrial habitats. Furthermore, some strains are capable of converting atmospheric nitrogen into a form which is accessible to eukaryotic organisms such as phytoplankton, macroalgae and higher plants.

Cyanobacterial communities in open habitats exposed to direct sunlight appear to be highly vulnerable to the effects of changing solar ultraviolet radiation. Exposure to UV can cause cytotoxic lesions in cyanobacteria such as DNA base dimers and photosystem II inactivation. It also has certain general debilitating effects associated with protein damage and pigment photooxidation. These responses are a function of wavelength, intensity and duration of exposure. Cyanobacteria appear to have four lines of defense against UV exposure: avoidance, screening, quenching, and repair (Vincent and Quesada 1994). However, there are marked differences among species in their ability to cope with UV. Intense solar radiation can impair some of these defense mechanisms.

According to Vincent and Quesada (1994), the UV-B flux over Antarctica is not likely to cause an abrupt decline in productivity in these microbial ecosystems, but is likely to alter community structure. The dispersal and primary colonization phases in Antarctic terrestrial environments may be especially vulnerable to UV-B radiation during the period of spring ozone depletion because at that time the microorganisms are still frozen and their biosynthetic repair mechanisms may not be functional. Any photochemical damage incurred during this period may remain unchecked until cellular metabolism resumes as temperatures rise (Vincent and Quesada 1994).

Donkor et al. (1993) studied the effects of tropical solar radiation on the motility of certain filamentous cyanobacteria. The percentage of motile filaments of *Anabaena variabilis, Oscillatoria tenuis* and two strains of *Phormidium uncinatum* were drastically reduced by unfiltered solar radiation. These in situ experiments were conducted in Ghana.

Some work done at Banaras Hindu University during the last few years is outlined below. In the freshwater *Nostoc muscorum*, UV-B (5 W m^{-2}) causes drastic imbalances in various physiological processes of the organism. Nitrogenase activity was inhibited instantly and complete inactivation occurred with a 20-min treatment of UV-B. $^{14}CO_2$ uptake was also abolished within 30 min of exposure to UV-B. However, a phycoerythrin-rich strain of *Nostoc spongiaforme* showed a higher tolerance level towards UV-B. Unlike its blue-green strain (phycocyanin rich), the brown form (phycoerythrin rich) did not show abrupt loss in pigment content. Furthermore, there was stimulation in nitrate reductase activity of the brown form (Tyagi et al. 1992). Strains containing phycoerythrin may be more highly tolerant to UV-B, probably because of their inherent property of adapting to a variety of light conditions.

Possible protection of damage caused by UV-B has been studied by simultaneous exposure of *Nostoc muscorum* to fluorescent light (Tyagi et al. 1992). Exposure of cultures to UV-B (5 W m^{-2}) in the presence of fluorescent light elicited significant levels of nitrogenase and $^{14}CO_2$ activity. A dose of 1 W m^{-2} UV-B did not cause any inhibitory effect when the cultures were simultaneously exposed to fluorescent light (Tyagi et al. 1992). Kumar et al. (1996) have shown a shielding (protective) role of ferric iron and certain cyanobacterial sheath pigments against UV-B-induced damage in the filamentous nitrogen-fixing cyanobacterium *Nostoc muscorum*, a rice field cyanobacterium.

Several strains of cyanobacteria isolated from habitats exposed to strong solar radiation contain one or more water-soluble, UV-absorbing mycosporine amino acid (MAA)-like compounds. The UV absorption spectra of MAAs complement that of the extracellular sunscreen pigment scytonemin, which many of the strains also produce (Garcia-Pichel and Castenholz 1991; 1993). Even though the specific MAA content was variable among strains, they were invariably higher when the cultures were grown with UV radiation than when it was absent. In five strains, the MAA complement accumulated as a solute in the cytoplasmic cell fraction. Significant, but not complete, protection from UV photodamage can be gained from the possession of either MAA or scytonemin but especially from simultaneous screening by both types of compounds (Garcia-Pichel and Castenholz 1993). Garcia-Pichel et al. (1993) studied the UV sunscreen role commonly ascribed to MAAs in a terrestrial cyanobacterium *Gloeocapsa* which accumulates intracellularly an MAA with an absorbance maximum at 326 nm but produces no extracellular sunscreen compound (i.e., scytonemin). The MAA prevented 3 out of 10 photons from hitting potential cytoplasmic targets. High contents of

MAA in the cells correlated with increased resistance to UV radiation. However, when resistance was measured under conditions of desiccation, with inoperative physiological photoprotective and repair mechanisms, cells with high MAA specific contents were only 20 to 25% more resistant. Some of the compounds were identical in several strains. In all, 13 distinct compounds were found. In general, the specific contents of MAA found in these isolates were between 0.16 and 0.84% of the total biomass. MAAs present a characteristic, single absorption band in the UV spectrum between 230 and 400 nm, with maxima ranging from 310 to 360 nm.

5.9.2
Effects on Amphibians

Blaustein et al. (1994) reported that the populations of many amphibian species, in widely scattered habitats, are declining whereas some other amphibians show no such declines. The widespread distribution of amphibians suggests involvement of global agents such as increased UV-B radiation. To test this hypothesis, Blaustein et al. focused on species-specific differences in the abilities of eggs to repair UV radiation damage to DNA and differential hatching success of embryos exposed to solar radiation at natural oviposition sites. Quantitative comparisons of activities of a key UV-damage-specific repair enzyme, photolyase, among oocytes and eggs from ten amphibian species were found to be characteristic for a given species but varied enormously among the species. Photolyase levels correlated with expected exposure of eggs to sunlight. Among some frog and toad species studied, the highest activity was shown by *Hyla regilla* whose populations are not known to be in decline. The toad *Bufo boreas* and the frog *Rana cascadae* whose populations have greatly declined showed much lower photolyase levels. In field experiments, the hatching success of embryos exposed to UV radiation was significantly greater in *H. regilla* than in *R. cascadae* and *B. boreas*. Moreover, in *R. cascadae* and *B. boreas*, hatching success was greater in regimes shielded from UV radiation as compared to regimes that allowed UV radiation. These observations are thus consistent with the UV-sensitivity hypothesis (Blaustein et al. 1994).

5.10
Polymer Degradation

In materials exposed outdoors, the level and nature of the UV radiation is critical. The outdoor lifetime of most polymeric materials is relatively short. However, this can be even shortened by the adverse effects of UV on materials. With increasing availability of more durable and stable materials, long-term variations of light exposure will have to be evaluated and taken into account.

Only very small amounts of radiation shorter than about 295 nm reach the Earth's surface except at high altitudes. The UV-B is responsible for most of the polymer photodegradation, although some polymers are sensitive to longer wavelength radiation (315–400 nm) as well. Any changes in ozone levels affect mostly the UV-B because ozone has essentially no absorbance at wavelengths longer than 330 nm.

5.11
Effect of UV-B on Air Quality

The photodissociation of ozone by atmospheric ultraviolet radiation is a key reaction that controls urban and regional oxidants, the self-cleaning capacity of the troposphere and the atmospheric lifetimes of many natural and anthropogenic gases. Recent work has brought out the need for some revision of the quantum efficiency for this process, with longer UV wavelengths playing a larger role than hitherto assumed. This is likely to have some impact on predicted geographical and seasonal distributions of gases such as methane and carbon monoxide.

The linkage between stratospheric UV transmission and tropospheric composition has been established by studying the records of surface CH_4 and CO concentrations. Temporary increases in tropical CH_4 and CO concentrations were observed in the second half of 1991, in phase with high levels of stratospheric sulfur dioxide and sulfate aerosols that resulted from the June 1991 eruption of Mt. Pinatubo. The reduced stratospheric UV transmission appears to have lowered the rate of CH_4 and CO removal, due to decreased tropospheric ozone photodissociation and hence decreased hydroxyl radical formation.

The atmospheric chemistry of CFC substitutes and the possible build-up of their breakdown products (particularly trifluoroacetic acid, TFA) has attracted much attention. The background atmospheric concentrations of HFC-134a (a principal precursor of TFA) have been increasing rapidly, according to measurements at Cape Grim, Tasmania (41°S) between 1978 and 1995, and Mace Head, Ireland (53°N) between July 1994 and May 1995. Model estimates suggest that HFC-134a emissions have risen rapidly from about 250 tons in 1991 to about 8000 tons in 1994.

TFA concentrations in several air and water samples collected in Europe in 1995 have been found to contain surprisingly high levels of TFA comparable to concentrations predicted by models for the year 2010. This suggests that there may exist an additional, yet unknown, source of TFA. One recent three-dimensional model shows that the deposition fluxes of TFA will reach maximum levels of 2 μmol m^{-2} yr^{-1} (about 1 nmol l^{-1} in rainwater) in the year 2020, with maximum deposition occurring in tropical areas. However, other modeling studies have indicated that several factors can enhance local concentrations of TFA. The TFA concentrations in precipitation of arid and semi-arid regions may be 2–4 times greater than the global mean and the source strength of TFA may be substantially enhanced in urban air where high OH levels lead to an increase in the rate of TFA formation. Therefore, local rainwater TFA concentration of 2 to 20 μg l^{-1} seems therefore plausible. Water bodies characterized by little or no outflow and high evaporation rates may be susceptible to accumulation of rainborne TFA. The model calculations predict that a concentration of 100 ng l^{-1} could be achieved in as short a period as three decades even when seepage is as high as 10%.

References

Anonymous (1992) Solar and ultraviolet radiation. Vol. 55: IARC Monograph on the evolution of carcinogenic risk to humans. WHO/Internat. Agency for Research on Cancer, Lyon, France

Blaustein AR, Hoffman PD, Hokit DG, Kiesecker JM, Walls SC, Hays JB (1994) UV repair and resistance to solar UV-B in amphibian eggs: a link to population declines? Proc. Natl. Acad. Sci. USA 91:1791–1795

Bothwell ML, Sherbot P, Roberge AC, Daley RJ (1993) Influence of natural ultraviolet radiation on lotic periphytic diatom community. Growth, biomass accrual, and species composition: short-term versus long-term effects. J Phycol 29:24–35

Brühl C, Crutzen PJ (1989) On the disproportionate role of tropospheric ozone as a filter against solar UV-B radiation. Geophys Res Lett 16: 703–706

Büchel C, Wilhelm C (1993) In vivo analysis of slow chlorophyll fluorescence induction kinetics in algae: progress, problems and perspectives. Photochemistry and Photobiology 58:137–148

Bukata RP (1993) Radiation attenuation in aquatic ecosystems. Abst. Workshop on the Effects of Increased UV-B Radiation. April 1993. Environment Canada Report No. 93-009, Toronto, pp 5–6

Caldwell MM, Teramura AH, Tevini M, Bornman JF, Björn LO, Kulandaivelu G (1995) Effects of increased solar ultraviolet radiation on terrestrial plants. Ambio 24:166–173

Coohill TP (1991) Action spectra again? Photochem. Photobiol. 54:859–70

Critchley C, Russell AW (1994) Photoinhibition of photosynthesis in vivo: the role of protein turnover in photosystem II. Physiologia Plantarum 92:188–196

Cullen JJ, Neale PJ, Lesser MP (1992) Biological weighting function for the inhibition of phytoplankton photosynthesis by ultraviolet radiation. Science 258:646–50

Dey DB, Damkaer DM, Heron GA (1988) UV-B dose/dose-rate responses of seasonally abundant copepods of Puget Sound. Oecologia 76:321–29

Donkor VA, Amewowor DHAK, Häder D-P (1993) Effects of tropical solar radiation on the motility of filamentous cyanobacteria. FEMS Microbiol. Ecol. 12:143–148.

Dring MJ, Makarov V, Schoshina E, Lorenz M, Lüning K (1996) Influence of ultraviolet radiation on chlorophyll fluorescence and growth in different life history stages of three species of *Laminaria* (Phaeophyta). Marine Biology 126:183–191

Ekelund NGA (1990) Effects of UV-B radiation on growth and motility of four phytoplankton species. Physiol Plant 78:590–594

Ekelund NGA (1991) The effects of UV-B radiation on dinoflagellates. J Plant Physiol 138:274–78

Fleischmann EM (1989) The measurement and penetration of ultraviolet radiation into tropical marine water. Limnol Oceanogr 38:1623–1629

Frederick JE (1990) Trends in atmospheric ozone and ultraviolet radiation: mechanisms and observations for the Northern Hemisphere. Photochem Photobiol 51:757–763

Frederick JE (1993) Ultraviolet sunlight reaching the Earth's surface: a review of recent research. Photochem Photobiol 57:175–78

Frederick JE, Koob AE, Alberts AD, Weatherhead EC (1993) Empirical studies of tropospheric transmission in ultraviolet broadband measurements. J Appl Meterol 32:1883–92

Frederick JE, Snell HE, Haywood EK (1989) Solar ultraviolet radiation at the Earth's surface. Photochem Photobiol 50:443–50

Frederick JE, Snell HE (1990) Tropospheric influence on solar ultraviolet radiation: the role of clouds. J Climate 3:373–381

Frederick JE, Weatherhead EC (1992) Temporal changes in surface ultraviolet radiation: a study of the Robertson–Berger meter and Dobson data records. Photochem Photobiol 56:123–31

Garcia-Pichel F, Castenholz RW (1991) Characterization and biological implications of scytonemin, a cyanobacterial sheath pigment. J Phycol 27:395–409

Garcia-Pichel F, Castenholz RW (1993) Occurrence of UV-absorbing, mycosporine-like compounds among cyanobacterial isolates and an estimate of their screening capacity. Appl Environ Microbiol 59:163–169

Garcia-Pichel F, Wingard CE, Castenholz RW (1993) Evidence regarding the UV sunscreen role of a mycosporine-like compound in the cyanobacterium *Gloeocapsa* sp. Appl Environ Microbiol 59:170–176

Gleason DF, Wellington GM (1993) Ultraviolet radiation and coral bleaching. Nature 365:836–838

Gómez I, Wiencke C (1996) Photosynthesis, dark respiration and pigment contents of gametophytes and sporophytes of the Antarctic brown alga *Desmarestia menziessi*. Botanica Marina 39:149–157

Gómez I, Wiencke C, Weykman G (1995) Seasonal photosynthetic characteristics of the brown alga *Ascoseira mirabilis* from King George Island (Antarctica). Marine Biology 123:167–172

Grobe CW, Murphy TM (1994) Inhibition of growth of *Ulva expansa* (Chlorophyta) by ultraviolet-B radiation. J Phycol 30:783–790

Häder D-P (1997) Penetration and effects of solar UV-B on phytoplankton and macroalgae. Plant Ecology 128:4–13

Häder D-P, Figueroa FL (1997) Photoecophysiology of marine macroalgae. Photochemistry and Photobiology 66:1–14

Häder D-P, Schäfer J (1994) In-situ measurement of photosynthetic oxygen production in the water column. Environmental Monitoring and Assessment 32:259–268

Häder D-P, Worrest RC (1991) Effects of enhanced solar ultraviolet radiation on aquatic ecosystems. Photochem Photobiol 53:717–725

Häder D-P, Worrest RC (1997) Consequences of the effects of increased solar ultraviolet radiation on aquatic ecosystems. In: Häder, D-P: The effects of ozone depletion on aquatic ecosystems. Ch. 3. Environmental Intelligence Unit. Academic Press and R.G. Landes Comp., Austin, pp 11–30

Häder D-P, Worrest RC, Kumar HD, Smith RC (1995) Effects of increased solar ultraviolet radiation on aquatic ecosystems. Ambio 24:174–180

Häder DP, Worrest RC, Kumar HD, Smith RC (1994) Aquatic ecosystems. In UNEP. Environmental Effects of Ozone Depletion. 1994 Assessment. UNEP, Nairobi

Hanelt D, Jaramillo MJ, Nultsch W, Senger S, Westermeir R (1994) Photoinhibition as a regulatory mechanism of photosynthesis in marine algae of Antarctica. Serie Cient Inst Antarct, Chil 44:67–77

Herrmann H, Ghetti F, Scheuerlein R, Häder D-P (1995) Photosynthetic oxygen and fluorescence measurements in *Ulva laetevirens* affected by solar irradiation. Journal of Plant Physiology 145:221–227

Herndl GJ, Muller-Niklas G, Frick J (1993) Major role of ultraviolet-B in controlling bacterioplankton growth in the surface layer of the ocean. Nature 361: 717–719

Karentz D (1994) Ultraviolet tolerance mechanisms in Antarctic marine organisms. In: Weiler CS, Penhale PA (eds) Ultraviolet radiation in Antarctica: measurement and biological effects. Vol. 62. Antarctic Research Series, pp 93–1010

Kerr JB, McElroy CT (1993) Evidence for large upward trends of ultraviolet-B radiation linked to ozone depletion. Science 262:1032–34

Koller LR (1965) Ultraviolet radiation. Wiley, New York, pp 105–53

Krupa SV, Kickert RN (1989) The greenhouse effect: impacts of ultraviolet-B, carbon dioxide, and ozone on vegetation. Environ Pollut 61:263–93

Kumar A, Tyagi MB, Srinivas G, Singh N, Kumar HD, Häder D-P (1996) UV-B shielding role of $FeCl_3$ and certain cyanobacterial pigments. Photochem Photobiol 64:321–325

Larkum AWD, Woo WF (1993) The effect of UV-B radiation on photosynthesis and respiration of phytoplankton, benthic macroalgae and seagrasses. Photosynth Res 36:17–23

Lesser MP, Cullen JJ, Neale PJ (1994) Carbon uptake in a marine diatom during acute exposure to ultraviolet B. J Phycol 30:183–192

Liu SC, McKeen SA, Madronich S (1991) Effect of anthropogenic aerosols on biologically active ultraviolet radiation. Geophys Res Lett 18:2265–68

Lloyd RE, Larson RA, Adair TL, Tuveson RW (1991) Cu(II) sensitizes pBR322 plasmid DNA to inactivation by UV-B (280–315 nm). Photochem Photobiol 57:1011–1017

Longstreth JD, de Gruijl FR, Kripke ML, Takizawa Y, van der Leun JC (1995) Effects of increased solar ultraviolet radiation on human health. Ambio 24:153–165

Lubin D, Jensen EH (1995) Effects of clouds and stratospheric ozone depletion on ultraviolet radiation trends. Nature 377:710–13

Lüning K (1990) Seaweeds. Their environment, biogeography and ecophysiology. Wiley, New York, p 527

Madronich S (1991) Implications of recent total atmospheric ozone measurements for biologically active radiation reaching the Earth's surface. Geophys Res Lett 18:2269–72

Madronich S (1995) The radiation equation. Nature 377:682–83

Mann KH, Chapman ARO (1975) Primary production of marine macrophytes. In: Cooper JP (ed) Photosynthesis and productivity in different environments. Cambridge Univ. Press, pp 207–223

Markager S, Sand-Jensen K (1994) The physiology and ecology of light-grown relationship in macroalgae. In: Round FE, Chapman DJ (eds) Progress in phycological research Vol 10. Biopress Ltd, Bristol, pp 209–298

Mopper K, Zhou X (1990) Hydroxyl radical photoproduction in the sea and its potential impact on marine processes. Science 250:661–64

Niu X, Frederick JE, Stein M, Tiao GC (1992) Trends in column ozone based on TOMS data: dependence on month, latitude, and longitude. J Geophys Res 97:14661–69

Pickett JE (1994) Effect of stratospheric ozone depletion on terrestrial ultraviolet radiation: a review and analysis in relation to polymer photodegradation. Polymer Degradation and Stability 43:353–62

Porst M, Herrmann H, Schäfer J, Santas R, Häder D-P (1997) Photoinhibition in the Mediterranean green alga *Acetabularia mediterranea* measured in the field under solar irradiation. J Plant Phys 151:25–32

Runeckles VC, Krupa SV (1993) The impact of UV-B and biologically effective dose-rates in natural waters. Photochem Photobiol 29:311–323

Schreiber U, Schliwa U, Bilger W (1986) Continuous recording of photochemical and non-photochemical chlorophyll fluorescence quenching with a new type of modulation fluorometer. Photosynthesis Research 10:51–62

Smith RC, Baker KS (1979) Penetration of UV-B and biologically effective dose-rates in natural waters. Photochem Photobiol 29:311–323

Smith RC, Prezelin BB, Baker KS, Bidigare RR, Boucher NP, Coley T, Karentz D, MacIntyre S, Matlick HA, Menzies D, Ondrusek M, Wan Z, Waters KJ (1992) Ozone depletion: ultraviolet radiation and phytoplankton biology in Antarctic waters. Science 255:952–958

Stolarski RS, Bloomfield P, McPeters R, Herman JR (1991) Total ozone trends deduced from Nimbus 7 TOMS data. Geophys Res Lett 18:1015–1018

Tyagi R, Srinivas G, Vyas D, Kumar A, Kumar HD (1992) Differential effects of ultraviolet-B radiation on certain metabolic processes in a chromatically adapting *Nostoc*. Photochem Photobiol 55:401–407

UNEP (1989) Action on ozone. United Nations Environment Programme, Nairobi

UNEP (1994) Environmental effects of ozone depletion. 1994 Assessment. United Nations Environment Programm, Nairobi

Vincent WF, Quesada A (1994) Ultraviolet radiation effects on cyanobacteria: Implications for Antarctic microbial ecosystems. In: Weiler CS, Penhale PA (eds) Ultraviolet Radiation in Antarctica: Measurements and Biological Effects. Antarctic Research Series. American Geophysical Union, Washington DC 62, pp 111–124

Vincent WF, Roy S (1993) Solar ultraviolet-B radiation and aquatic primary production: damage, protection and recovery. Environ Res 1:1–12

Wood WF (1987) Effect of solar ultraviolet radiation on the kelp *Ecklonia radiata*. Mar Biol (Berl) 96:143–150

Worrest RC (1983) Impact of solar ultraviolet-B radiation (290–320 nm) upon marine microalgae. Physiol Plant 58:428–434

Zölzer F, Kiefer J (1993) Risk estimates for UV-B enhanced solar radiation. Naturwissenschaften 80:462–465

Index